P9-BYG-632

TREATISE ON ANALYTICAL CHEMISTRY

A comprehensive account in three parts

PART I

THEORY AND PRACTICE

PART II

ANALYTICAL CHEMISTRY OF
INORGANIC AND ORGANIC COMPOUNDS

PART III

ANALYTICAL CHEMISTRY IN INDUSTRY

TREATISE ON ANALYTICAL CHEMISTRY

Edited by I. M. KOLTHOFF
School of Chemistry, University of Minnesota

and PHILIP J. ELVING
Department of Chemistry, University of Michigan

PART II

ANALYTICAL CHEMISTRY OF INORGANIC AND ORGANIC COMPOUNDS

VOLUME *14*

WILEY—INTERSCIENCE

a division of John Wiley & Sons, Inc., New York–London–Sydney–Toronto

Library of Congress Catalog Card Number: 59-12439
ISBN 0-471-50005-4

PRINTED IN THE UNITED STATES OF AMERICA

10 9 8 7 6 5 4 3 2 1

TREATISE ON ANALYTICAL CHEMISTRY

PART II

ANALYTICAL CHEMISTRY OF INORGANIC AND ORGANIC COMPOUNDS

VOLUME 14

With the cooperation of Edward W. D. Huffman, *Huffman Laboratories, Inc.* and John Mitchell, Jr., *E. I. du Pont de Nemours & Co.*, as section advisors.

AUTHORS OF VOLUME 14

ADAM S. INGLIS STANLEY T. HIROZAWA
EDWARD C. OLSON ROBERT T. HALL
ROBERT D. MAIR

Acknowledgment

In view of the comprehensive nature of the Treatise, the editors have felt it desirable to consult with experts in specialized fields of analytical chemistry. For the sections of Part II dealing with "Organic Analysis," they have been fortunate in securing the cooperation of Dr. Edward W. D. Huffman and of Mr. John Mitchell, Jr. The competence of the authors of the individual chapters in the sections on organic analysis, combined with the Messrs. Huffman's and Mitchell's broad knowledge of the area, has resulted in a succinct and critical treatment of elemental and functional organic analysis as well as of certain areas basic to organic analysis. The constructive help of the authors and of the Messrs. Huffman and Mitchell in the preparation of Volume 11 and the following volumes is acknowledged with gratitude.

AUTHORS OF VOLUME 14

Robert T. Hall

Research Center, Hercules Incorporated Research Center, Wilmington, Delaware

Stanley T. Hirozawa

BASF-Wyandotte Corporation, Wyandotte, Michigan

Adam S. Inglis

Division of Protein Chemistry, Wool Research Laboratories, C.S.I.R.O., Victoria, Australia

Robert D. Mair

Research Center, Hercules Incorporated Research Center, Wilmington, Delaware

Edward C. Olson

Biochemical Research Division, The Upjohn Company, Kalamazoo, Michigan

Foreword

The chapter on the determination of chlorine, bromine, and iodine in organic compounds, with which the present volume starts, properly belonged at the end of Volume 12 or at the beginning of Volume 13 of Part II of the *Treatise* to complete the section dealing with the determination of the elements that most commonly occur in organic compounds. Unfortunately the author who originally undertook the preparation of the chapter was not able to fulfill this commitment and another expert accepted the Editors' invitation to write the chapter. Meanwhile, in fairness to the authors and users of the *Treatise,* the Editors decided not to delay publication of Volume 13.

PART II. ANALYTICAL CHEMISTRY OF INORGANIC AND ORGANIC COMPOUNDS

CONTENTS VOLUME 14

SECTION B-1. Organic Analysis I. The Elements

Chlorine, Bromine, and Iodine. By *Edward C. Olson*

SECTION B-2. Organic Analysis II. Funtional Groups

Unsaturation. By *Stanley T. Hirozawa*

Determination of Acyl Groups. By *Adam S. Inglis*

O-Alkyl, *N*-Alkyl, and *S*-Alkyl. By *Adam S. Inglis*

Determination of Ethers and Epoxides. By *Robert T. Hall* and *Robert D. Mair*

Part II
Section B-1

ORGANIC ANALYSIS: CHLORINE, BROMINE, AND IODINE

BY EDWARD C. OLSON, *Biochemical Research Division, The Upjohn Company, Kalamazoo, Michigan*

Contents

Contents (*continued*)

I. INTRODUCTION

Halogen compounds are included in several major classes of organic compounds. They frequently occur as refrigerants, plastics, insecticides, pesticides, herbicides, solvents, antiknock additives in gasoline, and pharmaceuticals. Many organic halogen compounds are toxic and it is therefore often necessary to demonstrate that these materials are substantially absent from products intended for human or animal consumption. Thus the determination of trace quantities of organic halogen compounds is often necessary. The extensive use of organic halogen compounds in agriculture as insecticides and pesticides involves the organic analyst in the whole field of residue analysis and its specialized techniques (89). Since the determinations of the various halogens can be carried out with great sensitivity as well as with precision and accuracy, methods for the determination of the halogens are useful in determining not only the purity of organic halogen compounds but also the presence of small amounts of halogen-containing impurities, i.e., solvents, intermediates, and residues.

Organic halogen compounds present, in many cases, some unique sample-handling problems; for example, many amine hydrohalides and quaternary amine halides are often extremely hygroscopic, whereas acid halides may be very reactive. Thus in many cases these types of compound must be rigorously protected from contact with traces of atmospheric moisture if a satisfactory analysis is to be achieved. Many of the lower molecular weight organic halogen compounds (particularly those containing fluorine) are either low boiling liquids or gases at room temperature and atmospheric pressure, so that special sample handling and weighing techniques must be applied; e.g., the use of sealed capillaries, vacuum transfer systems, and cold traps.

Often a knowledge of the structure and/or the origin of a sample will enable the analyst to select the best method for more rapid and accurate determination of any halogen present; for example, in many cases it is pointless to combust an amine halide salt. Such valuable structural information, however, all too frequently is not transmitted to the analyst.

The large number of published methods for the determination of the halogens in organic compounds attests to the difficulties which can be encountered in the determination of chlorine, bromine, and iodine. In general, these difficulties arise in the decomposition of the organic halogen compounds rather than in the determinations of the halogen following decomposition of the sample. For this reason, methods of decomposition will receive major emphasis in this chapter. Many decomposition methods which give satisfactory results on a macro and semimicro scale have been found totally unsatisfactory or difficult to apply on a micro scale. Unfortunately, the converse is also true.

II. DETECTION AND COMPOUND IDENTIFICATION

The presence of one of the halogens in an organic material is usually demonstrated by the precipitation of the corresponding silver halide upon the addition of silver nitrate solution to the acidified aqueous solution resulting from a sodium fusion of the organic compound. For the well-equipped and well-staffed microanalytical laboratory, however, it is frequently easier to carry out a quantitative determination of the halogen in question.

The well-known Beilstein (4) test is extremely sensitive, and often traces of impurities will give positive results. Thus, a positive Beilstein test should always be confirmed by the silver nitrate test or by a quantitative halogen determination. Some organic compounds containing no halogen have been reported to interfere in the Beilstein test (25,80). Feigl reports a method which is applicable only to nonvolatile organic halides that is not subject to interference from certain organic nitrogen-containing compounds (22).

Bromine and iodine may be identified by oxidation with hypochlorite and extraction into carbon tetrachloride. If both bromine and iodine are present, they may be detected simultaneously by the addition of excess hypochlorite to further oxidize the iodine. A change in color from the characteristic purple color of iodine to the red-brown of bromine will occur in the carbon tetrachloride layer if bromine is also present.

As with any organic compound, the physical properties of a pure material may be useful in the identification of that material. Selected properties of organic halogen compounds can be found in all standard reference works.

The physical chemical methods of organic structure determination are, of course, useful for the identification of specific compounds, with infrared spectroscopy (68), nuclear magnetic resonance (13), and mass spectrometry (51) being particularly useful. The general analytical chemistry and physical properties of the halogens and of many inorganic and organic halogen-containing compounds have been discussed earlier in the treatise (2).

III. SEPARATION OF THE HALOGENS

Occasionally, analyses of mixtures of two or of all three halogens are required. Here the analyst has a choice of separation of the halogens into single-component systems, selective methods, or a method which will determine each of the three simultaneously. In the author's opinion none of the published methods for the simultaneous titration of chlorine, bromine, and iodine is entirely satisfactory, particularly on the microchemical scale. The most useful separations are those based on the ease with which the various halides are oxidized to the free halogen, which is then isolated and determined by any applicable method (65,14).

Mixtures containing chlorine plus either bromine or iodine may be readily analyzed without separation by first measuring the total halogen argentimetrically as halide and then determining either bromine or iodine iodometrically (on a separate aliquot) after oxidation to bromate or iodate.

IV. DECOMPOSITION METHODS

The most important methods for the decomposition of organic compounds are summarized in Table I, together with their useful range of application and pertinent references.

A. THE OXYGEN FLASK METHOD

The oxygen flask combustion method is probably the method most widely used today for the destruction of organic halogen compounds prior to the determination of their halogen content. When applicable, the method is rapid, simple, and easy to apply. However, there are many organic halogen compounds which are only partially decomposed by the

TABLE I

Decomposition Methods for the Determination of the Halogens[a]

Method	Most useful range	Ref.
Carius	Micro, semimicro	12, 35, 72
Oxygen flask	Micro, semimicro	47, 52, 63
Oxygen (catalytic)	Micro	6, 11, 15, 17, 42, 57, 72
Oxygen (empty tube)	Micro	7, 9
Catalytic hydrogenation	Micro	38, 76
Metallic fusion	Micro, semimicro	21
Alkali metals in liquid systems	Semimicro, macro	61, 67, 78, 79
Sodium in ethanol	Semimicro, macro	69
Sodium in liquid ammonia	Semimicro, macro	81
Organic alkali in solution	Semimicro, macro	10, 43

[a]This table is by no means a complete listing of all decomposition methods applicable to organohalogen compounds.

oxygen flask combustion method. It is, in addition, difficult to handle volatile and/or reactive materials by this method. Certain highly halogenated materials tend to distil out of the flame with only partial combustion. Interestingly, considerably more difficulty has been encountered in the application of the oxygen flask combustion method to bromine compounds than to chlorine compounds. It has been established that these problems arise from incomplete decomposition of the sample rather than from the formation of higher oxidation states of bromine (30).

The oxygen flask method was originated in 1893 by Hempel (28), who applied it to the determination of sulfur in macro (0.5 g) samples using 10-liter combustion flasks. The method, on a macro scale, was applied to the determination of chlorine (49) but gained little attention until it was modified by Mikl and Pech (52) to be used on a semimicro scale, and by Schöniger (63), who reduced the method to the microchemical scale of operation and successfully applied it to the determination of the halogens and sulfur. Lysyj and Zarembro (47) later suggested the use of capsules; this permitted the technique to be applied to some liquid samples.

In the oxygen flask method the sample is wrapped in filter paper (free of the element to be determined), fixed in a container which is attached to the stopper of the combustion flask, ignited, and burned in a flask (250–2000 ml) which has been flushed with oxygen and contains an appropriate absorbing solution. The absorbing solution may be varied to suit the particular finish to be used. The small volume of solution thus obtained is, in general, free from interfering ions and the determination of the halogens may be completed by any suitable method.

B. THE OXYGEN TUBE

1. Catalytic

Among the earliest (and most satisfactory) methods for the decomposition of organic halogen compounds was the method of Pregl (57), which was based upon the earlier work of Dennstedt (17). In this method, the sample is burned in a slow (4–5 ml/min) stream of oxygen and passed over platinum which is maintained at about 700°C. Platinum cylinders, gauze, or platinized quartz have all been proposed for this purpose.

A large variety of solutions have been proposed for absorbing the halogen from the exit gas from the catalytic combustion tube, depending largely upon what end measurement was to be employed. Among these absorbing solutions have been sodium hydroxide (42), sodium carbonate (57), and bisulfite solution (57). Solid absorbers have also been proposed; for example, Bobranski (11) used a boat of barium carbonate, whereas Coulson (16) employed both calcium oxide and sodium carbonate. Excellent results have been obtained using the method given by Belcher and Godbert (6), although care must be taken that the flow rate of the gas through the system does not exceed the recommended value; if it does, halogen will be lost.

Alternatively, in the absence of sulfur and phosphorus, the halogen can be absorbed directly on a silver gauze, the increase in weight being a direct measure of the halogen present in the sample (39,40,53).

2. Empty

In 1943 Belcher and Spooner (9) published the first account of the empty tube combustion method for the rapid decomposition of organic compounds. They used an empty quartz tube maintained at a temperature of approximately 800°C and an oxygen flow rate of about 50 ml/min. Great care was required to assure complete decomposition of the sample. Later, Belcher and Ingram (7) described a tube in which the combustion chamber contained baffle plates to assure proper mixing of the sample with the large excess of oxygen. They maintained the temperature in the combustion chamber at 900°C and found, with a 50-ml/min flow rate, that the combustion and absorption were complete in 10–12 min for micro samples.

C. CATALYTIC HYDROGENATION

The catalytic hydrogenation method for the determination of the halogens, in spite of its ease and general applicability, has never become widely used. terMeulen (76), who pioneered catalytic hydrogenation methods, employed a nickel catalyst and recommended a mixture of hydrogen and ammonia in preference to pure hydrogen as a carrier gas. However, Lacourt (38), who reduced the method to the micro scale of operation, used pure hydrogen and recommended nickel chromate as a catalyst and absorption of the hydrogen halide in alkaline solution rather than on barium carbonate (76). A major advantage of this method is that all of the halogens are obtained in the halide form. Sulfur is, of course, reduced to hydrogen sulfide and must be removed from solution prior to the determination of the halogens.

D. WET COMBUSTION METHODS

1. Carius Method

The Carius digestion (12) is among the most reliable and most universally applicable procedures for the decomposition of organic compounds. The basis of the Carius halogen method is the digestion of the organic halogen compound with fuming nitric acid in the presence of silver nitrate. The reaction is carried out in a sealed, heavy-walled tube at 250°C for a period of 7–8 hr. Such treatment produces the corresponding silver halide which is then determined gravimetrically. Some attempts have been made to complete the determination volumetrically (35), but these methods have not gained wide acceptance.

2. Zacherl and Krainick Method

An interesting microchemical method for the determination of chlorine and bromine was described by Zacherl and Krainick (88), who determined the halogens via alkalimetric titration following their liberation as the free halogen from a mixture of sulfuric acid, potassium dichromate, and silver dichromate. The halogen is absorbed in a standard solution of NaOH–H_2O_2 in which it is reduced to the sodium halide, consuming hydroxide according to equation (1).

$$Cl_2 + 2NaOH + H_2O_2 \rightarrow 2NaCl + 2H_2O + O_2 \quad (1)$$

Titration of the excess base can be carried out by any convenient method or, if desired, the halide ion generated may be directly determined by any of the usual argentimetric end measurements. In a similar procedure, Vieböck (83) utilized silver persulfate as a catalyst for the oxidation of the halides to free halogen. In both procedures, iodine remains in solution as iodate and thus is not determined. This fact, however, affords a convenient method for the separation of chlorine and bromine from iodine.

E. FUSION METHODS

The use of fusion methods, both peroxide fusion and alkali metal fusion, has declined markedly with the advent of some of the more rapid and less demanding decomposition methods. These methods offer definite advantages in that very few organic halogen compounds cannot be successfully decomposed by at least one of the fusion techniques. The fusion methods are also readily applied to samples which boil below room temperature or exhibit a high vapor pressure. Here the samples may be transferred by any suitable method into a soft glass capillary, which is then sealed and weighed. The capillary containing the sample is sealed into the bomb and the analysis carried out in the usual fashion, taking care to add sufficient reagent to react with the capillary as well as the sample.

1. Peroxide Fusion

Nearly all organic halogen compounds can be successfully decomposed by fusion with sodium peroxide. Many investigators (3,20) prefer to use an "accelerator," which is generally a mixture of potassium nitrate and sucrose. However, it is equally as effective to heat only the sample with sodium peroxide at temperatures near 600°C for about 2 hr in a muffle furnace. Under these conditions, chloride only is obtained from chlorine compounds, and a mixture of bromide and bromate (which must, of course, be reduced prior to the determination of bromine (73)) is formed with bromine compounds. Iodate is quantitatively formed from iodine-containing samples permitting direct iodometric determination of iodine.

Bombs for this purpose should be constructed of nickel and consist of a cup in which the fusion is carried out, a cap, a washer, and a suitable means of keeping the assembly gastight during the reaction. Satisfactory bombs are readily available from the usual commercial sources.

2. Alkali Metal Fusion

In certain cases, particularly with perhalogenated organics and/or those containing a high percentage of fluorine, complete decomposition can often be difficult to achieve. In these cases, fusion of the sample with potassium metal at 600°C for 2–4 hr in a nickel bomb (46) will nearly always result in total destruction of the sample. By such treatment, all of the halogens are obtained as the halide, facilitating the completion of the analyses.

The same bombs used for peroxide fusions may be used for the alkali metal fusions. Of course, great care must be taken in working with potassium metal and in assuring its complete destruction after opening the bomb.

As an alternative, heavy-walled Pyrex, Vycor, or silica tubes may be used. This procedure offers a definite advantage in that, short of an explosion, the sample cannot escape from the tube prior to reaction. Elving and Ligett (21) recommended that the tube be dried with anhydrous ether and evacuated prior to sealing. Under these conditions there is virtually no danger of explosion.

F. ALKALI METAL SUSPENSION, SOLUTIONS, AND ADDUCTS

The use of sodium metal suspended in ethanol as a reagent for the reduction of organic halogen compounds to the corresponding halide was originally proposed by Stepanow (69) in 1906. Many solvents (61,67,79) and metals other than sodium (78) have been proposed by later workers. Rauscher proposed the use of a mixed solvent (ethanolamine–dioxane) at reflux temperature in the most successful modification of this decomposition procedure (59).

Others have proposed the use of sodium biphenyl (43) or sodium naphthalene (10) as reagents for the decomposition of organic halogen compounds. Sodium in liquid ammonia has also been used (81). In general these methods are slow, are applicable to only a limited number of materials, and are not generally satisfactory on a micro scale. Thus, none has gained wide acceptance for the quantitative determination of the halogens in organic compounds.

V. METHODS FOR THE DETERMINATION OF THE HALOGENS

A. GRAVIMETRIC METHODS

All the halogens can be determined gravimetrically as their silver salts, with the determination of iodine being somewhat less satisfactory than that of chlorine and bromine because of the smaller gravimetric factor. Collaborative studies reported by Steyermark (70,71,73,74) have confirmed that the precision of the gravimetric micro method following the Carius digestion is more satisfactory for chlorine and bromine than for iodine. Procedures for the gravimetric determination of the halogens can be found in nearly all organic quantitative analysis textbooks (6,54,57,72). The gravimetric method may be applied on the macro, semimicro, and micro scales with excellent results.

B. TITRIMETRIC METHODS

Titrimetric methods are, in the opinion of the author, to be preferred wherever they can be applied with a precision and accuracy comparable to those attainable by gravimetric methods. In general, they are more rapid and require considerably less exacting technique than gravimetric methods to obtain results as satisfactory.

There exists a large variety of excellent titrimetric methods for the determination of the halogens. The method of choice must be selected after taking into consideration such facts as the scale of operation, possible impurities, other constituents of the sample, apparatus available, and the skill of the individual carrying out the analysis.

1. Argentimetric Titrations

a. Indicator Methods

In general, few of the indicator methods have found acceptance on the microchemical scale and thus are rarely used in organic analysis. However, if one can carry out the analysis on the semimicro or macro scale, and chooses to do so, a number of excellent indicator methods are available for the titration of the halogens.

b. The Volhard Method

Undoubtedly one of the most satisfactory argentimetric titration methods employing an indicator is the Volhard titration (5,27,43,46,50,77) in which an excess of silver nitrate is added. The silver chloride is removed by filtration, and the excess silver ion is back-titrated with a standard ammonium thiocyanate solution using a saturated ferric alum indicator solution. The endpoint is indicated by the first permanent formation of a buff color. Alternatively, the silver chloride may be coagulated by the addition of nitrobenzene and the titration carried out without separation of the silver chloride. Bromide and iodide may be similarly determined except that in the case of these elements the analysis may be completed without filtration or addition of nitrobenzene.

c. Absorption Indicators

Eosin (11), fluorescein (7,11,32), and dichlorofluorescein (7,32) have all been used as absorption indicators for the titration of halide ions liberated from organic compounds. However, absorption indicator methods are generally less satisfactory than other methods because of the adverse effect of neutral salts. Thus, care must be taken to work at a reasonably constant ionic strength. Therefore, absorption indicator methods should not be applied folllwing any of the fusion methods of decomposition.

d. Other Argentimetric Methods

The well-known Mohr method which is in common use in inorganic analyses is rarely applied in organic analysis, possibly because it is not applicable to the micro or semimicro scales of operation.

Among other titrimetric methods are the bromphenol blue method (64,67) for chlorine and the brilliant yellow method (82) for bromide. Kirsten and Alperowicz (35,36) obtained excellent results on the micro and semimicro scale by titrating an acidic alcoholic (ethanol) solution of the halide with ethanolic silver nitrate using mercuric chloride or bromide and diphenylcarbazide as an indicator.

2. Electrometric Methods

a. Potentiometric Titration

Perhaps the most satisfactory methods for the titration of the halogens are the electroanalytical methods. Of these, the most widely used methods involve potentiometric detection of the equivalence point. The indicating electrode may be either a silver or a silver–silver halide electrode, while the reference may be any suitable reference electrode. The mercury–mercurous sulfate electrode is excellent for this purpose since it need not be isolated from the solution by a bridge as is necessary when any of the usual calomel reference electrodes are employed. The details of one such system have been given by Huffman (29), who chose to carry out the titration in aqueous solution. A much sharper potential break at the equivalence point is obtained, particularly with chloride, if one carries out the titration in a solution containing 50–80% acetone (44) or alcohol (56).

Numerous automatic devices have been described for potentiometric titration of the halides with standard silver nitrate solutions. Huffman (29) has described a system based upon potentiometric titration to the apparent equivalence point as described by Kolthoff and Kuroda (37). Instruments to carry out automatic potentiometric titrations are available from several commercial sources.

b. Amperometric Titration

The amperometric titration of the halogens with a standard solution of silver nitrate offers an excellent, rapid, reliable, and (in the opinion of the author) insufficiently used method for the determination of the halogens. The most satisfactory conditions employ a pair of silver wire electrodes polarized by the application of approximately 0.25 V between the two electrodes. Under these conditions, oxygen need not be removed from solution. The endpoint can be located by the usual method of plotting points in amperometric titrations (45), by the dead stop method (23), or by automatic recording of the amperometric titration curve (55).

3. Methods Utilizing Mercury

a. Mercuric Oxycyanide Method

Chloride (7,83), bromide (83), and iodide (31) have all been determined by the mercuric oxycyanide method. The method, which is based on reaction (2)

$$2NaX + 2HgOCN \rightarrow Hg(CN)_2 + HgX_2 + 2NaOH \qquad (2)$$

can be applied very accurately at all levels of operation. The alkali released may, of course, be titrated in any convenient manner with standard acid. The modified method of Belcher (8) has been shown to give excellent results and is to be highly recommended.

b. Mercurimetric Methods

Many procedures for the titration of chloride and bromide with mercury have been published (7,18,34,35). However, these methods appear to offer little or no advantage over argentimetric methods (34,35).

4. Coulometric Titration

The coulometric generation of a silver ion offers a convenient method for the addition of silver ions to a solution containing any of the halide ions. Any convenient (amperometric or potentiometric) endpoint detection method can be applied. Lingane (44), Sundberg et al. (75), and Olson and Krivis (56) have studied this system extensively.

Equally as precise is the coulometric titration of the halogens with electrolytically generated mercurous ion, proposed by Przybylowicz and Rogers (58). They use amalgamated silver or gold electrodes, which when freshly prepared behave as pure mercury anodes and are considerably more convenient to use than the classical mercury pool electrodes. The equivalence point is determined potentiometrically using a mercury indicating electrode. It is necessary that the titration of chloride and bromide be carried out in 80% methanol to decrease the solubilities of the mercurous halides. The relative error of the method was approximately ±1% for all of the halogens.

C. AMPLIFICATION METHODS

1. Iodine

Iodine, while it may be determined either gravimetrically or by argentimetric titration, is best determined by oxidation to iodate followed by the liberation of iodine upon the addition of iodide to the acidified iodate solution. The iodine is then determined by titration with standard thiosulfate solution.

Bromine is the oxidizing agent of choice (41), although chlorine water (26), ozone (87), and permanganate (24) have also been used. The oxidation with bromine is best carried out in an acetate buffer (33) and the excess bromine removed by the addition of formic acid (84). The method is satisfactory at all levels of operation and has been applied to the determination of traces of iodine by employing a spectrophotometric finish (66).

An excellent procedure for the determination of trace quantities of iodine was recently published by Malmstadt (48), who modified the earlier work of Sandell and Kolthoff (62). The method utilizes the catalytic effect of iodine (either as iodide or iodine) upon the reaction between cerium(IV) and arsenic(III) (3)

$$2Ce(IV) + As(III) \xrightarrow[\text{catalyst}]{\text{Iodine}} 2Ce(III) + As(V) \quad (3)$$

The rate of disappearance of cerium(IV) ion is measured spectrophotometrically and iodine is determined from a standard curve. As little as 0.01 μg of iodine may be determined with a relative error of about $\pm 2\%$. The method is rapid and ideally suited to running large numbers of routine analyses.

2. Bromine

Bromine can be very precisely and conveniently determined by oxidation to bromate (1,85,86) utilizing hypochlorite or chlorine water as the oxidizing agent. Chlorine, of course, does not interfere, but iodine is quantitatively determined. Thus, the method is of considerable use in the determination of bromine (or iodine) and chlorine in the same sample—the sum may be determined argentimetrically and bromine (or iodine) only by the iodometric method after oxidation with hypochlorite; chlorine may then be determined by difference.

D. RECOMMENDED DECOMPOSITION PROCEDURES

Where applicable, the oxygen flask combustion method is to be recommended above all others because of its ease and simplicity and also because a small volume of solution of known composition is obtained for analysis. For the determination of the halogens, 10 ml of 0.1N sodium hydroxide has been found to be a satisfactory absorbing solution in all cases.

Care must be taken to provide sufficient oxygen for complete combustion of the sample. The author recommends the use of 1000-ml flasks for ordinary micro samples (i.e., 5-mg samples). For larger samples (50–100 mg) or for liquids burned in capsules, 1000- or even 2000-ml flasks are recommended. The flasks may be filled with oxygen by simply passing a brisk stream of oxygen through the flask for an appropriate time or, more reliably, by the displacement of distilled water from the flask by oxygen. In the latter method, care must be taken to avoid contamination with halogen ions.

Remote combustion systems are commercially available to circumvent the potential hazard of explosion. As a precaution, it is advisable to carry out the combustion behind a safety shield. However, since oxygen is consumed and the combustion products are absorbed, there is very little danger of explosion. Only two explosions have occurred in nearly ten years of experience with the method in this laboratory, and both of these resulted from an organic liquid being inadvertently added to the combustion flask instead of the aqueous absorbing solution.

The sample is transferred to the filter paper holder using a platinum weighing scoop or a microchemical weighing stick. The paper, folded as shown in Figure 1, is inserted into the holder attached to the stopper of the combustion flask. The paper wick is lighted in a burner and the stopper is rapidly inserted into the oxygen-filled flask. The flask is inverted and held tightly together. When combustion is complete, the contents of the flask are shaken briskly to wash down the holder. The flask is then set aside to permit complete absorption of the combustion products. Some investigators prefer continuous mechanical shaking during the absorption period. After about 10 to 15 min, the flask is opened and the stopper and the sample holder are carefully washed, allowing the wash solution to run into the flask. The solution is then transferred to a container suitable for the completion of the determination. Following this transfer, the combustion flask is clean and ready for reuse. It need not, and indeed should not, be washed further, since this only increases the possibility of contaminating the combustion flask for the next determination.

If the determination is to be completed by titration as the halide ion, it is necessary that care be taken to reduce any higher oxidation states of bromine and iodine formed during the combustion. Many procedures call for the addition of hydrogen peroxide to the absorbing solution (or following

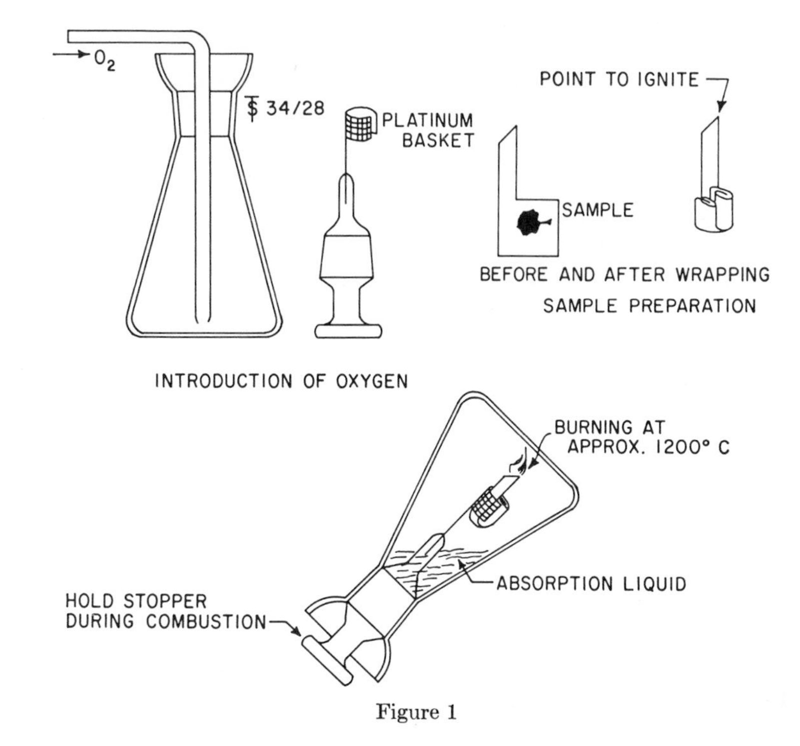

Figure 1

the combustion) in the determination of chlorine. This has been found to be completely unnecessary and, in fact, undesirable (56). Elimination of the addition of hydrogen peroxide in the chloride determination makes unnecessary the boiling to destroy the hydrogen peroxide prior to completion of the titration. Thus, the procedure is both shortened and made more precise. Alkaline hydrogen peroxide (63) and hydrazine sulfate (19) solutions are often recommended for the reduction of the higher oxidation states of bromine or iodine to the halide. However, the author favors the use of an alkaline bisulfite solution to accomplish this reduction (63,73).

E. RECOMMENDED METHODS FOR THE DETERMINATION OF THE HALOGENS

The determination of the halogens, once decomposition of the sample has been achieved, is relatively straightforward. Few interferences occur except in the relatively rare cases of mercury and silver compounds containing one of the halogens and those compounds containing more than one halogen. A small volume of relatively pure halide solution results from the usual microchemical decomposition methods. Thus, nearly any method suitable for use on a microchemical scale may be applied to the determination of the halogens. The reader is referred to Part II, Section A, Volume 7, p. 335 of this Treatise for additional methods (2) which have been employed chiefly in inorganic analyses, but can, in many cases, equally well be applied to the solutions resulting from the decomposition of organic compounds.

1. The Carius Method

In recent years, the Carius method, which was once in nearly universal use, has largely been replaced by the more rapid and less demanding oxygen flask combustion as the method of choice for the decomposition of the sample.

However, the gravimetric Carius method should be considered whenever a referee technique is called for. In the hands of a skilled microchemical technician, the method is capable of yielding excellent results on nearly all types of organic halogen compounds. This nearly universal applicability and the fact that all the halogens are determined in the same fashion are among the major advantages of the procedure.

For the experimental details of the Carius method, the reader is referred to a book by Steyermark (72) which includes an excellent, detailed, and copiously referenced treatment of the method.

2. Potentiometric Titration

Potentiometric titration procedures can be applied to the halide solutions resulting from any of the previously discussed methods of sample decomposition.

Procedure

Transfer the reduced sample solution from the oxygen combustion flask to a 150-ml beaker containing a Teflon-covered magnetic stirring bar, with 65 ml of denatured ethyl alcohol or acetone. Add 5 ml of 1*M* nitric or perchloric acid and a few drops of nonionic detergent (or gelatine solution). Insert a silver or silver halide indicating electrode and a mercury–mercurous sulfate reference electrode (if a saturated calomel electrode is employed, it should be isolated from the solution by an agar–potassium nitrate bridge). Stir the solution magnetically and titrate with standard silver nitrate solution. The endpoint of the titration is taken at the point at which $\Delta E/\Delta$ ml is a maximum from the plot of E vs. ml of silver nitrate added.

Automatic determination of the equivalence point is, of course, possible if suitable equipment is available.

Calculations

$$\% \text{ X} = \frac{\text{ml AgNO}_3 \times N \times \text{meq. wt. Hal} \times 100}{\text{mg sample}}$$

where Hal = Cl, Br, or I.

3. Constant-Current Coulometric Titration

The recommended procedure employs potentiometric detection of the equivalence point (44,56). For systems using amperometric endpoint detection, see Cotlove et al. (15) and Richter (60).

Any constant-current source capable of delivering between 10 and 30 mA is suitable. The generating anode is a spiral of silver wire and the cathode is a bright platinum electrode isolated from the sample solution by a medium glass frit (electrolyte should flow slowly from the cathode chamber into the solution). The reference pair consists of a silver indicating electrode and a saturated calomel reference electrode isolated from the sample solution by an agar–potassium nitrate bridge. The solution is stirred magnetically.

Transfer a suitable reduced aliquot of the sample solution or the contents of the combustion flask to a 150-ml beaker containing a Teflon-covered magnetic stirring bar with 65 ml of denatured ethyl alcohol. Add 5.0 ml of dilute nitric acid (1:10) and titrate the sample using a generation current of 20 mA and a silver–saturated calomel indicating pair. The cutoff potential is selected by running a standard sample of the halogen being determined and plotting the potential between the indicating electrodes as a function of time. The inflection point is chosen as the cutoff potential.

Titrate a reagent blank prepared in exactly the same fashion.

$$\% \text{ X} = \frac{(t_s - t_B)i \times \dfrac{(\text{atomic wt. of X})}{F} \times 100}{\text{Sample weight, mg}}$$

where t_s = time in seconds for sample titration
t_B = time in seconds for blank titration
i = generating current in milliamperes
F = the Faraday

4. Amperometric Titration

Transfer a suitable aliquot of the reduced sample solution to a 100-ml beaker which contains a magnetic stirring bar. Adjust to pH 3–4 and then add 1 ml of 0.1% gelatine (or a suitable nonionic detergent) and 50 ml of denatured alcohol or acetone. Insert a pair of silver electrodes polarized by about 0.2 V and titrate with standard silver nitrate solution. The endpoint is found by plotting i vs. milliliters of silver nitrate and extrapolating the straight-line segments of the curve until they intersect.

Calculation

$$\% \text{ Halogen} = \frac{\text{ml AgNO}_3 \times N \times \text{meq. wt. Hal} \times 100}{\text{Sample weight, mg}}$$

5. Iodometric Determination of Bromine

Add a few drops of aqueous methyl red solution to the solution in the oxygen combustion flask (or transfer a suitable aliquot of the solution resulting from any suitable sample decomposition) and neutralize using $2N$ sulfuric or hydrochloric acid. Buffer the solution by adding 5 ml of 20% potassium dihydrogen phosphate solution and then add 5 ml of 5–6% sodium hypochlorite solution (Clorox). Cover loosely and heat at 90–95°C (do not boil) for 8–10 min. Remove the flask from the hot plate, add 5 ml of 50% sodium formate solution, and allow the flask to cool with occasional swirling to remove all traces of free chlorine. Add 5 ml of $10N$ sulfuric acid, a few drops of 5% ammonium molybdate solution (catalyst), and about 0.5 g of potassium iodide. Stopper the flask tightly and titrate the liberated iodine after 5–10 min using $0.01N$ sodium thiosulfate solution. Starch or Thyodene may be added near the endpoint as an indicator. Since commercial hypochlorite and hypochlorite solutions always contain a small amount of bromine, it is imperative that a blank be carried out in exactly the same manner.

Bromine is calculated as follows:

$$\% \text{ Br} = \frac{\text{ml Na}_2\text{S}_2\text{O}_3 \times N \times 13.318 \times 100}{\text{mg sample}}$$

6. Iodometric Determination of Iodine

Add a few drops of aqueous methyl red solution to the solution in the oxygen combustion flask (or transfer a suitable aliquot of the solution resulting from any suitable sample decomposition) and neutralize using

2N sulfuric acid. Add 1 ml of a saturated aqueous solution of bromine (or bromine in acetic acid) and allow the stoppered flask to stand for 5–10 min. Destroy the excess bromine by the dropwise addition of 90% formic acid. Check for complete removal of bromine by the addition of a drop of methyl red solution. Add 5 ml of 2N sulfuric acid and 0.2–0.3 g of potassium iodide. Titrate the liberated iodine with 0.01N sodium thiosulfate solution, adding starch indicator (or Thyodene) as the end-point of the titration is approached. Correct for a blank determined in the same manner and calculate the percent iodine as follows:

$$\% \text{ I} = \frac{\text{ml } Na_2S_2O_3 \times N \times 21.151 \times 100}{\text{mg sample}}$$

7. Analysis of Chlorine–Bromine and Chlorine–Iodine Mixtures

Combust two samples by any suitable method or take measured aliquots of a single sample. Argentimetrically titrate one sample for total halogen by any of the previously mentioned methods. Determine bromine (or iodine) via the iodometric method in the usual fashion.

The calculations are as follow:

$$\% \text{ Br} = \frac{\text{ml } Na_2S_2O_3 \times N \times 13.318 \times 100}{\text{mg sample}}$$

$$\% \text{ I} = \frac{\text{ml } Na_2S_2O_3 \times N \times 21.151 \times 100}{\text{mg sample}}$$

$$\text{meq Br (or I)} = \frac{\text{mg Br (or I)}}{\text{meq Br (or I)}}$$

$$\text{Total meq halogen} = \text{ml titrant} \times \text{N}$$

$$\text{meq Cl} = \text{Total meq} - \text{meq Br (or I)}$$

$$\% \text{ Cl} = \frac{\text{meq} \times 35.457 \times 100}{\text{Sample weight (mg)}}$$

8. Analysis of Samples Containing Bromine–Iodine

Transfer two suitable aliquots of the sample solution resulting from the reduction of the solution obtained from the oxygen flask combustion. Treat one aliquot with bromine as in the iodometric determination of iodine and calculate the milliequivalents of iodine in the sample. Treat the other aliquot with hypochlorite as in the iodometric determination of bromine and determine the total milliequivalents of bromine and iodine. Bromine may then be calculated in the usual manner.

REFERENCES

1. Alicino, J. F., A. Crickenberger, and B. Reynolds, *Anal. Chem.*, **21**, 755 (1949).
2. Armstrong, G. W., H. H. Gill, and R. F. Rolf, "The Halogens," in *Treatise on Analytical Chemistry*, Part II, Vol. 7, I. M. Kolthoff and P. J. Elving, Eds., Interscience, New York, 1961, p. 335.
3. Beamish, F. E., *Ind. Eng. Chem., Anal. Ed.*, **5**, 348 (1933); **6**, 352 (1934).
4. Beilstein, F., *Chem. Ber.*, **5**, 620 (1872).
5. Belcher, R., J. E. Fildes, and A. M. G. Macdonald, *Chem. Ind. (London)*, **1955**, 1402.
6. Belcher, R., and A. L. Godbert, *Semimicro Quantitative Organic Analysis*, 2nd ed., Longmans Green, London, 1954.
7. Belcher, R., and G. Ingram, *Anal. Chim. Acta*. **7**, 319 (1952).
8. Belcher, R., A. M. G. Macdonald, and A. J. Nutten, *Mikrochim. Acta*, **1954**, 104.
9. Belcher, R., and C. E. Spooner, *J. Chem. Soc.*, **1943**, 313.
10. Benton, F. L., and W. H. Hamill, *Anal. Chem.*, **20**, 269 (1948).
11. Bobranski, B., *Zh. Anal. Chem.*, **84**, 225 (1931); **99**, 108 (1934).
12. Carius, L., *Ann.*, **116**, 1 (1860); **136**, 129 (1865); *Chem. Ber.*, **3**, 697 (1870).
13. Chamberlain, N. F., "Nuclear Magnetic Resonance and Electron Paramagnetic Resonance," in *Treatise on Analytical Chemistry*, Part I, Vol. 4, I. M. Kolthoff and P. J. Elving, Eds., Interscience, New York, 1963, p. 1885.
14. Conway, E. J., *Microdiffusion Analysis and Volumetric Error*, 4th ed., Lockwood, London, 1957, Chapters XLI–XLV.
15. Cotlove, E., H. V. Trantham, and R. L. Bowman, *J. Lab. Clin. Med.*, **51**, 461 (1958).
16. Coulson, A. F., *Analyst.*, **67**, 47 (1942).
17. Dennstedt, M., *Anleitung zur Vereinfachten Elementaranalyse*, Meisner, Hamburg, 1919.
18. Dubský, J. V., and J. Trtílek, *Mikrochemie*, **12**, 315 (1933); **15**, 95 (1934).
19. Elek, A., and R. A. Harte, *Ind. Eng. Chem., Anal. Ed.*, **9**, 502 (1937).
20. Elek, A., and D. W. Hill, *J. Am. Chem. Soc.*, **55**, 2550 (1933); **55**, 3479 (1933).
21. Elving, P. J., and W. B. Ligett, *Ind. Eng. Chem., Anal. Ed.*, **14**, 449 (1942).
22. Feigl, F., *Spot Tests in Organic Analysis*, 5th ed., Elsevier, Amsterdam, 1956, p. 81.
23. Foulk, C. W., and A. T. Bawden, *J. Am. Chem. Soc.*, **48**, 2045 (1926).
24. Gautier, J. A., *Bull. Soc. Chim. France*, **4**, 219 (1937).
25. Gilman, H., and J. E. Kirby, *J. Am. Chem. Soc.*, **51**, 1571 (1929).
26. Groák, B., *Biochem. Z.*, **175**, 455 (1926).
27. Gysel, H., *Helv. Chim. Acta*, **24**, 128 (1941).
28. Hempel, W., *Angew. Chem.*, **13**, 393 (1893).
29. Huffman, E. W. D., *Microchem. J., Symp. Ser.*, **2**, 811 (1962).
30. Huffman, E. W. D., and E. C. Olson, unpublished results.
31. Ingram, G., *Analyst*, **69**, 265 (1944).
32. Ingram, G., *Mikrochem., Mikrochim. Acta*, **36/37**, 690 (1951).
33. Kirk, P. L., and K. Dod, *Mikrochemie*, **18**, 179 (1935).
34. Kirsten, W., *Mikrochem., Mikrochim. Acta*, **34**, 149 (1949).
35. Kirsten, W., *Anal. Chem.*, **25**, 74 (1953).
36. Kirsten, W., and I. Alperowicz, *Mikrochem., Mikrochim. Acta*, **39**, 234 (1952).
37. Kolthoff, I. M., and P. K. Kuroda, *Anal. Chem.*, **23**, 1304 (1951).
38. Lacourt, A., *Mikrochemie*, **23**, 308 (1938).
39. Lacourt, A., Ch. T. Chang, and R. Vervoort, *Bull. Soc. Chim. Belges*, **50**, 67 (1941).

40. Lacourt, A., and Ch. T. Chang, *Bull. Soc. Chim. Belges,* **50,** 115 (1941); **52,** 175 (1943).
41. Leipert, T., *Mikrochem. Pregl Festschr.,* 266 (1929).
42. Leipert, T., and O. Watzlawek, *Zh. Anal. Chem.,* **98,** 113 (1934).
43. Liggett, L. M., *Anal. Chem.,* **26,** 748 (1954).
44. Lingane, J. J., *Anal. Chem.,* **26,** 622 (1954).
45. Lingane, J. J., *Electroanalytical Chemistry,* 2nd ed., Interscience, New York, 1958, Chapter XII.
46. Lohr, L. J., T. E. Bonstein, and L. J. Frauenfelder, *Anal. Chem.,* **25,** 1115 (1953).
47. Lysyj, I., and J. E. Zarembo, *Anal. Chem.,* **30,** 428 (1958).
48. Malmstadt, H. V., and T. P. Hadjiioannou, *Anal. Chem.,* **35,** 2157 (1963).
49. Marcusson, J., and H. Döscher, *Chem.-Ztg.,* **34,** 417 (1910).
50. McNevin, W. M., and G. H. Brown, *Ind. Eng. Chem., Anal. Ed.,* **14,** 908 (1942).
51. Melpolder, F. W., and R. A. Brown, "Mass Spectrometry," in *Treatise on Analytical Chemistry,* Part I, Vol. 4, I. M. Kolthoff and P. J. Elving, Eds., Interscience, New York, 1963, p. 1959.
52. Mikl, O., and J. Pech, *Chem. Listy,* **46,** 382 (1952); **47,** 904 (1953).
53. Mitsui, T., and H. Sata, *Mikrochim. Acta,* **1956,** 1603.
54. Niederl, J. B., and V. Niederl, *Micromethods of Quantitative Organic Analysis,* Wiley, New York, 1942.
55. Olson, E. C., and E. W. D. Huffman, unpublished results.
56. Olson, E. C., and A. F. Krivis, *Microchem. J.,* **4,** 181 (1960).
57. Pregl, F., *Quantitative Organic Microanalysis,* 5th ed., (J. Grant, Ed.) Churchill, London, 1946.
58. Przybylowicz, E. P., and L. B. Rogers, *Anal. Chem.,* **28,** 799 (1956).
59. Rauscher, W. H., *Ind. Eng. Chem., Anal. Ed.,* **9,** 296 (1937).
60. Richter, H. L., Jr., *Anal. Chem.,* **27,** 1526 (1955).
61. Ruzhentseva, A. K., and V. S. Letina, *Zh. Anal. Khim.,* **3,** 139 (1948).
62. Sandell, E. B., and I. M. Kolthoff, *Mikrochim. Acta,* **1,** 9 (1937).
63. Schöniger, W., *Mikrochim. Acta,* **1955,** 123; **1956,** 869.
64. Schulz, F., *Collection Czech. Chem. Commun.,* **3,** 281 (1931); **4,** 21 (1932).
65. Scott, W. W., *Standard Methods of Chemical Analysis,* 5th ed., Van Nostrand, New York, 1939.
66. Shahrokh, B. K., and R. M. Chesbro, *Anal. Chem.,* **21,** 1003 (1949).
67. Sisido, K., and H. Yagi, *Anal. Chem.,* **20,** 677 (1948).
68. Smith, A. L., "Infrared Spectroscopy," in *Treatise on Analytical Chemistry,* Part I, Vol. 6, I. M. Kolthoff and P. J. Elving, Eds., Interscience, New York, 1965, p. 3535.
69. Stepanow, A., *Chem. Ber.,* **39,** 4056 (1906).
70. Steyermark, A., *J. Assoc. Office Agr. Chemists,* **40,** 381 (1957).
71. Steyermark, A., *J. Assoc. Office Agr. Chemists,* **41,** 297 (1958).
72. Steyermark, A., *Quantitative Organic Microanalysis,* 2nd ed., Academic Press, New York, 1961.
73. Steyermark, A., and M. B. Faulkner, *J. Assoc. Office Agr. Chemists,* **35,** 291 (1952).
74. Steyermark, A., and M. W. Garner, *J. Assoc. Office Agr. Chemists,* **36,** 319 (1953).
75. Sundberg, O. E., H. C. Craig, and J. S. Parsons, *Anal. Chem.,* **30,** 1842 (1958).
76. terMeulen, H., *Rec. Trav. Chim.,* **41,** 112 (1922); **53,** 118 (1934).
77. Thompson, J. J., and U. O. Oakdale, *J. Am. Chem. Soc.,* **52,** 1195 (1930); **52,** 3466 (1930).
78. Tomicek, O., and K. Petak, *Casopis Ceskoslov. Lekarnictva,* **17,** 309 (1937).

79. Umhoefer, R. R., *Ind. Eng. Chem., Anal. Ed.*, **15**, 383 (1943).
80. vanAlphen, J., *Rec. Trav. Chim.*, **52**, 567 (1933).
81. Vaughn, T. H., and J. A. Nieuwland, *Ind. Eng. Chem., Anal. Ed.*, **3**, 274 (1931).
82. Vecera, M., and J. Bulusek, *Mikrochim. Acta*, **1958**, 41.
83. Vieböck, F., *Chem. Ber.*, **65B**, 493, 586 (1932).
84. Vieböck, F., and C. Brecher, *Chem. Ber.*, **63B**, 3207 (1930).
85. White, L. M., and M. D. Kilpatrick, *Anal. Chem.*, **22**, 1049 (1950).
86. Willard, H. H., and A. H. A. Heyn, *Ind. Eng. Chem., Anal. Ed.*, **15**, 321 (1943).
87. Willard, H. H., and L. L. Merritt, Jr., *Ind. Eng. Chem., Anal. Ed.*, **14**, 489 (1942).
88. Zacherl, M. K., and H. G. Krainick, *Mikrochemie*, **11**, 61 (1932).
89. Zweig, G., Ed., *Analytical Methods for Pesticides, Plant Growth Regulators, and Food Additives*, Academic Press, New York, 1963.

Part II
Section B-2

UNSATURATION

By Stanley T. Hirozawa, *BASF-Wyandotte Corporation, Wyandotte, Michigan*

Contents

Contents (*continued*)

I. INTRODUCTION

Organic compounds of the homologous series with the general formula C_nH_{2n} are called alkenes or olefins. They are frequently called unsaturated hydrocarbons because they do not contain the maximum number of hydrogen atoms found in the saturated hydrocarbons. Unsaturated hydrocarbons may be deficient of more than two hydrogen atoms per mole, e.g., acetylenes, cumulenes, conjugated polyenes, and isolated polyenes. The simplest member of this family of compounds is ethylene, whose structure is written:

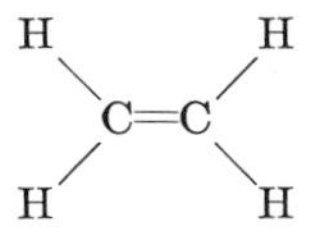

Replacement of one or more of the hydrogen atoms with alkyl and other functional groups gives rise to a large number of unsaturated compounds.

The reactivity of the double bond makes this family of compounds very important industrially and from the standpoint of chemical synthesis. A naturally occurring class of unsaturated compounds contains fats and oils derived from plants and animals. They are used as food, and for the manufacture of paints, varnishes, soaps, and detergents. Alkenes are produced by the cracking of natural gas and petroleum hydrocarbons. The alkenes are used in the manufacture of gasolines, rubber, plastics, and numerous basic chemicals.

The reactivity of unsaturated compounds is determined by the electronic property of the group or groups attached to the olefinic carbons. Thus, this family of compounds possesses a very wide range of reactivities. Their importance and variable reactivity account for the fact that about 3000 papers have been published on analytical methods for unsaturated compounds (34). The complete coverage of this vast literature is beyond the scope of this chapter, which will deal with selected methods which are believed to be most useful to the analyst. For a comprehensive treatment of this subject, see reference 271. Determination of unsaturation by chemical methods is covered in references 322 and 340.

II. TOXICOLOGY

Unsaturated hydrocarbons are considered moderately toxic when inhaled (306). Ethylene can be used as an anesthetic, and when inhaled in sufficient quantity, it can be an asphyxiant. Prolonged or repeated exposure to high concentrations of various olefins has caused certain toxic effects in animals, such as liver damage and hyperplasia of the bone marrow (due to 2-butene). However, no corresponding effects have been discovered in human beings as due to industrial exposure. The diolefins, butadiene and isoprene, are more irritating than paraffins or monoolefins of the same volatility. In general, the olefins are comparatively innocuous materials.

The greatest hazard from the handling of olefins is the danger of fire and explosion. Volatile olefins should be handled under adequate ventilation.

III. PROPERTIES

A. PHYSICAL PROPERTIES

The general physical properties of the alkenes are much the same as those of the corresponding alkanes (258). Alkenes of four carbons or less exist as gases while those with more than 18 carbons are solid under normal laboratory conditions. The solubility of the lower alkenes in water, though slight, is considerably greater than that of the alkanes, because the higher concentration of electrons in the double bond leads to greater attraction for the dipolar water molecules. The solubility is greatly enhanced by dilute acids because of the increased attraction of the positively charged hydronium ions by the double bonds.

The refractive indices of alkanes, alkenes, and conjugated olefins differ quite markedly. Refractometric properties have been used in the characterization and quantitative analysis of mixtures of hydrocarbons, vegetable oils, and synthetic rubber (271).

B. CHEMICAL PROPERTIES

1. Reactions

The unsaturated hydrocarbons are very reactive compared to the saturated hydrocarbons. This reactivity can be attributed to the nature of the carbon–carbon double bond. The usual graphical representation of a double bond shows two equal bonds. Actually the double bond consists of a σ-bond which is similar to a single carbon–carbon bond (80 cal/mol) and a weaker π-bond (60 cal/mol) (258,271). The π-electrons are held less firmly between the carbon nuclei and are therefore more readily available to electrophilic (cationoid) substances. Thus, unsaturated aliphatic compounds react readily with halogens, ozone, other oxidizing agents, and strong acids. Nucleophilic (anionoid) substances such as weak acid and ammonia generally do not react (271). Nearly all the chemical methods for the determination of unsaturation are based on the reactivity of the π-electrons.

Some of the characteristic reactions of unsaturated compounds are addition of halogens, hydrogen, bisulfite, and morpholine; oxidation by peracid, ozone, and permanganate; mercuration; sulfonation; and nitration. These reactions are given in Section VI.

Another characteristic reaction, polymerization, is useful industrially but may be detrimental analytically. Polymerization of unsaturated monomers in the sample itself or in the reaction medium can lead to low results. This phenomenon has been observed in our laboratory with certain unsaturated compounds in 90:10 acetic acid–water medium.

2. Inductive Effects

The reactivity of unsaturated compounds is related to the availability of the π-electrons. Thus, when the hydrogen atoms of ethylene are substituted by more electropositive groups—e.g., phenyl or methyl—the rate of electrophilic addition is increased because of the increase in electron density at the double bond. Conversely, when the substituted groups are more electronegative than hydrogen—e.g., halogen or carbonyl—the rate of addition is decreased because of decrease in electron density at the double bond. This effect is illustrated by the relative rate of bromination of the following substituted ethylenes determined in CH_2Cl_2 solution at $-78°C$ (22):

$(CH_3)_2C{=}C(CH_3)_2$	$(CH_3)_2C{=}CH_2$	$C_6H_5CH{=}CH_2$	$CH_3CH{=}CH_2$	$CH_2{=}CH_2$
14.0	5.5	3.4	2.0	1.0

$CH_3CH{=}CHCOOH$	$CH_2{=}CHCOOH$	$CH_2{=}CHBr$
0.3	small	small

The electron-withdrawing power of the carboxylic acids is reversed by neutralizing with a base. Thus the $CH_2{=}CHCOO^-$ ion is about 300 times more reactive than the undissociated acid (371).

The quantitative addition of halogen to conjugated diolefins is difficult. The first mole of halogen adds readily, predominantly in the 1,4 position and partly in the 1,2 position. The second mole adds on very slowly because of the inductive effect of the halogen atoms on the carbons adjacent to the remaining double bond. A similar effect is shown by acetylene.

Strongly electronegative groups such as carbonyl can deplete the electron density of the double bond sufficiently to change its bond character from nucleophilic to electrophilic. These electrophilic unsaturated compounds react readily with nucleophilic substances such as bisulfite and morpholine. The inductive effect of electronegative groups on ethylene is shown by the sequence of reactivity with morpholine (322):

$$H_2C{=}CH{-}C{\equiv}N \gtrsim H_2C{=}CHCOOR \gg H_2C{=}CHCOOH > H_2C{=}CHCOR \gtrsim H_2C{=}CHCOONa$$

Substitution of alkyl groups on the α- and β-carbons increases the electron density of the double bond, thereby retarding the reaction with nucleophilic reagents such as morpholine.

When the double bond is shielded from an electronegative group by one or more carbons, the inductive effect becomes less apparent. For example, allyl cyanide can be determined by bromination (322). Allyl cyanide can also be determined by the morpholine method because it isomerizes to the corresponding α,β-unsaturated compound under the conditions of the method.

Thus it appears that the choice between electrophilic or nucleophilic reagent depends on the electronegativity of the substituent(s) and its (their) proximity to the double bond.

3. Steric Effects

The double bond in alkenes restricts the rotation of the molecule about this bond. All alkenes in which one hydrogen on each carbon of ethylene is substituted by some other group could exist as *cis* and *trans* isomers.

cis-alkene *trans*-alkene

The initial attack by an electrophilic reagent occurs at the π-bond. The reagent approaches in a direction perpendicular to the double bond (348). In the *cis* isomer, the approach to one side of the molecule is unhindered, while in the *trans* isomer, both sides are partially hindered. Thus hydrogenation and bromination of *cis* isomers are faster than for the corresponding *trans* isomer. Another example is that the addition of mercuric acetate to terminal unsaturation, to cyclic alkenes, and to internal unsaturation with the *cis* configuration is rapid, while the mercuration of the *trans* isomer is very slow.

In chromatographic analysis, where the interaction of the π-bonds with an electrophilic agent on the column is involved, the *cis* isomers are held more tightly and therefore move more slowly than the *trans* isomers. Certain solid materials with controlled pore size can separate the skeletal isomers; for example, molecular sieves can absorb the normal olefins, whereas the branched olefins are passed on because they are too bulky to enter the pores.

4. Catalysis

Because of the extremely wide range of reactivities among unsaturated compounds, numerous reagents must be used, each covering a certain section of this range. A specific reagent for a group of similar unsaturated compounds reacts at a different rate for each compound because of slight variations in the π-electron density. Catalysis is applied in many methods to hasten the slow reaction.

The catalysis of halogenation of double bonds by mercury(II) is well known. Various metal and metal oxide catalysts are absolutely essential in the hydrogenation of unsaturation. These and other examples of catalysis will be discussed in the appropriate sections.

C. ELECTROCHEMICAL PROPERTIES

The polarographic behavior of unsaturated compounds depends on their electronic structure. Isolated double bonds are not reducible, while double or triple bonds that are either conjugated or twinned with another double or triple bond, a benzene ring, carbonyl, carboxyl, or ester groups are reducible. The polarography of unsaturated compounds has been covered thoroughly by Kolthoff and Lingane (200) and Brezina and Zuman (59a,380) and reviewed by Elving (118) and Page (264).

D. SPECTRAL PROPERTIES

1. Infrared

The absorption bands of unsaturated compounds in the infrared region result principally from the C═C stretching vibrations and the C═C—H stretching and deformation vibrations. The relationship between bond structure and vibration frequencies is given in Table I. The intensity of the absorption band of the C═C stretching vibration which occurs around 6.0–6.1 μ is dependent on the structure of the molecule. Since infrared absorption depends on changes in dipole moment, the absorption band is strong for unsymmetrical olefins, such as α-olefins, and very weak for symmetrical olefins such as ethylene and *trans*-2-butene.

Conjugated diolefins absorb strongly near 6.25 μ and the allenes absorb strongly at 5.1 μ due to unsymmetrical stretching of the paired double bonds.

Olefins with at least one hydrogen atom on a double-bonded carbon absorb in the infrared region due to C—H vibrations. The C—H stretching vibrations in the 3.2–3.5 μ region are intense and characteristic, but overlap in part the C—H stretching modes of saturated alkyl groups. The absorption bands due to the in-plane bending vibrations of the C═C—H which occur in the 7–12 μ region are weak and lie in the range of the C—C stretching vibration absorptions. The out-of-plane bending vibrations of the C═C—H absorbs intensely in the 10–15 μ region. These bands are

TABLE I

Relationship Between Bond Structure and Vibration Frequencies (35)

C═C Stretching Vibrations		
Nonconjugated	1680–1620 cm^{-1} (intensity very variable)	
Phenyl conjugated	Near 1625 cm^{-1} (intensity enhanced)	
CO or C═C conjugated	Near 1600 cm^{-1} (intensity enhanced)	
C—H Stretching and Deformation Vibrations		
—CH═CH-(*trans*)	3040–3010 cm^{-1} (m)	CH stretching
	970–960 cm^{-1} (s)	CH out-of-plane deformation
	1310–1295 cm^{-1} (s–w)	CH in-plane deformation
—CH═CH-(*cis*)	3040–3010 cm^{-1} (m)	CH stretching
	Near 4675 cm^{-1}	Combination band (135)[a]
	Near 690 cm^{-1}	CH out-of-plane deformation (correlation uncertain)
—CH═CH_2 (Vinyl)	3040–3010 cm^{-1} (m)	CH stretching
	3095–3075 cm^{-1} (m)	CH stretching
	6220–6100 cm^{-1}	Overtone of CH stretching (135)[a]
	4770 cm^{-1}	Combination band (135)[a]
	995–985 cm^{-1} (s)	CH out-of-plane deformation
	915–905 cm^{-1} (s)	CH_2 out-of-plane deformation
	1856–1800 cm^{-1} (m)	Possible overtone of the above
	1420–1410 cm^{-1} (s)	CH_2 in-plane deformation
	1300–1290 cm^{-1} (sw)	CH in-plane deformation
CR_1R_2═CH_2	3095–3075 cm^{-1} (m)	CH stretching
	895–885 cm^{-1} (s)	Out-of-plane deformation
	1800–1750 cm^{-1} (m)	Possible overtone of the above
	1420–1410 cm^{-1} (s)	CH_2 in-plane deformation
CR_1R_2═CHR_3	3040–3010 cm^{-1} (m)	CH stretching
	840–790 cm^{-1} (s)	CH out-of-plane deformation

[a]All these data were obtained from reference 35 with the exception of the data for the combination and overtone bands occurring in the near-infrared region.

highly specific for some types of olefins in spite of the fact that many C—C stretching and CH_3 rocking vibrations occur in this region. This region is most useful for the determination of the olefin type. Anderson and Seyfreid (23) reported the characteristic bands for the C═C—H out-of-plane deformation vibrations for five classes of olefins (R, R′, R″ are alkyl groups):

Class	Absorption band, μ
RCH═CH_2	10.05 and 10.98 ± 0.2
RR′C═CH_2	11.24 ± 0.02
trans-RCH═CHR′	10.36 ± 0.02
cis-RCH═CHR′	14.0–14.6 (variable)
RR′C═CHR″	11.9–12.7 (variable)

The first three classes give characteristic strong bands while the latter two classes give weaker and variable bands.

The characteristic absorption bands discussed above may be shifted to other wavelengths when one or more of the substituents is not an alkyl group.

The correlations between infrared spectra and structure of olefins have been described by numerous investigators (23,122,191,237,279,320,342), and the characteristic frequencies of olefinic and unsaturated hydrocarbons have been discussed by Bellamy (35) and reviewed by Sheppard and Simpson (319).

The near-infrared region includes wavelengths from 0.7 to 3.0 μ. With a few exceptions, all the absorption bands in the near-infrared region arise from overtones and combinations involving hydrogenic stretching vibrations. Because of the smaller absorptivities of combination and overtone bands, sample paths must be longer than those required for fundamental bands. The fundamental and overtone bands of hydrogenic vibrations can be assigned with little ambiguity because they are highly characteristic, while the identification of combination bands is generally more difficult.

Olefinic compounds yield overtone and combination bands in the near-infrared region (see Table I). The first overtone of the C—H stretching vibrations at 1.61–1.64 μ (6220–6100 cm^{-1}) in the near-infrared region is useful in the determination of vinyl unsaturation (135a). The combination band at 2.14 μ (4675 cm^{-1}) has been used in the determination of *cis*-olefins (135) while the combination band at 2.10 μ (4770 cm^{-1}) has been applied to vinyl unsaturation (135).

The near-infrared region has been reviewed in several publications (134a,187a,187b,367a).

2. Raman Spectroscopy

Raman lines originate from the displacement in frequency of monochromatic ultraviolet or visible light which is scattered by organic molecules. These displacements are usually expressed in cm^{-1} and correspond to the vibration frequencies of the atoms in a molecule. Therefore, Raman spectra are very similar to infrared spectra. However, the strength of a fundamental frequency usually is not equivalent in the two techniques because Raman lines arise principally from changes in polarizability resulting from vibrations, while infrared absorptions arise from changes in dipole moment resulting from vibrations. For example, the infrared absorption due to the C=C stretching vibration is very weak for centrosymmetric molecules. The Raman line arising from the C=C stretching vibrations is intense for all olefins. The Raman frequency shifts occur in the 1650 cm^{-1} region.

The principles and applications of Raman spectroscopy have been described by Hibben (153).

3. Ultraviolet Absorption

Absorptions in the ultraviolet and visible regions are due to excitation of the electrons of organic molecules. The greater the availability, mobility, and resonance of the electrons, the lower is the frequency of radiation absorbed. This trend is shown by the unsaturated compounds in Table II (271). Thus, an isolated double bond absorbs in the 170–200 mμ region while conjugated dienes absorb strongly in the 210–250 mμ region. The highly conjugated carotenes absorb in the visible region.

TABLE II

Effect of Conjugation on the Ultraviolet Absorption of Olefins[a]

Structure	Absorption maxima, m μ
C=C	170–200
$(C{=}C{-})_2$	210–240
$(C{=}C{-})_{10}$, α-carotene	477, 509
$(C{=}C{-})_{11}$, β-carotene	485, 520
$C{=}C{-}C_6H_5$, styrene	244
C=C—C(=O)—	210–240, 300–330
$(C{=}C{-})_2COOH$	233
$(C{=}C{-})_3COOH$	268
$(C{=}C{-})_4COOH$	315

[a]Additional information on absorption maxima of olefins is given in Part II, Volume 13 of this Treatise (249a).

4. Nuclear Magnetic Resonance

Approximately half the known isotopes have nuclei which behave like spinning magnets. When placed in a magnetic field, these nuclei can be caused to absorb radiofrequency energy. The most useful isotopes are ^{1}H, ^{19}F, and ^{31}P because of their high nuclear magnetic strength, relative natural abundance, and spherical distribution of the nuclear charge. ^{1}H is the most useful isotope to the organic chemists and the term *proton magnetic resonance* (PMR) is commonly used.

The general technique is to place a pure sample or a solution of a solid or viscous material in a strong, homogeneous magnetic field. The sample is irradiated with a radiofrequency signal and the intensity of the absorption of the signal is plotted as a function of increasing magnetic field strength.

The magnetic field strength is expressed in three ways: ν (hertz), δ or τ. The relationship among these three ways of expression is as follows. The position of the tetramethylsilane (TMS) absorption peak is arbitrarily selected as the zero point. The majority of organic compounds absorb downfield or at lower magnetic field strength than TMS. The position of any absorption peak is the difference between the absorption frequencies of the unknown and the reference. The negative sign of the absorption frequency is omitted because there are very few protons that absorb upfield from TMS. δ and τ are calculated from ν, the absorption frequency, and the radiofrequency. For example, if $\nu = 120$ Hz and a radiofrequency of 60 MHz is used,

$$\delta = \frac{(120 \text{ Hz})(10^6)}{60 \times 10^6 \text{ Hz}} = 2 \text{ ppm}$$

and

$$\tau = 10 - \delta = 8 \text{ ppm}$$

δ and τ are independent of the radiofrequency while all three will vary with the reference standard.

The position of the absorption peak relative to the reference is commonly called the chemical shift. The chemical shift is used analytically as the primary means of identification of functional groups containing or affecting the observed proton. It also indicates the relative spatial positions of the functional groups. A convenient chart of the characteristic chemical shifts (at 60 MHz with reference to TMS) of various types of organic protons is given on page 16 of reference 42. The positions of the chemical shifts of different types of protons on unsaturated carbons are listed in Table III.

TABLE III

Characteristic Positions of NMR Signals of Protons Attached to Unsaturated Carbons (42)

Proton type	Chemical shift, Hz
$CH_2=$	278–295
$CH=$	302–333
$CH=$, conjugated	350–384
(cyclohexene ring)—H	318–355
$HC\equiv CY$	
$Y = CH_3$	108
$Y = COH$	147–165
$Y = C_6H_5$	183

TABLE IV

Spin-Spin Coupling Constants for Protons on Multiple Bonds (42)

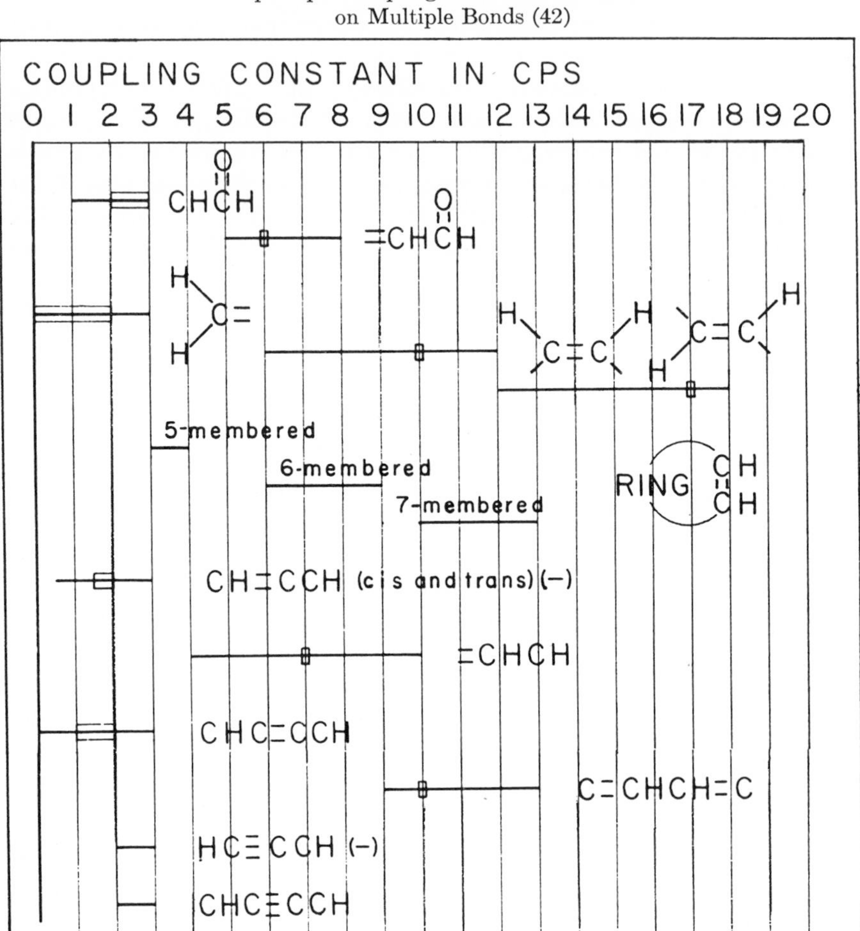

Chemically equivalent protons, which are shielded from other types of protons, e.g., TMS, resonate at a single frequency and give a single sharp band. In the presence of neighboring protons, the band is split by spin–spin interaction between the two groups of neighboring protons. The distance between peaks in a band is called the *coupling constant*. The spin coupling constants for protons on multiply bonded carbons are given in Table IV. If the difference in the chemical shift between two groups of protons is at least six times as great as the coupling constant, (*1*) the number of peaks in a band due to a group of equivalent protons is equal to the number of neighboring protons plus one, (*2*) the peaks are symmetrically arranged about the chemical shifts of the groups, and (*3*) the ratio of the intensities of the peaks correspond to the coefficients of the expansion of $(R + 1)^N$, where N is the number of neighboring protons. Thus for 1, 2, 3, and 4 neighboring protons, the approximate peak intensity ratios are 1:1, 1:2:1, 1:3:3:1, and 1:4:6:4:1, respectively.

The coupling constants for proton–proton interaction ranges from 20 to less than 1 Hz. Therefore, very high resolution is required to observe the multiplets. The constant for coupling between vicinal protons or geminal protons is a function of the angle between the carbon–hydrogen bonds involved and of the electronegativity of adjacent groups (75). Whereas $\Delta\nu$ is directly proportional to the frequency used in the measurement, the coupling constant is independent of the frequency.

Although chemical shifts provide information about the functional groups that are present, spin–spin multiplets supply valuable information concerning the spatial arrangements of functional groups in molecules. They give direct information about those groups that are most closely associated with each other and the number of equivalent protons in each of the groups. Under special circumstances, they provide information concerning bond lengths and bond angles. The absence of spin–spin multiplicity indicates that a group of equivalent protons is isolated from the other magnetic nuclei in the molecule. Information derived from spin–spin interaction of protons makes NMR one of the most powerful tools in the study of the structure of organic molecules.

Quantitative information is obtained from an NMR spectrum by integrating the area under the bands. The area under the absorption band is proportional to the number of protons responsible for the absorption. By a comparison of the areas under different absorption bands, the relative number of protons represented by each band can be determined. In order to assign absolute numbers of protons, a structure can be proposed for the compound or the molecular weight of the compound can be determined.

The preceding discussion of NMR was intentionally very superficial, because its purpose was to give the reader the barest essentials that should help him understand some of the terminologies used. Higher order spin patterns, multiple resonances, and problems which complicate the interpretation of NMR spectra have been omitted. For a more complete picture, refer to one of the comprehensive monographs (24,272). The chapter on NMR in this Treatise by Chamberlain (75) bridges the gap between theory and application and is recommended for analytical chemists. The empirical approach to the interpretation of NMR spectra by Bible (42) is very helpful.

IV. SEPARATION

The purpose in separating unsaturated compounds is (*1*) to purify aliphatic and alicyclic solvents, (*2*) to purify unsaturated compounds, and (*3*) to break down complex mixtures into homologous groups which simplifies analysis by various instrumental methods. In some cases the physical separation of unsaturated compounds is incidental to the method of analysis, e.g., gas–liquid chromatography and mass spectrometry. Such incidental separations are covered in Section VI-C.

A. ADSORPTION CHROMATOGRAPHY

By far the most important method of separating olefins from other homologs is adsorption chromatography. The principles and techniques of chromatography have been treated in several monographs (74,207,341, 377) and in Part I, Volume 3 of this Treatise (197). The removal of trace amounts of aromatics and olefins from aliphatic and alicyclic solvents by passing them through a column of activated silica gel is a well-known technique. The main purpose of such an application is the complete removal of the unsaturated impurities; no attempt is made to recover the unsaturates and part of the saturated components is also held up by the adsorbent.

When the quantitative recovery of one or more of the major components is desirable, the components must first be separated by development. Then the boundary between the components must be determined. The three methods of development are frontal, displacement, and development analysis. Frontal analysis, in which the sample solution is percolated continuously through the separating medium, is useful only for simple mixtures and is seldom used. Both displacement and development analyses have been very useful in the analysis of complex olefinic materials.

Adsorbents most commonly used are activated carbon, alumina, paper, and silica gel. On nonpolar adsorbents such as carbon, adsorption is a function of molecular weight; e.g., the sequence of emergence of hydrocarbons from a carbon column is methane, acetylene, ethylene, ethane, etc. On acid (electrophilic) adsorbents, such as alumina and silica gel, adsorption is a function of nucleophilic character and geometry of the molecules. Thus the sequence of emergence from a silica column is alkanes, alkenes, aromatic. For steric reasons, cyclic alkenes are more strongly adsorbed than straight-chain alkenes and *cis*-alkenes are more strongly adsorbed than *trans*-alkenes. Adsorbability is also increased by conjugation of a double bond by another double bond and by carboxylic and carbonyl groups.

Adsorption chromatography may be divided into column, paper, and thin-layer chromatography, depending on the form and type of adsorbent.

1. Column Chromatography

a. Displacement Analysis

In displacement analysis, a small amount of sample is placed on the top of the column and is followed by a substance that has greater affinity for the adsorbent than the sample. In lieu of a desorbing agent, heat may be used to move the sample through the column. The advantages are that displacement of the sample is quantitative and tailing is minimized. One disadvantage of the displacement method is that overlapping occurs to some extent even in optimum cases.

Various physical and chemical properties may be utilized to characterize the fractions emerging from the column, e.g., refractive index, density, thermal conductivity (gases), and bromine number. In one method of separation the effluent is collected in appropriate fractions and the physical or chemical property of each fraction is determined and plotted against the cumulative volume of the effluent. Figure 1 shows such a separation of catalytically cracked gasoline by displacement chromatography (125).

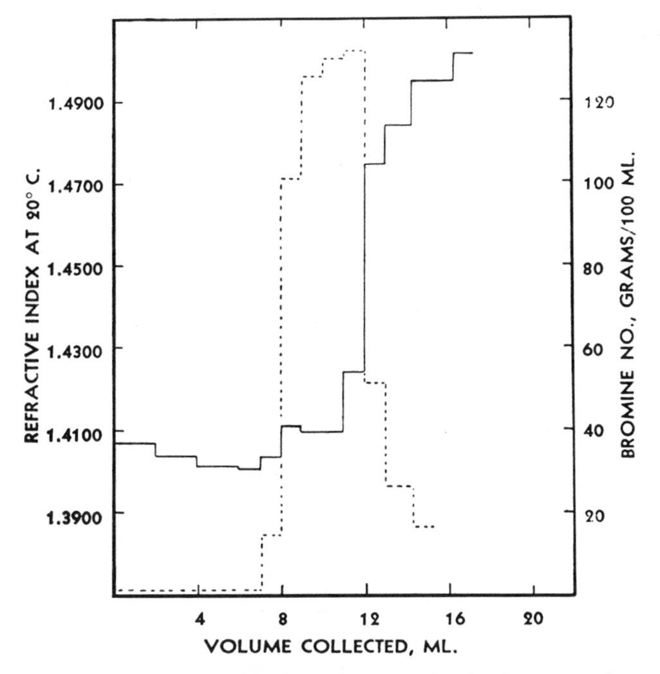

Fig. 1. Separation of a catalytically cracked gasoline by displacement chromatography (125). (Courtesy of *Analytical Chemistry.*) (---) Bromine No. (——) Refractive index. Column, 1 × 234 cm. Gel, Davison 100–200-mesh (125.5 g). Charge, 20 ml. Temperature, −40°C.

The solid line gives the refractive index of the fractions, while the broken line gives the bromine number of the fractions. The bromine number adsorptogram gives a more distinct separation of the three homologous groups. The first emerging group is the saturates with low bromine number and low refractive index. The second emerging group is the olefins with high bromine number and intermediate refractive index. The last group is the aromatic compounds, which are characterized by low bromine number and high refractive index. The overlapping of the boundaries of the three groups is quite evident.

The foregoing example of displacement chromatography pertained to a full-range cracked gasoline. More distinct adsorptograms can be obtained by using narrow boiling ranges (78). However, this technique has several disadvantages: precision is only moderate, procedure is too involved and time consuming, and the evaluation of adsorptogram is particularly difficult when the olefin content is less than 5%.

The separation of petroleum fractions into saturated, olefinic, and aromatic groups is simplified considerably by using dyes to mark the boundaries between the groups (81,88). It is essential that the volume of aromatics plus olefins should not exceed the capacity of the column. The ratio of the volume of aromatics plus olefins to the volume of silica gel should be equal to or less than 1:18. Therefore, it is convenient to know the aromatic and olefin content of the sample; the fluorescence indicator adsorption (FIA) method may be used for this purpose.

In the FIA method a special column is packed with 100–200 mesh silica gel. The column has four sections: charger, neck, separator, and analyzer. A small amount of gel containing Sudan III and other fluorescent dyes is placed in the middle of the separator section. Approximately 0.75 ml of sample is introduced to the charger section and is forced down the column with isopropanol under about 5 psi of air pressure. As the hydrocarbons are separated, the dyes are also separated so that under ultraviolet light, distinct zones of the hydrocarbon groups are evident. The forward extent of the saturates zone is marked by the "wet" front, olefins by the yellow front, aromatics by the blue front, and, finally, the alcohol zone follows the red Sudan III band. When the sample passes into the analyzer section of the column, the length of each zone is measured. The volume percent of each hydrocarbon type is calculated from the ratio of the length of a zone to the total length of the sample.

The details of the FIA method are given in A.S.T.M. Method D-1319-55T (14). The effectiveness of this method has been established by cooperators of the A.S.T.M. for gasoline, kerosene, and jet fuel (up to 315°C) boiling ranges. The accuracy of the method is approximately 2%. Heavily cracked samples tend to give indistinct olefinic boundaries.

The information derived from the FIA method is used in the bulk separation of cracked petroleum distillate fractions. The procedure is essentially the same as in the FIA method except that the scale of operation is much larger and the fractions are actually collected. For samples containing less than 1% olefins (for details see A.S.T.M. Method D-2002-64 (19)), the quantity of sample that may be charged is dependent upon the aromatic content of the sample, the column efficiency, and the gel capacity of the column. The ratio of volume of aromatics to volume of gel should not exceed 1:12. The saturates fraction is collected until the yellow color is about $^1/_4$ in. above the column tip. Cooperative tests have established that a representative saturates fraction will be obtained if the percentage by volume of saturates recovered is not less than 90% of the saturates in the sample as determined by sulfonation (see Section IV-B-1).

For samples containing 1–35% olefins, a detailed separation procedure is given in A.S.T.M. Method D-2003-64 (20). This method is similar to the one just discussed except that the ratio of volume of aromatics plus olefins to volume of silica gel should not exceed 1:18. Using 2-propanol as desorbent under 10 psi nitrogen pressure, about 4–5 hr is required to obtain 10 ml of saturates from a 30-ml charge. The isolated saturates will be representative if the percentage of saturates recovered is within ±5% relative of the percentage determined by the FIA method.

Snyder applied FIA and linear elution adsorption chromatography to 21 gasolines of various types (334) and petroleum oil boiling in the range 350 to 1000°F (333). The olefins are detected by the ultraviolet absorption of their iodine complex. The method is useful in determining trace amounts of olefins in the range 0.05–2.0%.

Tailing of olefins into the aromatic fraction and consistent low olefin contents found in FIA analysis have been overcome by the use of silica gel which has been partially deactivated by the addition of 4% water to dry silica gel (259).

Displacement analysis on silica gel was used to investigate the separation of terpenes (134), of saturated from unsaturated acids (77), and the separation of C_4–C_{22} saturated and unsaturated acids and their glycerides (1,145,161).

Displacement analysis on carbon was used to separate various unsaturated acids (159,160) and dibasic acids with isolated and conjugated double bonds (119).

Zakupra and co-workers (376) found that freshly crushed silica gels containing fine particles separated alkanes from alkenes more easily but it catalyzed the isomerization and polymerization of the alkenes. They pretreated the silica gel with hydrochloric acid and hydrogen peroxide to prevent isomerization and polymerization of the alkenes.

Gaseous olefins may be adsorbed at low temperatures and fractionally desorbed at controlled partial vacuums and at different fixed temperatures. Such methods are slow and inconvenient and lack selectivity. This technique was improved by using a moving hot zone to displace the sample. Using a carbon column and thermal conductivity detection, the method was rapid and convenient but the alkenes were incompletely separated from the alkanes of the same carbon number (146). Working at elevated temperature and using strongly adsorbed displacing agents saturated in nitrogen, separation of substances boiling as high as 150°C was possible. Some of the displacement agents used were simple esters of acetic acid (76,266), bromobenzene (169), and mercury vapor (351,352).

b. Development (Elution) Analysis

In development analysis a small amount of sample is moved by a substance which is less strongly adsorbed than the sample. When the fractionation is complete, the adsorbent column may be sectioned for recovery of the separated components. If the development is continued until the fractions leave the column, the procedure is referred to as elution analysis.

The adsorbents and detection methods used in development analysis is the same as in displacement analysis. The big difference is that in development analysis, the eluent is less strongly adsorbed than any component of the sample. Therefore, elution of the components is often very slow, especially when their adsorption affinities vary widely. In such cases it is common practice to complete the desorption with successively stronger developers (281). Another undesirable characteristic of development analysis is "tailing." This phenomenon can be suppressed by using an eluent with continuously increasing desorptive strength (6).

For the fractionation of petroleum distillates silica gel is used almost exclusively. The saturates are eluted with pentane. When the refractive index of the effluent reaches that of pentane, the eluent is switched to pentene in order to hasten the desorption of the olefins. After the complete desorption of olefins, the aromatic substances are eluted with a polar compound such as ether, acetone, or alcohol. The separation of individual components in the three hydrocarbon groups is negligible. The eluents should boil outside the range of the fraction to facilitate recovery. Polar eluents can usually be washed out with water. This technique is not suitable for high boiling petroleum fractions.

Spence and Vahrman (339) separated paraffins from olefins in low temperature tars on a silica gel column after reacting the sample with 80% excess of iodine chloride. The halogenated compounds were more strongly adsorbed, thereby improving the separation. The paraffins were eluted with a light petroleum (40–60°C) until the purple-brown front of the halogenated olefin derivatives approached the bottom of the column. The halogenated olefins were recovered by displacing with ethanol. The olefins are regenerated by refluxing the halogenated derivatives with ethanol and excess sodium iodide. There is a small loss of olefins during the separation. The authors did not detect any isomerization of the olefins as confirmed by the infrared absorption spectra of the original sample and the regenerated olefins.

Elution chromatography was used to separate various saturated and unsaturated acids (112,151,181,187), esters (113,286), and glycerides of linseed oil (360,361) and of soybean oil (283). Improved separation of C_{18} acids and esters can be achieved by using eluents of intermediate adsorptivity (204). Phenol red indicator impregnated on a magnesium oxide column was used to follow the separation of stearic and oleic acids (139). Addition products of unsaturated fatty acids with 2,4-dinitrobenzenesulfenyl chloride (179) were separated effectively on a magnesium sulfate column using 5% ethyl ether in benzene as developer (329).

Low-boiling olefins may be trapped on the adsorbent at low temperature and developed with gaseous eluents such as carbon dioxide, nitrogen, hydrogen, and air. To desorb higher boiling components the temperature may be raised to 200–250°C or a moving furnace may be used. Numerous applications of gas adorption chromatography have been reported for the separation of low-boiling olefins (85–87,171,172,265,309) and for the fractionation of a high-boiling organic acid fraction (152). The separation of C_1 and C_2 hydrocarbons by Patton and co-workers (265) is shown by the chromatogram in Fig. 2. Gas adsorption chromatography is the fore-

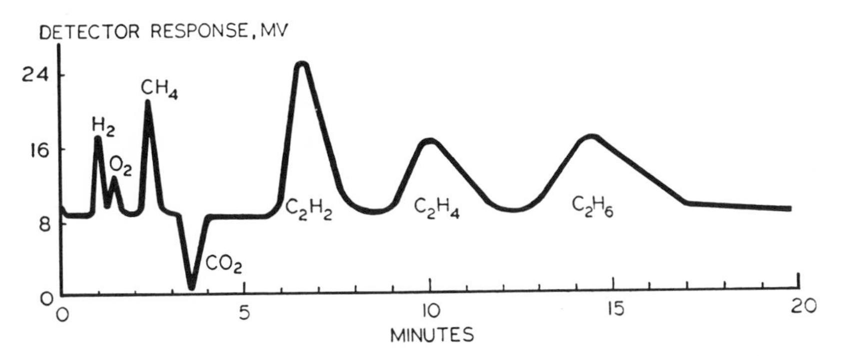

Fig. 2. Separation of gas mixtures by gas elution chromatography (265). (Courtesy of *Analytical Chemistry.*) Column, active carbon. Developer, nitrogen. Detector, thermal conductivity cell.

runner of gas–liquid partition chromatography which has undergone a tremendous development in the last decade. Because of the latter's importance in the quantitative determination of olefins, it will be discussed separately in Section VI-C.

Martin and Smart (230) converted the separated components to carbon dioxide and water by passing them over copper oxide heated to 600°C. Significant advantages were the increased sensitivity and uniform response by all hydrocarbons in the thermal conductivity cell.

Scott and Phillips (315) were able to determine 10 ppb of 1-hexene and 1-heptene in heptane by preliminary concentration of the olefins. The sample was passed through a column of aluminum oxide impregnated with silver nitrate and the column was freed of liquid. The last traces of heptane was removed by eluting with nitrogen. The olefins were finally displaced by a stream of nitrogen saturated with 1-octene. Duffield and Rogers (109) showed that between 92 and 235°C, silver nitrate supported on Chromosorb-W or Chromosorb-P retains olefins but not saturated hydrocarbons. The olefins are desorbed at 250–300°C with argon carrier gas. Using polysiloxane resin adsorbent and helium carrier gas, Hachenberg and Gutberlet (144) separated *n*-olefins and isoolefins of each molecular weight from C_8 to C_{18} cracked wax. Temperature programming over 80–175°C greatly improved resolution over isothermal adsorption, but branched and internally double-bonded olefins were not separable.

In the usual gas chromatographic technique, very small sample sizes are used. Preparative-scale gas chromatographs may be used to prepare relatively large amounts (1–2 g in ideal cases) of a pure component which can serve as a reference standard for other methods. It should be cautioned, however, that isomerization and structural changes may occur during the separation process.

2. Thin-Layer Chromatography

Adsorption chromatography, in which the adsorbent is prepared as a thin coating on glass or plastic, is commonly referred to as thin-layer chromatography (TLC). This technique has become quite popular in recent years and several books have been published on this topic (48,339a). The principles involved are the same as in adsorption column chromatography but the sample size is much smaller and the development of the chromatogram is faster.

Kirchner, Miller, and Keller applied this technique for the separation of lemonene, α-pinene, pulegonene, and other terpene-type compounds (190). The positions of the developed spots were revealed by spraying with suitable reagents. Prey and co-workers (275) separated C_2–C_{12} olefins as the mercury addition compounds by chromatography on thin silica gel layers (or paper). With two-dimensional development, they separated 12 olefins, which were detected with diphenylcarbazide. Den Boer (99) separated closely related unsaturated compounds which differ in number, position, or environment of the double bonds by using thin layers of silica gel treated with silver nitrate.

3. Paper Chromatography

Chromatography in which thin sheets of paper are used as adsorbent is called paper chromatography. The principles and techniques are similar to TLC but either ascending or descending development may be used because of the flexibility of the paper.

Huber (163) determined olefins and diolefins by paper chromatography after converting them to nonvolatile compounds by treatment with mercuric acetate. The spots were developed with lower alcohols containing ammonium hydroxide, ammonium carbonate, and water and detected by exposure to hydrochloric acid vapor and spraying with ammonium sulfide or dithizone. The error with photometric measurement of the spot is $\pm 7.5\%$.

Pairs of fatty acids having identical R_f values, e.g., palmitic and oleic acids, and myristic and linoleic acids, were resolved by reverse-phase paper chromatography by converting the unsaturated acids into their hydroxyiodo derivatives (278). The latter have higher R_f values than the corresponding unsaturated acids. The hydroxyiodo acids yielded two spots, probably due to the formation of diastereoisomers.

Frequently the paper adsorbent is impregnated with stationary liquid phase. Boldingh (50) achieved good separation of the ethyl ester of oleic and erucic acids from C_{12}–C_{18} saturated acids on paper impregnated with rubber latex using a methanol–acetone mixture as the developing agent. McHale and co-workers (235) were able to determine the number of double bonds. The stationary phase was the paper which was impregnated with zinc carbonate containing a small amount of sodium fluorescein and with liquid paraffin. The mobile phase was aqueous ethanol. They used this technique in the study of the intermediates formed during the hydrogenation of some natural products.

B. CHEMICAL METHODS

1. Absorption in Sulfuric Acid

Simple olefins react with sulfuric acid to yield mainly the acid sulfate and, with fuming sulfuric acid, some sulfonate addition products (271).

$$>C=C< + H—OSO_2OH \rightarrow H—\overset{|}{\underset{|}{C}}—\overset{|}{\underset{|}{C}}—OSO_2OH \qquad (1)$$

$$>C=C< + HOO_2S—OSO_2OH \rightarrow HOO_2S—\overset{|}{\underset{|}{C}}—\overset{|}{\underset{|}{C}}—OSO_2OH \qquad (2)$$

However, depending on the concentration of sulfuric acid and the nature of the olefin, side reactions occur, resulting in the formation of esters, polymers, alcohols, and other oxidation products.

The ease of addition of sulfuric acid depends on the number of alkyl substituents at the double bond. For example, ethylene reacts slowly at room temperature with concentrated sulfuric acid, requiring a catalyst such as silver sulfate for rapid addition. Propylene reacts with 85% sulfuric acid in the absence of a catalyst, while isobutylene reacts with 65% sulfuric acid at room temperature. These differences in reactivity are utilized in separating mixtures of olefins (258).

The best reagent for the complete removal of olefins is sulfuric acid containing mercuric sulfate and magnesium sulfate. The latter suppresses the reaction of the reagent with carbon monoxide. Silver catalysts are said to catalyze the reaction with carbon monoxide and the polymerization of propylene and butylenes (271).

The separation of olefins from aromatic compound by sulfuric acid absorption is not very practical except for certain simple known mixtures. However, the olefins plus aromatics can be separated from the saturated compounds using concentrated sulfuric acid containing phosphorus pentoxide. Kattwinkel (180) originally used 14% phosphorus pentoxide, but Mills and co-workers (249) found that a reagent containing 30% phosphorus pentoxide was most satisfactory. This latter method was adopted as standard by the American Society for Testing and Materials (12). The details of the procedure are given in reference 12. The repeatability of the method is about 1% and the reproducibility (interlaboratory) about 2%. The FIA method (Section IV-A-1-a) is preferred over the acid absorption method because it provides simultaneous determination of the saturates, olefins, and aromatics. The acid absorption method is useful, however, in checking the accuracy of the chromatographic separation of petroleum distillates described in Section IV-A-1-a.

2. Nitration

Olefins react with nitrogen tetroxide to give the dinitro and the nitronitrite addition products (eq. 3) (271). The nitrito group may be

$$-CH{=}CH_2 + N_2O_4 \rightarrow -\underset{\substack{|\\ }}{\overset{NO_2}{\overset{|}{C}}}H{-}\underset{\substack{|\\ NO_2}}{C}H_2 + -\overset{ONO}{\overset{|}{C}}H{-}\underset{\substack{|\\ NO_2}}{C}H_2 \quad (3)$$

oxidized to the nitrate or hydrolyzed to give an alcohol (eqs. 4 and 5).

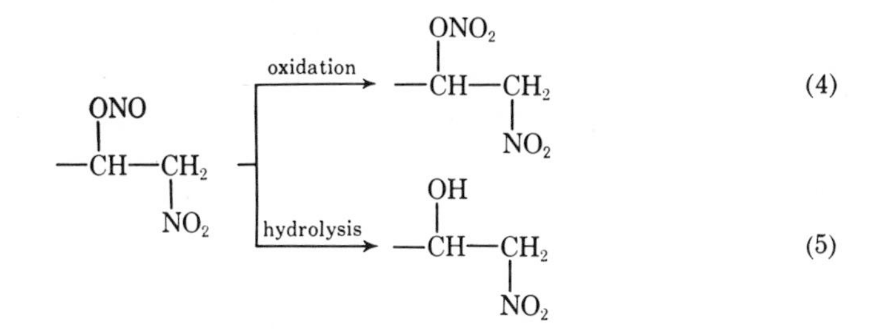

The nitration products have a much lower volatility than the hydrocarbons. This property was utilized by Bond (52) in the separation of the nonreacted fraction by steam distillation. In a variation of this procedure he solubilized the nitration products so that the volume of the unreacted hydrocarbons could be measured directly in the reaction flask. The details of the method are given by Bond and co-workers (52,307). Outlines of the two methods are given below (307).

Method A. Steam Distillation Procedure

Gaseous nitrogen tetroxide is passed into the chilled sample to react with the olefins. After the completion of the reaction, an aqueous solution of urea is added to decompose the excess nitrogen tetroxide. The unreacted hydrocarbons are steam-distilled from the mixture. The olefin content is determined from the difference in volume of the unreacted hydrocarbons and the volume of the original sample. The unreacted fraction is essentially olefin free and may be used for the determination of other hydrocarbon types.

Method B. Direct Volumetric Procedure

Gaseous nitrogen tetroxide is passed into the chilled sample to react with the olefins. The nitration products are made soluble in aqueous alcohol by reacting with alcoholic potassium sulfide solution, permitting the measurement of the unreacted portion of the test sample directly in the reaction mixture. The olefin content is determined from the difference in volume of the unreacted hydrocarbons and the original volume of the sample.

This method is applicable where only the total olefin content of the sample is required.

The two nitration methods for the separation of olefins from other hydrocarbons were evaluated by the cooperators of the American Society for Testing and Materials (307). The standard deviation for the four synthetic blends was 1.2% for both methods. When only the volume percentage of the olefins is required, method B is preferred because of its relative simplicity and rapidity. The authors preferred the nitration method over the bromination method (standard deviation 1.1%) in the determination of volume percent olefins because the molecular weight and density of the olefins are required in the bromination method.

There are several disadvantages to the nitration method. Highly branched olefins react with difficulty and the nitration products formed are less readily separated from the unreacted hydrocarbons. Two possible hazards are the toxicity of nitrogen tetroxide and the thermal instability of some of the reaction products. Nitration products are subject to rapid decomposition if heated much above 120°C. Violent reactions have occurred where nitrogen tetroxide has been left in contact with hydrocarbons for a long period of time.

Other nitration agents used in the determination of olefins are nitrogen trioxide, nitric oxide, nitrosyl chloride, nitrosobenzene, and tetranitromethane (271).

V. DETECTION

Unsaturation in organic compounds may be detected by means of any of the properties discussed in Section III. Convenient chemical methods are the decolorization of bromine water and the transformation of the purple color of permanganate to brown. However, both bromine and permanganate are strong oxidizing agents and are not specific for olefinic unsaturation.

Sharefkin and Sulzberg (316) detected olefins by the wine red color of the 2,4-dinitrophenylhydrazones of methyl ketones formed by Friedel-Crafts acetylation. The degree of olefinic character was estimated by employing various Friedel-Crafts catalysts of different electrophilic strengths.

Okuda (263) detected unsaturated organic compounds by their color reaction with benzaldehyde and sulfuric acid. A drop of benzaldehyde and then a drop of sulfuric acid were added to a crystal or a drop of sample. The color appeared in 30 sec for double and triple bonds with at least one hydrogen on the unsaturated carbons. Saturated compounds, compounds with a carbonyl group adjacent to the unsaturated carbon, and aromatic and heterocyclic compounds did not give the color reaction.

VI. DETERMINATION OF UNSATURATION

A. CHEMICAL METHODS

Halogenation methods are most commonly used for the determination of nucleophilic olefins, whereas bisulfite and morpholine addition methods are best suited for the electrophilic α,β-unsaturated compounds. Hydrogenation is generally applicable but it is not suitable for routine work because the hydrogenation rate varies with structure and the method is much slower than others. Special methods are available for specific olefin types: mercuration for vinyl unsaturation, treatment with maleic anhydride for conjugated diolefins, and hydration for acetylenic unsaturation. Methods involving oxidation are particularly useful in the elucidation of the structure of complex unsaturated compounds.

1. Halogenation

Halogenation methods have several disadvantages arising from the high oxidizing power of the halogens. Any reducing agent oxidizable by the halogens will give high results. Another common interference is substitution. Because of the wide range of reactivities of the olefins, there are numerous halogenation methods, and moreover, each method has special modifications for specific compounds, e.g., reagent concentrations, reagent excess, time of reaction, type and amount of catalyst, etc. The problems involved in the application of halogenation methods to complex mixtures of unsaturated compounds are quite apparent. In spite of these disadvantages halogenation methods are used in commerce to set "unsaturation" specifications for various commodities because of their convenience and rapidity. There are also many nearly pure compounds or simple mixtures in commerce which can be assayed simply and rapidly by halogenation methods.

a. General

(1) Reactions

The standard potentials of halogens, interhalogens and the pseudohalogen, thiocyanogen, are given in Table V (206). The reactivity of

TABLE V

Standard Potentials of Some Halogenating Agents (206)

Half-cell reactions	E^0, V
$I_2 + 2e \rightleftharpoons 2\,I^-$	0.5355
$(SCN)_2 + 2e \rightleftharpoons 2\,SCN^-$	0.77
$IBr + 2e \rightleftharpoons I^- + Br^-$	1.02
$Br_2 + 2e \rightleftharpoons 2Br^-$	1.087
$ICl + 2e \rightleftharpoons I^- + Cl^-$	1.19
$BrCl + 2e \rightleftharpoons Br^- + Cl^-$	1.2
$Cl_2 + 2e \rightleftharpoons 2Cl^-$	1.3595

these compounds toward olefins is closely related to their oxidizing power. Iodine is too nonreactive, whereas chlorine is too reactive and is often accompanied by severe substitution. The most useful and thoroughly tested reagents are bromine, iodine chloride, and iodine bromide. The usefulness of bromine chloride has been recognized only recently (34,69, 311,312) although undoubtedly this reagent must have been generated unintentionally by earlier workers.

The complexity of halogenation reactions is probably due in part to the fact that the halogens can react with the solvent to form various active species; for example, Sidgwick (321) calculated that bromine equilibrates in water to give the following species at the indicated concentrations in millimoles per liter:

Br_2	HOBr	HBr	Br_3^-	Br_5^-
211.8	1.92	1.92	0.26	0.0021

Bromine in chloroform yields HBr_3 and HBr_5 (130).

The addition of bromine to a double bond in a polar solvent is believed to take place in two steps. The first step involves the electrophilic attack by the partially polarized bromine molecule resulting in the formation of the intermediate π-complex (reaction (6)) (27a, 258). In the next step

$$\text{RHC}{=}\text{CHR} + {:}\overset{\delta+}{\ddot{\text{Br}}}\,{:}\overset{\delta-}{\ddot{\text{Br}}}{:} \longrightarrow \left[\text{RHC}{-}\text{CHR}\ \cdots\ {:}\overset{+}{\ddot{\text{Br}}}\right] + \text{Br}^- \qquad (6)$$

the bromonium π-complex is attacked by any nucleophilic agent which happens to be near the side opposite from the complexed bromonium ion, resulting in the *trans* addition product (reactions (7)–(9)). The extent of each reaction and the stereospecificity are dependent on the composition of the titration medium (268a). It is doubtful that the bromonium ion can exist as shown in reaction (8); it is probably associated or chemically combined with a polar solvent molecule (41). The bromonium ion can combine with a bromide ion to form bromine or react with another mole of olefin. Therefore, reaction (8) does not affect the stoichiometry. In reaction (9), A is any negative portion of a solvent, e.g., acetate ion in acetic acid, hydroxide in water, alkoxy in ether. It is evident that reaction (9) also does not affect the stoichiometry of the desired reaction.

The formation of the π-complex by the interhalogens is favored because of their polarity (e.g., see reaction (10)). Again the final step involves the

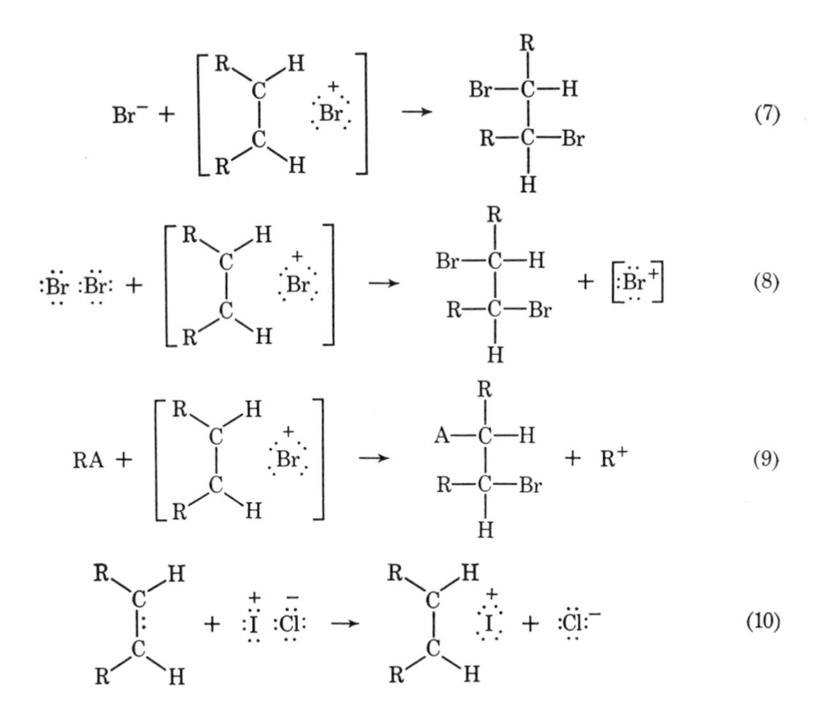

nucleophilic attack by an anion or a negatively charged fragment of the solvent similar to reactions (7)–(9).

Reactions that affect the stoichiometry of halogen addition to double bonds are (*1*) substitution at positions other than the unsaturated carbons, (*2*) elimination of hydrogen bromide or bromine from a reaction product, and (*3*) the addition of hydrogen bromide to the double bond (271). Dewar and Fahey (101) have shown that, unlike bromine, deuterium bromide and deuterium chloride add predominantly *cis* to acenaphthylene and indene. These results rule out the π-complex mechanism in the polar addition of hydrogen halides to olefins. This apparent anomaly can be explained by the results obtained by Cook and co-workers (82) in their freezing point diagram studies on the molecular interaction of hydrogen chloride with π-orbital donor molecules. They found that alkenes formed complexes with 2 moles of hydrochloric acid while alkynes formed complexes with 4 moles of acid per mole. Complexes of this type were not formed by alkanes. The bromonium ion was able to form the cyclic three-membered intermediate in equation 1 by using the π-electron pair to complete its octet; electronic considerations preclude an analogous reaction for the proton. Thus an initial electrophilic attack of the proton at one of the unsaturated carbons, rather than perpendicular to the π-bond, could explain the predominantly *cis* addition of hydrogen halides to alkenes.

McIlhiney (236) attempted to apply corrections for substitution by determining the acid formed during brominations in carbon tetrachloride. However, Buckwalter and Wagner (68) showed that acid correction for substitution leads to generally low results and even to negative results in some cases. With increasing understanding of the conditions which are conducive to substitution, it is much easier to control this side reaction rather than applying uncertain corrections.

(*a*) *Control of Substitution*

Highly branched olefins are most susceptible to substitution. Some of the conditions that promote substitution are high temperature, light, high halogen concentration, and catalysis by mercury. Thus, this side reaction may be minimized by titrating at low temperature in the absence of light. Reagent excess should be kept at a minimum and mercury catalyst should be used only when necessary. In spite of these precautionary measures, it is not possible to eliminate substitution completely. Substitution reactions (excluding aromatic compounds, e.g., phenol and aniline) are usually slower than addition reactions. Therefore, it is possible to correct for this side reaction by extrapolation methods described in Sections VI-A-1-a-(2) and VI-A-1-b-(2).

(*b*) *Solvent Effects*

The kinetics of bromination and the general course of the reaction are greatly influenced by the solvent system. Carbon tetrachloride, which is a useful solvent for the bromination of rubber-type compounds, is very prone to free radical chain addition to the double bonds which presents many unknown features. Such reactions may be initiated by peroxides or light. Rabilloud (276) found that irradiation of carbon tetrachloride solutions of olefins caused addition of carbon tetrachloride to the double bond.

$$CCl_4 \xrightarrow{\text{light}} Cl\cdot + \cdot CCl_3 \tag{11}$$

$$RCH{=}CH_2 + \cdot CCl_3 \rightarrow R\dot{C}HCH_2CCl_3 \tag{12}$$

$$R\dot{C}HCH_2CCl_3 + CCl_4 \rightarrow RCHClCH_2CCl_3 + CCl_3 \tag{13}$$

Telomerization,

$$R\dot{C}HCH_2CCl_3 + RCH{=}CH_2 \rightarrow R\dot{C}HCH_2CH[R]CH_2CCl_3 \tag{14}$$

was suppressed by using a large excess of carbon tetrachloride. The order of reactivity was $RR'C{=}CH_2 > RCH{=}CH_2$ with $RCH{=}CHR$ being nonreactive, probably due to steric hindrance. Asscher and Vofsi (26) reported that copper(I) or -(II) and iron(II) or -(III) chlorides catalyzed the addition of carbon tetrachloride. They proposed a redox-transfer mechanism which, for copper, is given by reactions (15) and (16).

$$Cl_3CCH_2CHR^{\cdot} + CuCl^+ \rightarrow Cl_3CCH_2CHClR + Cu^+ \qquad (15)$$

$$Cu^+ + CCl_4 \rightarrow CuCl^+ + \cdot CCl_3, \text{ etc.} \qquad (16)$$

Reactions (11)–(16) would obviously lead to low results.

The desirable course is halogen addition by the ionic mechanism described earlier. This course is followed by the majority of methods which employ polar or ionizing solvent. If nonpolar solvents such as carbon tetrachloride must be used for solubility reasons, addition of a small amount of polar material may successfully conduct the reaction along ionic lines (271). Exposure to light and peroxide should be avoided.

(2) Titration Methods

Volumetric titration is by far the most common technique used in the analytical halogenation of olefins. The application of coulometry to such analyses has increased in recent years.

(a) Volumetric Titration

There are two techniques commonly employed in volumetric titrations. In the back-titration method, an excess of halogen reagent is added to the sample and after a precisely controlled reaction time, the analysis is finished iodometrically with standard thiosulfate solution using the starch–iodine endpoint.

$$X_2 + 2I^- \rightarrow 2X^- + I_2 \qquad (17)$$

$$I_2 + 2S_2O_3^{2-} \rightarrow 2I^- + S_4O_6^{2-} \qquad (18)$$

The preparation of standard thiosulfate and starch solution is given in any textbook on quantitative analysis (e.g., reference 201). Starch containing a high amylose fraction gives a very sharp endpoint. Duke and Maselli (110) used the dead-stop technique to detect the endpoint in the back-titration with thiosulfate, while Knowles and Lowden (193) titrated the thiosulfate amperometrically with iodate in a medium containing iodide ions.

Arsenic(III) oxide solution has also been used to back-titrate the excess bromine. This method is less convenient than iodometry because it is recommended (293) that excess arsenic(III) should be added and back-titrated with the bromine reagent.

For direct volumetric titrations the most commonly used endpoint detection method is the electrometric dead-stop method (56,106). Kolthoff and Bovey (196) determined the endpoint amperometrically, using a rotating platinum electrode, while Sweetser and Bricker (344) determined the endpoint spectrophotometrically.

The halogenating reagent may be a solution of a halogen or interhalogen in a suitable solvent. Frequently, however, halogenating reagents are not very stable. To avoid this difficulty, the halogenating agent may be generated chemically *in situ* by solutions of stable, highly pure oxidizing materials, such as bromate, in the presence of halides. Cerium(IV) has also been used for this purpose (178,330).

The nature of the generated halogenating species is dependent on the acidity and on the concentration of the various halide ions (312). Thus, in the presence of stoichiometric concentration of bromide ions, bromine is formed according to

$$BrO_3^- + 5Br^- + 6H^+ \rightarrow 3Br_2 + 3H_2O \tag{19}$$

while in the presence of excess bromide, the tribromide ion is formed:

$$BrO_3^- + 8Br^- + 6H^+ \rightarrow 3Br_3^- + 3H_2O \tag{20}$$

In the presence of a controlled limited concentration of bromide ions and an excess of hydrochloric acid, bromine chloride is formed:

$$BrO_3^{2-} + 2Br^- + 3Cl^- + 6H^+ \rightarrow 3BrCl + 3H_2O \tag{21}$$

Bromine was not detectable even in a concentration of hydrochloric acid as low as 0.08M. In the absence of bromide ions, 1 mole of bromine chloride and 2 moles of chlorine are formed:

$$BrO_3^- + 5Cl^- + 6H^+ \rightarrow BrCl + 2Cl_2 + 3H_2O \tag{22}$$

It is apparent that the titration medium must be carefully controlled in volumetric titrations in which the active halogenating reagent is generated chemically.

(b) *Coulometric Titration*

Another method of producing halogenating agents *in situ* is by constant-current electrical generation, e.g.,

$$3Br^- \rightarrow Br_3^- + 2e \tag{23}$$

The quantity of reagent consumed is proportional to the quantity of electricity used to generate the reagent. However, since a constant current is used, the measured quantity is time. In some of the commercial coulometric power supplies, the current is conveniently calibrated in microequivalents per second.

Since the bromide–bromine association constant is quite small (206), an appreciable fraction of the oxidized bromide could exist as the bromine molecule.

$$Br_2 + Br^- \rightleftarrows Br_3^- \qquad K = 17 \tag{24}$$

In a 0.1*M* bromide solution containing 10^{-4}–$10^{-5}M$ total bromine, roughly 40% would exist as bromine molecules. Bromine is more reactive than tribromide because the negative charge on the latter diminishes its electrophilic strength. Therefore, excessive concentration of bromide should be avoided to keep reaction (24) as far to the left as possible. However, the bromide concentration cannot be cut back too much because the current efficiency of reaction (23) is dependent on a certain minimum bromide concentration.

The increasing popularity of the coulometric method in the determination of unsaturation is due to several advantages of this method: (*1*) No standard reagent is required, (*2*) substitution and other errors are minimized although not completely eliminated, because the maximum halogen concentration is usually kept at a low level, (*3*) the method is precise and accurate because the control of current and the measurement of time are much easier than the measurement of volume, (*4*) the method is especially well suited to low-level titrations, and (*5*) the method is more readily automated.

Bratzler and Kleemann (57) were the first to apply electrically generated bromine to determine unsaturation. They generated bromine in a flowing system and collected the overflow in a vessel containing potassium iodide and starch. A gas sample containing 2% ethylene was analyzed with a precision of ±5%.

Leisey and Grutsch (212) used the automatic dead-stop method to determine the endpoint. A polarized platinum–platinum electrode system was used with a meter relay placed in series. By adjustment of the set point on the meter relay, the generation current could be switched off at any predetermined "diffusion" current, hence at any predetermined concentration of the halogen. A manual reset button was provided to "unlock" the meter relay so that further increments of bromine could be added to the slower reactions. They obtained better than 0.5% agreement with theoretical values for synthetic samples over the bromine index range 0–180. With the use of mercuric acetate catalyst, they found it to be rapid, accurate, and well suited to routine use.

Roberts (287) used an automatic potentiometric endpoint detector which was ideal for slow reactions. The bromine generation was terminated when a predetermined potential was reached. The potential "decayed" as the bromine was depleted by the slow reaction, and successive increments of bromine were generated until an arbitrary "decay slope" was reached. The method had greater accuracy and could handle smaller sample sizes than the conventional bromate–bromide techniques. Roberts and Brejcha (288) later described an apparatus for terminating coulometric titrations in which both amperometric and potentiometric principles are used.

Baumann and Gilbert (31) used a constant-current potentiometric endpoint in the coulometric titration of unsaturation in various petroleum products with bromine numbers ranging from 0.006 to 300. Preanodization and precathodization in 10% (by volume) sulfuric acid is required to obtain reproducible results. The relative standard deviation was ±2%.

A spectrophotometric endpoint was used by Miller and co-workers (246,248) in the coulometric titration of olefins of various structures. The absorbance at 360 mμ was automatically recorded as bromine was generated at a constant rate. A break in the curve occurs after most of the olefins are reacted and the bromine concentration increases. The straight horizontal and rising portions of the titration curve are extrapolated to obtain the endpoint.

Simple olefins such as 1-octene gave ideal titration curves in which the excess reagent portion was straight and steep. For other compounds, e.g., 2,4,4-trimethyl-1-pentene, the excess-reagent-line was not as steep as that of the ideal samples, and moreover the slope started to decrease before the usual upper absorbance level was reached. The nonattainment of the normal slope would indicate a slow reaction of the sample or a side reaction, while the decrease in slope at higher levels of bromine concentration would indicate a definite acceleration of the side reaction. Since the volume does not change during a coulometric titration, the slope of the excess-reagent line should be the same as that for a blank titration unless the reaction is slow or side reactions are introduced by the sample. This last statement applies to any extrapolated titration in which the physical or chemical property of bromine is plotted, e.g., amperometric titration.

The importance of olefins in photochemical reactions in polluted atmospheres has stimulated the development of several methods for the determination of low concentrations of olefins. Among these methods are two coulometric methods. Nicksic and Rostenbach (257) developed an instrument to determine olefins in the atmosphere and automobile exhausts in the 0–4 and 0–100 ppm ranges. The gas samples are passed through a brominating solution and the bromine is replenished by intermittent coulometric generation. Altshuller and Sleva (10) used the same instrument to analyze artificial mixtures and automobile exhausts containing

20–1200 ppm olefins. They found that sulfur dioxide and hydrogen sulfide gave positive errors while nitrogen dioxide gave negative errors. The method is applicable after removal of the interferences with Ascarite. The coulometric results were about 50% higher than results obtained by the colorimetric dimethylaminobenzaldehyde method (7,9) but considering the complexity of the mixtures and the low level of the olefins (about 20 ppm), the agreement was considered satisfactory.

Austin (27) described an instrument in which the electrically generated bromine reacted with the olefins at 380–450°C. The depletion of the bromine was measured as the effluent gas passed through an electrolytic sensing cell. Since the reaction is not stoichiometric or uniform for all olefins, the instrument must be calibrated against known olefins. Carbonyl compounds interfere and should be filtered out of the input gas.

The difficulties involved in coulometric titrations in which a property of the generated titrant is extrapolated have been discussed above. An additional disadvantage is that bromine consumed by side reactions is included in the results, thus leading to positive bias. Most coulometric titrations are carried out to a preselected reference point and the endpoint is accepted when it holds for an arbitrarily specified length of time. Results from such techniques would also include the bromine consumed by side reactions and, moreover, the operator must select the proper endpoint, which is often a difficult decision to make for slow reactions. Corrections for background and side reactions become increasingly important as the sample size becomes smaller, because the relative error becomes larger. The blank corrections applied in chemical titrations can often lead to false results because the concentration of the halogen in the blank is greater than that in the sample, and, also, the sample can introduce side reactions which naturally cannot be corrected by the blank. Belcher and Fleet (34) have reported such a difficulty in the microdetermination of unsaturation using bromine chloride.

The above problems can be overcome by cumulative coulographic titrimetry, which is discussed below.

(c) *Cumulative Coulographic Titration*

Cumulative coulographic titrimetry (CCT) is based on the following principles (155). Basically, it is the same as previous coulometric methods in that the haolgen is generated electrically at a constant rate and the generating current is cut off automatically at a preselected indicator current. The indicator circuit consists of the usual platinum–platinum electrode system polarized with about 200–300 mV. The appearance of excess halogen is indicated by the rise in the indicator current. The halogen concentration is kept constant at a predetermined maximum throughout a titration, but the reference concentration may be varied simply by changing the set point on the meter relay. As the bromine is consumed by slow reactions and side reactions, the needle on the meter relay drops below the set point. This causes the automatic intermittent generation of bromine, keeping the bromine concentration essentially constant.

The main difference between the regular coulometric titration and CCT is the manner in which the intermittent "constant current pulses" are handled. In CCT the current pulses are summed electromechanically (155) or electronically (156) and the cumulative sum of the electricity used in the generation of the reagent is recorded automatically against elapsed time by a servo recorder. The fact that the reagent concentration is kept constant leads to several advantages: (*1*) The quantity of electricity used at any time during the titration is equivalent to the sum of sample plus impurities that had reacted with the reagent up to that time; (*2*) the rates of background and side reactions are essentially constant so that corrections can be applied by extrapolation; (*3*) the kinetics of the reaction is simplified from the usual second-order to a pseudo-first-order reaction; and (*4*) each titration curve gives the kinetics of the reaction.

Typical CCT curves are given in Fig. 3. The solid curve represents the titration of a sample that gives no side reaction. Curve $\overline{ab}$ represents the bromination of any rapidly reducible impurities in the titration medium and the buildup of the bromine concentration to a level predetermined by the selected set point. The constant slope of $\overline{bc}$ gives the rates of consumption of bromine by the slow reaction of the solvent or an impurity and perhaps also through the volatilization of the bromine. The sample is added at point c. In the region $\overline{cd}$, the bromine generation rate is slower than the reaction rate, but at point d the generation rate becomes faster than the reaction rate and the intermittent generation of bromine starts at this point. At point e the reaction is complete and the straight segment $\overline{fe}$ is extrapolated to t_0 to obtain the corrected generation time $\overline{gc}$, thus eliminating error $\overline{he}$. The slopes of $\overline{bc}$ and $\overline{ef}$ should be equal if the sample does not introduce any side reaction. If side reactions, e.g., substitution, are introduced, the slope of the excess-reagent line $\overline{ij}$ would be greater than that of $\overline{bc}$. However, extrapolation of $\overline{ji}$ to t_0 should eliminate error $\overline{hi}$. Correction by extrapolation would not be valid for rapid side reactions such as the substitution of aromatic compound with electronegative substituents, e.g., phenol and aniline, or for certain conditions of mercury catalysis which promote vigorous substitution (see next section).

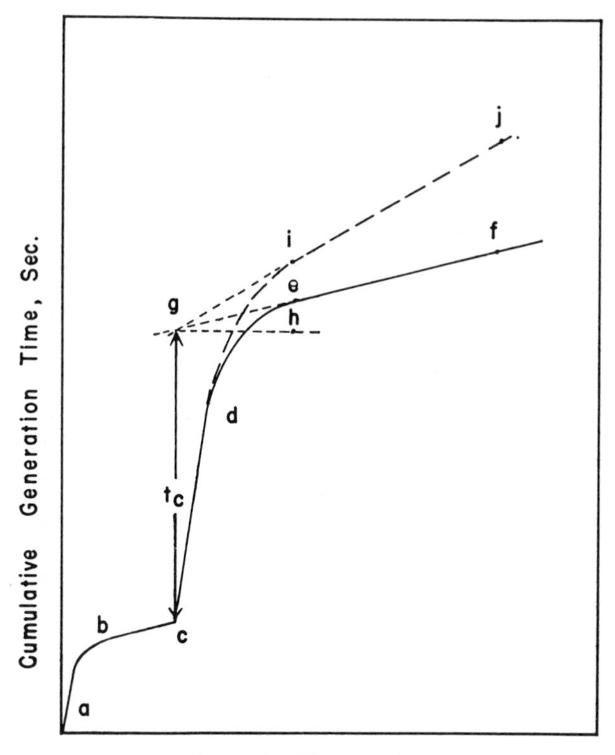

Fig. 3. Typical cumulative coulographic titration curves (155).

A schematic diagram of the cumulative coulographic titrator is given in Fig. 4. The coulometric power supply and the recorder are commercial components. The titration cell consists of a 100-ml electrolytic beaker containing two pairs of platinum–platinum electrodes. The working cathode is separated from the titration medium by a sintered glass tube. The meter relay is optically actuated and the switching is controlled entirely by solid state components. The heart of the current accumulator, which is placed in series with the coulometric power source, is the Solion integrator. The Solion output is amplified by a temperature-compensated transistor and the amplified output is fed to the recorder. The amplified output is attenuated so that the chart paper is calibrated for exactly 100 or 200 sec at current levels of 0.05, 0.1, 0.2, and 0.5 μeq/sec. The recorder is switched so that when the 200-sec mark is reached, the titration is stopped, the Solion is shorted out simultaneously, and a new ramp is started when the pen reaches the zero-current line. The only manual operation required is the addition of a known amount of sample when the

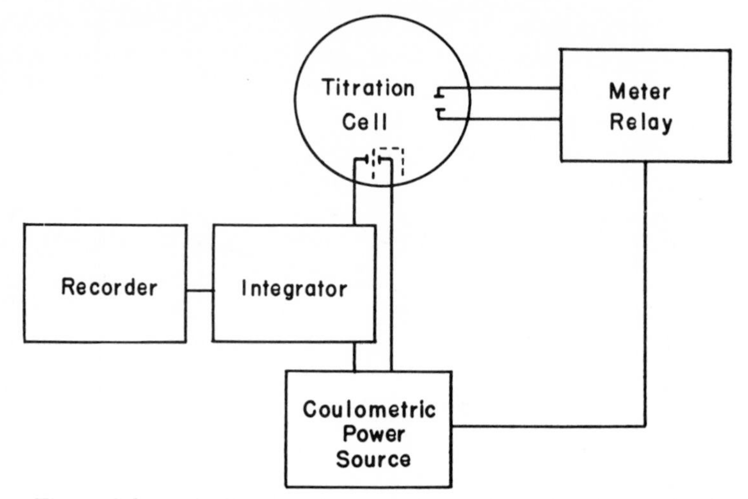

Fig. 4. Schematic diagram of cumulative coulographic titrator (155).

background becomes constant. Several samples may be added to the same titration medium provided there is no excessive side reaction of the samples. An example of consecutive titrations on the completely automated cumulative coulographic titrator is given in Fig. 5.

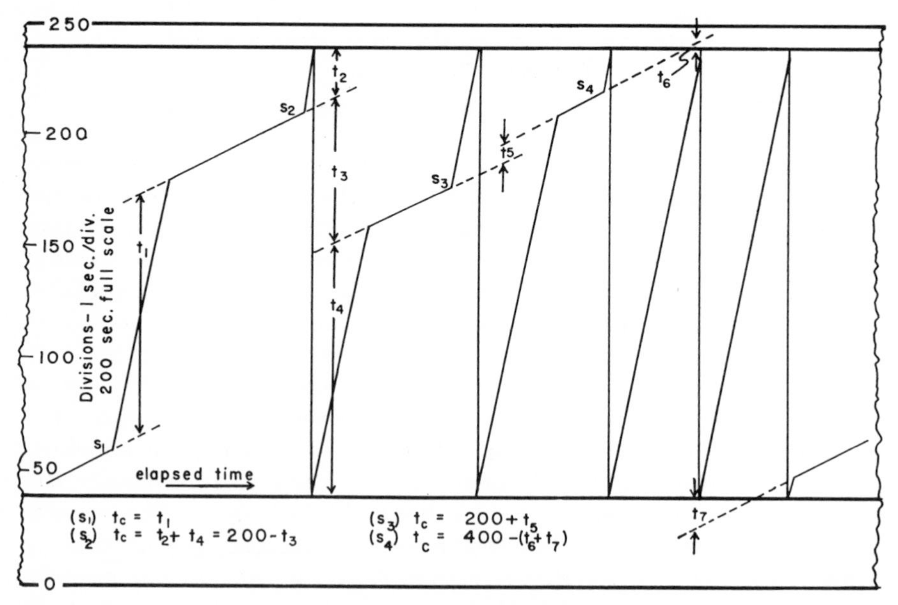

Fig. 5. Consecutive titrations of olefins with bromine on the automatically resetting cumulative coulographic titrator (155).

CCT is very useful in the study of the kinetics of halogenation reactions. Similar titrimetric techniques were applied by Walisch and co-workers (107,362,364) to determine the rate constants of various unsaturated compounds ranging between 50 and 10^8 liters $mole^{-1}$ min^{-1}. Their instrument was capable of maintaining a constant bromine concentration by intermittent generation of bromine. They followed the disappearance of the olefin by measuring the sum of the partial times of electrolysis by the constant current during the course of the total time of reaction. Walisch and Ashworth (363) applied this method to the microdetermination of unsaturated compounds and obtained a reproducibility of $\pm 0.3\%$ for monoolefins.

(3) Catalysis

The effect of electrolytes and the polarity of solvents on the rate and course of halogen addition to olefins have been discussed previously. Byrne and Johnson (71) found that the reaction between olefins and tribromide ion is accelerated by hydrochloric acid. Silver, nickel, and antimony salts have also been used to accelerate halogen addition. However, by far the most effective catalyst is mercury(II) ion. Hübl (164,165) first discovered the beneficial effect of mercuric chloride in the iodination of fatty materials. The beneficial effect was thought to be due to the formation of iodine chloride (120,370):

$$HgCl_2 + 2I_2 \rightarrow HgI_2 + 2ICl \tag{25}$$

Since the discovery of the catalytic effect of mercury(II) in the iodine addition, numerous applications have been published on the mercury-catalyzed addition of bromine, iodine chloride, iodine bromide, and bromine chloride. Chlorides, sulfates, and acetates of mercury(II) at various concentrations have been used. It is well known among analysts that acceleration of addition by mercury is accompanied by the acceleration of substitution. It is indeed difficult for the analyst to select an optimum catalytic condition for a specific application.

The first important contribution toward the systematization of mercury catalysis was made by Lucas and Pressman (220), who observed that the mole ratio of mercury(II) to bromide must be 1 or greater in order to have sufficient catalysis on the addition of bromine to alkynes and conjugated double bonds. However, because of the varying amounts of substitution for different compounds, it is difficult to ascribe a general accuracy and precision value to this procedure (322).

Schulek and Burger (311) showed that when mercuric chloride is added to a bromine solution, bromine chloride is formed:

$$HgCl_2 + Br_2 \rightarrow HgBrCl + BrCl \tag{26}$$

The absorption maximum of the bromine chloride was 348 mμ compared to 377 mμ for the bromine. The distillate from the solution containing mercuric chloride and bromine showed an absorption maximum of 343 mμ. Similar experiments confirmed the postulated formation of iodine chloride in equation (25).

Schulek and Burger (310) postulated that the catalytic function of mercury(II) in the addition of bromine chloride to olefins was to form the highly electrophilic bromonium ion.

$$BrCl + Hg^{2+} \rightarrow HgCl^+ + Br^+ \quad (27)$$

However, the participation of bromonium ion appears unlikely from the reactivity standpoint (34) and from other evidence (41). Belcher and Fleet (34) proposed that mercury(II) aids in the heterolysis of the bromine chloride during the π-complex formation step (reaction (28)). However,

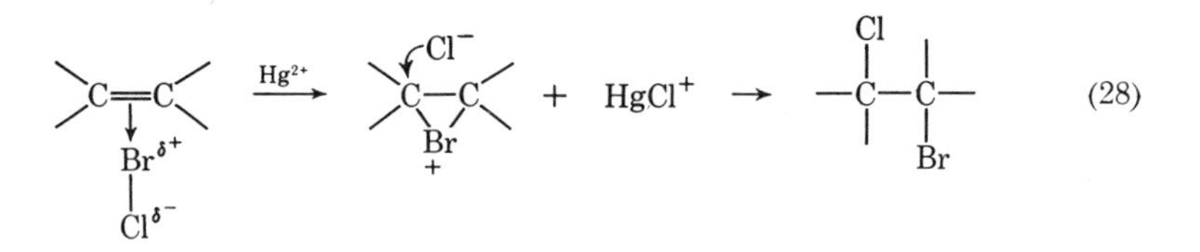

the mechanism is probably not this simple, because in the presence of twentyfold excess of chloride, the mercury(II) would exist as the trichloro and tetrachloro anions and the bromine chloride could be complexed by chloride ion to form the anion.

Because of the low concentration of halogen maintained in coulometric titrations, mercury catalysis is beneficial for many of the unsaturated compounds. To elucidate the nature of mercury catalysis in coulometric titrations, a systematic study was made using CCT (155). The kinetics of the halogenation of unsaturation in polyoxypropylene polyols was studied as a function of the type and concentration of mercury(II) salts and the type and concentration of halides. The reaction rates were pseudo first-order because the halogen concentration was kept constant throughout the titration. For the purpose of comparing reaction rates under different conditions, it would be convenient to assign five classes of rate curves as shown in Fig. 6. Classes I and II are ideal for routine applications, while III may be used with some loss in accuracy. Class IV is too slow for practicality, while V shows no visible reaction.

A summary of the data is given in Table VI. The principle oxidizable species were deduced from the concentration of the salts and the stability

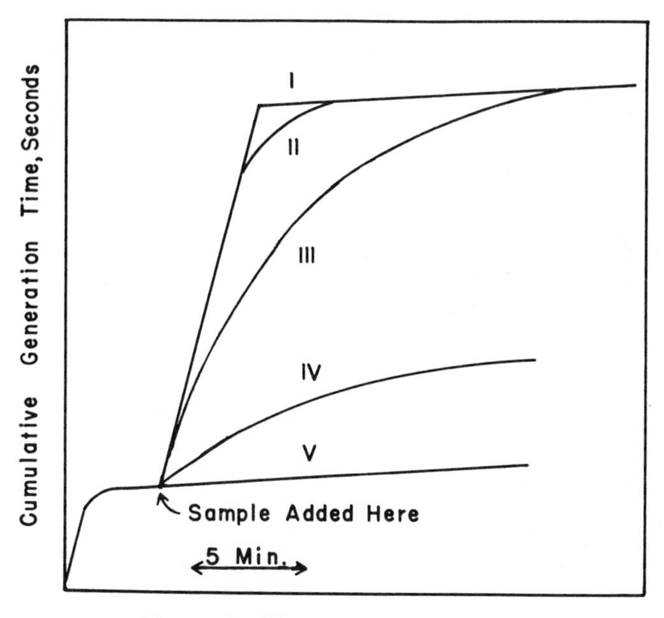

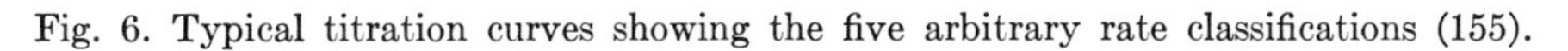
Fig. 6. Typical titration curves showing the five arbitrary rate classifications (155).

constants of the halides of mercury (47):

	K_1	K_2	K_3	K_4
Chloride	$10^{6.74}$	$10^{6.48}$	$10^{0.95}$	$10^{1.05}$
Bromide	$10^{9.40}$	$10^{8.28}$	$10^{2.41}$	$10^{1.4}$
Iodide	$10^{13.40}$	$10^{11.0}$	$10^{3.67}$	$10^{2.37}$

In the assignment of the structure of the oxidized species, the following assumptions were made: (*1*) The oxidizable species with highest negative charge is preferentially oxidized, and (*2*) for a given oxidizable species, the halide atoms that are oxidized are that pair that would give a halogen with the lowest oxidizing potential, e.g., in experiment 13:

$$\left[\begin{matrix}\text{Cl} \\ \text{Cl:}\ddot{\text{H}}\text{g:Br}\end{matrix}\right]^{-} \xrightarrow[\text{electrolysis}]{-2e} [\text{Cl:Hg} \leftarrow \text{ClBr}]^{+} \qquad (29)$$

and in experiment 12:

$$\left[\begin{matrix}\text{Cl} \\ \text{Cl:}\ddot{\text{H}}\text{g:Br} \\ \ddot{\text{Br}}\end{matrix}\right]^{2-} \xrightarrow[\text{electrolysis}]{-2e} \left[\begin{matrix}\text{Cl} \\ \text{Cl:}\dot{\text{H}}\text{g} \leftarrow \text{Br}_2\end{matrix}\right]^{0} \qquad (30)$$

TABLE VI

Effect of Halide Type and Concentration on the Mercury Catalysis of the Halogenation of Unsaturation in Polyols in 90:10 Acetic Acid–Water (155)

Expt. No.	Sodium halide	Mercury salt	Oxidizable species	Oxidized species	Rate[a]	Comment[b]
1	0.1*M* NaBr	None	Br^-	Br_3^-	IV	
2	0.1*M* NaBr	0.05*M* $HgBr_2$	Br^-, $HgBr_4^{2-}$	$HgBr_4^0$	III–IV	
3	0.1*M* NaBr	0.1*M* $HgBr_2$	$HgBr_2$, $HgBr_3^-$	$HgBr_3^+$	III	
4	None	0.1*M* $HgBr_2$	$HgBr_2$	$HgBr_2^{2+}$	II	1
5	None	0.1*M* HgO, 0.1*M* $HgBr_2$	$HgBr^+$	$[HgBr^{2+}]$	—	2
6	0.1*M* NaBr	0.033*M* $HgAc_2$	$HgBr_2$, $HgBr_3^-$	$HgBr_3^+$	III	
7	0.1*M* NaBr	0.05*M* $HgAc_2$	$HgBr_2$	$HgBr_2^{2+}$	II	1
8	0.1*M* NaBr	0.1*M* $HgAc_2$	$HgBr^+$	$[HgBr^{2+}]$	—	2
9	0.1*M* NaBr	0.05*M* $Hg(NO_3)_2$	$HgBr_2$	$HgBr_2^{2+}$	II	1
10	0.1*M* NaBr	0.1*M* $Hg(NO_3)_2$	$HgBr^+$	$[HgBr^{2+}]$	—	2
11	0.1*M* NaBr	0.037*M* $HgCl_2$	$HgBr_2Cl_2^{2-}$	$HgBr_2Cl_2^0$	III	
12	0.1*M* NaBr	0.074*M* $HgCl_2$	$HgBrCl_2^-$, $HgBr_2Cl_2^{2-}$	$HgBr_2Cl_2^0$	II	
13	0.1*M* NaBr	0.148*M* $HgCl_2$	$HgCl_2$, $HgCl_2Br^-$	$HgCl_2Br^+$	I	
14	0.1*M* NaBr	0.296*M* $HgCl_2$	$HgCl_2$, $HgCl_2Br^-$	$HgCl_2Br^+$	I	
15	0.1*M* NaBr, excess NaCl	0.165*M* $HgCl_2$	$HgCl_4^{2-}$, $HgCl_3Br^{2-}$	$HgCl_3Br^0$	I–II	
16	0.1*M* NaBr	0.05*M* $HgCl_2$, 0.05*M* HgO	HgBrCl	$HgBrCl^{2+}$	I	
17	0.03*M* KI, excess NaCl	Excess $HgCl_2$	Cl^-, $HgCl_4^{2-}$, $HgCl_3I^{2-}$	$HgCl_3I^0$	IV	4
18	0.03*M* KI, 0.2*M* NaBr	0.05*M* $HgBr_2$	Br^-, $HgBr_4^{2-}$, $HgBr_3I^{2-}$	$HgBr_3I^0$	V	5
19	0.1*M* NaCl	0.1*M* $HgCl_2$	$HgCl_3^-$	$HgCl_3^+$	I	3
20	0.1*M* NaBr, 0.1*M* NaCl	None	Cl^-, Br^-	Br_3^-	IV	

[a] See Fig. 6 for rate classification.

[b] Comments:

(1) Reaction is quite fast but there is a rather large side reaction between this active agent and the solvent system which makes it inapplicable.

(2) The side reaction with the titration medium was very large (~0.2 μeq/sec) while there was no indication of the halogenation of the polyol or even the much more reactive allyl alcohol.

(3) Desired reaction is very rapid but the solvent side reaction is also greater. However, this system is applicable if necessary, using CCT.

(4) Reacts slowly with polyol but rapidly with vinyl ether.

(5) Does not react with vinyl ether.

Whether the halogen and interhalogen complexes of mercury(II) remain in solution or dissociate has not been ascertained. The remarkable reactivity of the product in reaction (29) can be explained by the given structure in which the polarity of the bromine chloride is enhanced by the positive charge on the mercury:

$$\left[\mathrm{Cl}:\overset{+}{\mathrm{Hg}} \leftarrow \overset{\delta-}{\mathrm{Cl}}\text{—}\overset{\delta+}{\mathrm{Br}}\right]$$

If the halogen or interhalogen dissociates from the mercury complex, the rates of reaction can be explained according to the proposal of Belcher and Fleet with the argument that equation (8) will be driven to the right with a greater force the higher the average charge on the mercury complexes. In this discussion it will be assumed that the halogen or interhalogen remains associated to the mercury.

The following conclusions can be drawn from Table VI: (*1*) The catalytic effect of mercury is not dependent on the type and concentration of its salts but rather on the halo-mercury(II) complexes available for oxidation in the titration medium; (*2*) the rate of reaction is proportional to the positive charge of the oxidized complex (compare experiments 2 and 4, 14 and 15); (*3*) the bromine chloride complexes are more reactive than the corresponding bromine complexes; (*4*) experiments 1 and 20 show that it is probably not possible to generate bromine chloride by oxidizing a solution containing free chloride and bromide ions—the availability of bromide ions must be controlled by using the proper complexes containing not more than one bromide ion per mercury(II) cation; and (*5*) the average number of halide ions per cation should not be as low as 2 and preferably should be around 2.5 or higher to minimize solvent background and sample side reactions.

The $[HgBr]^{2+}$ complex reacts quite rapidly with the titration medium (about 0.2 μeq/sec) while it showed no indication of halogenating the polyol or even the much more reactive allyl alcohol. The reason for the instability is the presence of an unpaired electron in the bromine atom complex. Its reluctance to react with the double bonds in spite of its high electrophilic strength can be explained by the π-electron theory, which requires the release of a halonium ion by the active agent in the initial step to form the π-complex.

It is interesting to note that the one condition that must be satisfied to generate the bromine chloride complexes (experiments 13–16) is exactly the one determined by Lucas and Pressman (220) to be necessary to obtain maximum catalysis, namely that the mercury(II):bromide ratio must be greater than 1. They made an additional observation that the sample bleached the bromine until little or no color was perceptible; then, after the specified reaction time, they "liberated the bromine with sodium chloride" before iodometric titration with standard thiosulfate. From the results of our CCT studies, it appears that at the point at which Lucas and Pressman observed the bleaching of the bromine by the sample, the bromate had oxidized the bromides on all the complexes with two or more halide ions per mercury(II) ion. The acidic bromate is not capable of oxidizing bromomercurate(II) ion (this was proven in our laboratory) as this must yield the highly unstable bromine atom complex of mercury(II) (see experiments 5, 8, and 10). Therefore part of the bromate remained dormant throughout the reaction period and its oxidizing potential was harnessed after the addition of sodium chloride. Complexation of the chloride ions led to the formation of oxidizable species that yielded bromine chloride complexes of mercury(II). The occasional very high results may be attributed to the high reactivity of these complexes, especially at $10^{-2}M$ or greater concentration which was possible under their conditions. CCT has shown that $10^{-5}M$ $[Cl{:}Hg \leftarrow ClBr]^+$ has a solvent background only slightly higher than the tribromide ion while at $2 \times 10^{-5}M$ there is an appreciable increase and a large increase at the $10^{-4}M$ level.

The combination of CCT with controlled mercury catalysis is especially useful in the titration of compounds that are easily substituted. For

TABLE VII

Rate of Halogenation of Various Olefins by Tribromide and Bromine Chloride Complex of Chloromercurate(II) in 90:10 Acetic Acid–Water (155)

Compound	Bromination rate[a]	
	Br_3^-	$[ClHg \leftarrow ClBr]^+$
Allyl alcohol	III–IV	I–II
2-Butene-1,4-diol	III–IV	I–II
2-Butene-1-ol	I–II	I
Crotonaldehyde	II–III[b]	
Cyclohexene	I	
Diisobutylene	I–II	
2-Methyl-2-propene-1-ol	I	
Oleic acid		I
Propyl oleate	II	
Propylene trimer	II[c]	

[a]See Fig. 6 for the classification of rate curves.

[b]Tribromide in aqueous medium. Reaction is slow in acid medium.

[c]Propylene trimer must be titrated at ice temperature to minimize substitution. Mercury catalysis gives large positive error due to substitution.

example, each mole of 2,4,4-trimethyl-1-pentene consumed 1.49 moles of $[ClHg \leftarrow ClBr]^+$ while it consumed 1.00 mole of Br_3^- or $[Cl_2Hg \leftarrow Br_2]^0$. It is apparent that even an extrapolated titration cannot eliminate substitution which is greatly accelerated by catalysis. Hence, mercury catalysis should be used with discretion. The practice of forcing a difficult reaction by catalysis, high reagent concentration, and long reaction time to obtain the "theoretical" result should be discouraged, because it is obvious that any desired result can be obtained by such method. It would be preferable to select an electronically compatible reagent for any given sample.

The two most useful electrogenerated halogenating agents are the two extremes, Br_3^- and $[ClHg \leftarrow ClBr]^+$, which are convenient for the rapid spot-checking of unknown olefins. The rate classification of various olefins are given in Table VII for these two reagents.

Recently, Brand and co-workers (56a) used the electrogenerated bromine choride complex of mercury(II) in the amperometric titration of cinnamic acid. Of the several methods of endpoint detection studied, the biamperometric measurement of the excess reagent was found to be most satisfactory.

(4) Interferences

Since halogens are strong oxidizing agents, many reducible substances interfere. Interfering sulfur and nitrogen compounds are tabulated in Table VIII (13). Polycyclic aromatics, highly alkyl-substituted benzenes, *trans*-decalin, tetramethyllead, and tetraethyllead also interfere (13). Peroxides give low results in iodometric procedures because they oxidize iodide ions in the back-titration step.

(5) Definitions

Halogenation results are usually expressed as "iodine number" for fats and oils and "bromine number" for other olefinic compounds. These terms are defined as the grams of free halogen consumed by 100 g of samples:

$$\text{Iodine number} = \frac{(\text{Milliequivalents of } I_2)\ (12.69)}{(\text{Weight of sample, g})} \tag{30}$$

$$\text{Bromine number} = \frac{(\text{Milliequivalents of } Br_2)\ (7.99)}{(\text{Weight of sample, g})} \tag{31}$$

Unsaturation results may also be reported as millimoles of unsaturation

TABLE VIII

Compounds that Interfere in the Bromide–Bromate Method for Olefins (ASTM Designation 1159–64T) (13)

Compound	Purity,[a] %	Bromine number		
		Theory	Found	Deviation
Aromatics, Polycyclic				
Anthracene	—[b]	0.0	11.8	+11.8
Phenanthrene	—[b]	0.0	3.9	+3.9
Sulfur Compounds				
Ethanethiol (ethyl mercaptan)	99.95	0.0	209	+209
3-Thiapentane (ethyl sulfide)	99.94	0.0	184	+184
2,3-Dithiabutane (methyl disulfide)	99.97	0.0	1.1	+1.1
Thiacyclobutane (trimethylene sulfide)	99.95	0.0	214	+214
Thiophene	99.99	0.0	0.4	+0.4
Thiacyclopentane (tetrahydrothiophene)	99.95	0.0	183	+183
3,4-Dithiahexane (diethyl disulfide)	99.90	0.0	0.4	+0.4
2-Methyl-2-propanethiol (*tert*-butyl mercaptan)	99.92	0.0	141	+141
1-Pentanethiol (Amyl mercaptan)	99.92	0.0	83	+83
Nitrogen Compounds				
Pyrrolidine	99.85	0.0	11.8	+11.8
Pyridine	—[c]	0.0	1.4	+1.4
2-Methylpyridine	99.90	0.0	0.9	+0.9
4-Methylpyridine	99+[d]	0.0	1.7	+1.7
2,4,6-Trimethylpyridine	99+[d]	0.0	2.7	+2.7
2-(5-Nonyl)pyridine	—[e]	0.0	1.4	+1.4
Pyrrole	99.99	0.0	873	+873
2-Methylpyrrole	98+[f]	0.0	708	+708
2,4-Dimethylpyrrole	98+[f]	0.0	484	+484
2,5-Dimethylpyrrole	99.9[g]	0.0	869	+869
2,4-Dimethyl-3-ethylpyrrole	98+[f]	0.0	248	+248
1-(1-Butyl)pyrrole	98+[f]	0.0	472	+472

[a]API Standard Samples, unless otherwise noted.
[b]Eastman white label product.
[c]Meets ACS specifications.
[d]Purity estimated by spectra and GLC.
[e]Sample of best purity from Vigreaux distillation.
[f]Samples supplied by API Project 52.
[g]Purity estimated from freezing point.

per gram of sample:

$$\text{Unsaturation value} = \frac{\text{Milliequivalents of reagent}}{(\text{Weight of sample, g})\ (p)} \qquad (32)$$

where p is the number of milliequivalents of reagent required per millimole of unsaturation. Replacement of iodine number, bromine number, thiocyanogen number, hydrogen number, etc., with "unsaturation value" should greatly facilitate communication among chemists because all "unsaturation values" as defined by equation (32) should be directly comparable no matter what method was used to obtain the values.

The molecular weight of a pure olefin is given by

$$\text{Molecular weight} = \frac{1000}{(\text{Unsaturation value})\ (n)} \qquad (33)$$

where n is the number of double bonds per molecule.

The chemical industry has adopted the practice of reporting unsaturation results by the mercuration method for polyoxyalkylene polyols in milliequivalents per gram (17):

$$\text{Total unsaturation} = \frac{(\text{Milliequivalents KOH})}{(\text{Weight of sample, g})} - A \qquad (34)$$

where A is the milliequivalents of KOH used to titrate the acidity in 1 g of sample. Unfortunately, any chemist who is not aware of the fact that each millimole of unsaturation yields 1 meq of acid in the mercuration reaction [i.e., $p = 1$, in equation (32)] will erroneously divide the total unsaturation by 2 in the actual application of this value.

b. Bromine

Solutions of free bromine were the first halogenating agents used in the determination of olefins but these methods suffered from the instability of the reagents due to volatilization of the bromine. The most generally applicable solvents are acetic acid and carbon tetrachloride. These methods have largely been superseded by improved methods, but the bromine in carbon tetrachloride is still useful for samples that cannot be dissolved in the polar solvents used in the other methods.

(1) Bromine in Acetic Acid

Manchot and Oberhauser (224) used $0.1N$ bromine in acetic acid to determine unsaturation in vegetable oils by back-titrating the bromine with arsenious oxide. Uhrig and Levin (354) employed a direct titration to the visual orange-yellow endpoint using a $0.8N$ bromine solution. Mercuric chloride (373) and mercuric acetate (172) catalysts were used to determine unsaturation in vegetable oils.

Fritz and Wood (128a), who used a photometric endpoint, showed that in the absence of mercury(II) catalyst, they could titrate nucleoplilic olefins selectively in the presence of electrophilic olefins. Their data showed a slight positive bias due to the slow reaction of the electrophilic olefins with bromine. Application of CCT (see Section VI-A-1-a-(2)-(c)) can eliminate the bias due to slow side reactions.

The instability of the bromine in acetic acid was overcome by Rosenmund and co-workers (294,295) by making the solution 0.01*M* sulfuric acid and 0.0075*M* pyridine. The reagent reacted rapidly (2 min) with many fatty materials and it appeared that it did not undergo substitution. They attributed the good qualities of this reagent to a reaction product between bromine and pyridine. On the other hand, the effect of pyridine and sulfuric acid could have been simply catalytic since the mole ratio of bromine to pyridine was about 7. This reagent is commonly referred to as the Rosenmund-Kuhnhenn or the PSDB (pyridinium sulfate dibromide) reagent. The latter name may be a misnomer; a name such as pyridinium sulfate–bromine reagent would be more accurate since the bromine exists in the oxidized state.

The Rosenmund-Kuhnhenn reagent has been applied satisfactorily to sterols and to the unsaponifiable fraction of olive oil (51), to methallyl chloride (116), to polyallyl esters and ethers (55), to furan, sylvan, and furfuryl alcohol, but reacted only slightly with furfural, furoic acid, and furyl cyanide. Rowe and co-workers (297) found the reagent applicable to conjugated olefins in tall oil if catalyzed by mercuric acetate and the reaction period was 16 hr in 70% excess of the reagent. By using 10 ml of 2.5% mercuric acetate in acetic acid, Benham and Klee (36,192) cut the reaction time down to 1 min for isolated olefins and 30–120 min for conjugated oils. However, on polymerized oils, Lips (217) found that the apparent results increased with reaction time and reagent excess. This agrees with the observation by Wilson (372) that this reagent substitutes branched olefins but hardly substitutes straight-chain olefins.

Planck, Pack, and Goldblatt (269) found that the Benham and Klee procedure gave iodine values for conjugated oils that agreed very well with the catalytic hydrogenation values by using two modifications: (*1*) avoiding exposure to light during bromination and (*2*) reversing the order of addition of the catalyst, i.e., adding the catalyst to the sample before the reagent is added. Their results are given in Table IX. This modification has been adopted by ASTM for the determination of unsaturation in drying oils and the details of the procedure are given in reference 16.

The Rowe-Furnas-Bliss modification (297) of the Rosenmund-Kuhnhenn method (294) has been published by ASTM as a general method; the details

TABLE IX

Comparison of Iodine Values Obtained by the Planck-Pack-Goldblatt and the Catalytic Hydrogenation Methods (269)

Sample	Excess of reagent,[a] %	Value	Catalytic hydrogenation value
Tung oil			
domestic	222	231.9	232.3
	222	232.2	
Chinese	240	228.7	227.9
	259	228.6	
	266	228.1	
Chinese (3–5 yr old)	188	224.3	228.6
	298	225.9	
Nyasaland (*A. montana*)	260	215.9	216.0
	260	219.4	
	291	215.4	
isomerized (β)	235	230.7	229.2
	269	230.9	
	319	230.8	
α-Eleostearic acid, 100%	208	272.8	272.8
	243	274.4	
	247	272.3	
β-Eleostearic acid, 98%	216	273.2	270.2
	230	270.5	
	230	274.7	

[a]$100(B - S)/S$, where B and S represent the volumes of thiosulfate solution used for blank and sample titrations, respectively.

of the procedure are given in reference 21. This method has been applied to α-olefins, cyclic olefins, unsaturated aliphatic side chains on aromatic rings, unsaturated esters and alcohols, acrylic and methacrylic esters, acids, and salts. It is not satisfactory for diisobutylene, some tertiary substituted hydrocarbons, commercial propylene trimer and tetramer, butene dimer, and mixed nonenes, octenes, and heptenes.

(2) Bromine in Carbon Tetrachloride

McIlhiney (236) recommended bromine in carbon tetrachloride to replace the less stable and slowly reacting Hübl reagent ($I_2 + HgCl_2$). The sample, dissolved in carbon tetrachloride, is reacted with an excess of the reagent and the excess bromine is titrated iodometrically with thiosulfate to the starch endpoint in a two-phase system. The hydrogen bromide that develops during the titration is determined by adding an excess of iodate to liberate an equivalent amount of iodine which is again titrated with the thiosulfate.

$$6HBr + KIO_3 + 5KI \rightarrow 6KBr + 3I_2 \qquad (35)$$

McIlhiney attributed the hydrogen bromide formation to substitution and therefore subtracted twice the bromine substitution value from the first titration to obtain the bromine addition result. This method of correction appeared good for many compounds, but in general the results were low and negative in some cases.

Buckwalter and Wagner (68) pointed out that spontaneous cleavage of hydrogen bromide from the dibromide addition product occurs easily in the presence of water; this would have explained the overcorrection in the McIlhiney method. They suggested that both the bromine and hydrogen bromide be swept away from the addition product with nitrogen into a solution of potassium iodide. This modification was still not completely satisfactory.

In attempts to minimize these side reactions, it was found that high temperature increases substitution (68) and that protection from light is beneficial (73). Others tried direct titrations using visual bromine color endpoints (131,132,251) and dead-stop electrometric endpoint using mercuric chloride catalyst (56). The precision of the latter method on several monoolefins was about 3%.

Perhaps the best variation of the McIlhiney procedure was developed at the Shell Development Company (318). They combined many features to minimize substitution errors. The sample is reacted in the dark at ice temperature with an excess of bromine in carbon tetrachloride and the excess bromine is determined iodometrically. One or two more determinations are made at increased reaction times and the results are extrapolated to zero time if there is significant change in bromine absorption with time. This extrapolation method of correcting for substitution is considerably simpler than previous methods. It is based on the assumption that the substitution reaction is slower than the addition reaction and that after the completion of the addition reaction it remains approximately constant.

The results by the Shell method are given in Table X. The low results on the conjugated dienes agree with the usual reluctance of the second mole of bromine to add to such molecules. The high result for pinene was attributed to the rupturing of the bridge ring besides addition of the bromine to the double bond. The results are generally satisfactory but there are more convenient methods available today for most of the samples in Table X. However, this method is still very useful because many samples, e.g., rubber-type samples, are not soluble in the polar solvents

TABLE X

Analytical Data by the Extrapolation Method (Bromine in Carbon Tetrachloride) (318)

Compound	Extent of reaction, % of theor.
1-Pentene	101
2-Pentene	97
2-Methyl-2-butene	100
2,2-Dimethylpropene	100
1-Hexene	100
2,3-Dimethyl-2-butene	99
1-Heptene	99
4-Methyl-2-pentene	101
2,3,3-Trimethyl-1-butene	99
1-Octene	100
2,3,4-Trimethyl-2-pentene	96
Diisobutylene	101
1-Decene	99
Triisobutylene	99
1-Tetradecene	100
1-Hexadecene	97
1-Octadecene	99
Styrene	99
α-Methylstyrene	100
4-Phenyl-1-butene	97
Stilbene	ca. 60
Cyclohexene	99
3-Methylcyclohexene	99
Pinene	ca. 150
2,5-Dimethyl-1,5-hexadiene	101
d-Limonene	100
2-Methyl-1,3-butadiene	100
4-Methyl-1,3-pentadiene	54
2,3-Dimethyl-1,3-butadiene	54

used in the other methods. Therefore, the details of this method are given in Section VII-A.

c. Tribromide

The presence of bromide ions stabilizes bromine solutions against volatilization due to the formation of the tribromide ion (equation (24)). However, the bromine concentration can be appreciable because of the low stability of the tribromide ion. An excessive amount of bromide ions slows down bromine addition (39) due to the decrease in the more electrophilic free bromine. However, a large excess of bromide is not detrimental but in fact beneficial when catalyzed by hydrochloric acid (71). Even better than the tribromide reagent is the *in situ* generation of the tribromide by stable standard solutions of oxidizing agents which are available in very high purity or by the ultimate standard, electricity.

(1) Tribromide in Methanol

Bromine in absolute methanol which has been saturated with sodium bromide is commonly referred to as the Kaufmann reagent (182,183,186). This reagent was specifically developed for the determination of unsaturation in fatty oils. The iodine values for soybean oil obtained by this method deviated only slightly from those obtained by the Wijs and Hanus methods, but it required a long reaction time (2 hr) and a large excess (300%) of reagent (114).

Some controversial results have been obtained in the application of this reagent to olefinic hydrocarbons. Uhrig and Levin (354) reported that substitution of triisobutylene gave high results, while Wilson (372) reported that although up to about 90% of the absorption could take place by substitution, substitution and addition are mutually exclusive so that good results are obtained provided that corrections are made for peroxide.

Critchfield (89) applied 0.2*N* tribromide in methanol for the determination of α,β-unsaturated acids and esters. These compounds react very slowly with electrophilic reagents because of the depletion of the π-electrons by the strong electron-attracting power of the ester and carboxylic acid

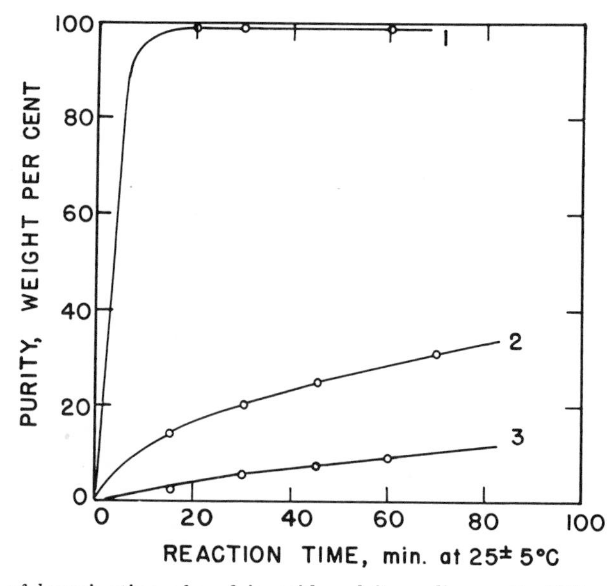

Fig. 7. Rate of bromination of maleic acid and its sodium salts (89). 1. Disodium salt; 2. Monosodium salt; 3. Free acid. (Courtesy of *Analytical Chemistry*.)

groups. Williams (371) showed that the rate of addition of bromine to acrylate ion is about 300 times faster than to the un-ionized acid. The effect of removing the protons on the rate of bromination of maleic acid is shown in Fig. 7 (89). It is evident that the acids must be neutralized in order to brominate the double bonds in a practical length of time. The esters were saponified with aqueous potassium hydroxide. After carefully neutralizing the potassium hydroxide, the unsaturation was determined by bromination. The results are shown in Table XI.

TABLE XI

Determination of Purity of α, β-Unsaturated Acids and Esters by Brominating Their Alkali Salts (89)

Compound	Average purity,[a] wt %	
	Bromination	Other
Acrylic acid	98.4 ± 0.2 (3)	98.5[b] (3)
Butyl acrylate	98.0 ± 0.1 (4)	98.4[b] (3)
Cellosolve acrylate	94.5 ± 0.1 (2)	95.4[b] (3)
Crotonic acid	99.7 ± 0.1 (4)	99.9[c] (2)
Dibutyl fumarate	99.8 ± 0.0 (2)	99.8[d]
Dibutyl maleate	96.5 ± 0.1 (2)	97.2[d]
Diethyl fumarate	99.4 ± 0.1 (2)	98.7[c] (2)
Diethyl maleate	98.3 ± 0.1 (2)	98.7[c] (2)
Ethyl acrylate	99.0[e]	99.0[b] (3)
Ethyl crotonate	97.6 ± 0.1 (2)	98.0[b] (3)
Fumaric acid	99.2 ± 0.1 (3)	99.9[f]
Maleic acid	98.6 ± 0.1 (3)	99.0[f]
Methyl methacrylate	99.5 ± 0.1 (3)	99.5[b] (3)

[a]Figures in parentheses represent number of determinations.
[b]Morpholine method.
[c]Sodium sulfite method.
[d]Saponification.
[e]Standard deviation for 15 degrees of freedom is 0.11.
[f]Total acidity.

(2) Tribromide in Water

Aqueous bromine containing excess potassium bromide was reported to have satisfactory stability and to be especially applicable to oxygen-containing olefinic compounds which react incompletely with some other halogenating agents (318). The normal procedure involves the shaking of the sample in the presence of ice with a 65–70% excess of reagent for 20 min and back-titration of the bromine iodometrically with sodium thiosulfate. To avoid loss of bromine, the reaction is carried out in an evacuated bottle (Procedure A). With difficultly reacting samples, the reaction is carried out at room temperature for 1 hr (Procedure B). Details of this procedure are found on page 241 of reference 271. Typical results by this procedure are given in Table XII.

The *in situ* generation of bromine by adding a bromide–bromate standard solution to an acidic medium is commonly referred to as the Francis (126) method although others employed this technique earlier. This reagent might have been classified under bromine instead of tribromide because only a small excess of bromide is used beyond the stoichiometric requirement in the generation of bromine. However, the trend today is

TABLE XII

Results by Aqueous Tribromide Method[a] (271)

Compound	Extent of reaction, % of theoretical	
	Proc. A	Proc. B
Allyl alcohol	99	—
β-Chloroallyl alcohol	98	100
Methallyl alcohol	102	—
Methylvinylcarbinol	97	—
Crotyl alcohol	100	—
4-Hydroxy-2-methyl-1-pentene	96	—
Acrolein	100	—
α-Chloroacrolein	10	92
Acrolein dimer	100	—
Methacrolein	100	—
Methyl isopropenyl ketone	99	—
Mesityl oxide	99	—
Acrolein diacetate	100	—
Methacrolein diacetate	99	—
1,1,3-Triallyloxypropane	101	—
Acrylic acid	86	97
Sodium acrylate	—	100
Methacrylic acid	99	99
Tetrahydrobenzoic acid	99	—
Vinyl acetate	100	—
Vinyl propionate	—	103
Vinyl butyrate	—	99
Vinyl crotonate	—	64
Vinyl oleate	—	100
Allyl acetate	98	—
Allyl propionate	100	—
Chloroallyl propionate	ca. 40	93
Allyl butyrate	101	—
Allyl caproate	98	—
Allyl caprylate	100	—

TABLE XII (*continued*)

Compounds	Extent of reaction, % of theoretical	
	Proc. A	Proc. B
Allyl hexahydrobenzoate	99	—
Diallyl adipate	91	97
Diallyl phthalate	ca. 60	101
Allyl crotonate	77	100
Methyl α-chloroacrylate	<1	66
Ethyl acrylate	10	101
α-Allyl glycerol ether	100	—
Allyl glycidyl ether	100	—
Allyl 1,3-dichloro-2-propyl ether	99	—
Allyl sulfolanyl ether	100	—
Dimethallyl ether	86	101
Vinyl chloride	100	—
Dichloroethylene	—	36
1,1-Dichloro-1-propene	—	(low)
1,3-Dichloro-1-propene	(low)	96
2,3-Dichloro-1-propene	(low)	99
3-Chloro-1-butene	2	103
1-Chloro-2-butene	101	—
1,4-Dichloro-2-butene	80	97
Allyl *N*-ethyl carbamate	100	—
Dibutyl allyl phosphate	100	—
2,4-Dimethyl-3-sulfolene	98	100

[a]Procedures A and B are defined in text.

to use direct titrations to minimize substitution. In direct titrations, the bromine concentration is kept very low and in the presence of the accumulated bromide, hydrochloric acid, mercuric chloride, etc., the bromine would probably be in equilibrium with halides, interhalogen, halogen halide, and interhalogen halide. Thus the classification of the Francis reagent under tribromide is justified.

The earlier applications of the Francis method was essentially an aqueous bromine method in which the sample was reacted with an excess of the reagent and the bromine was back-titrated iodometrically. Improvements and refinements by many workers (271) were combined into a procedure by Johnson and Clark which was incorporated into the ASTM Standards (11) (Designation: D-1158-55-T; this method is not included in the 1966 Book of ASTM Standards; for details of the procedure, see page 248 of reference 271). The reaction is carried out at ice temperature in 0.24N hydrochloric acid in acetic acid to which the 0.5N aqueous bromide–bromate reagent is added dropwise until a yellow color persists for at least

5 sec. Then 1 ml of the reagent is added, shaken for 40 sec, and back-titrated iodometrically with 0.1*N* sodium thiosulfate to the starch endpoint. The results by this method are good for straight-chain and cyclic isolated olefins, but the results for the branched olefins are generally too high.

DuBois and Skoog (106) used the same bromide–bromate reagent in the direct titration of olefins, using the dead stop "electric eye" endpoint and mercuric chloride catalyst to sharpen the endpoint. The titration was carried out at 0–5°C to minimize substitution. The titration medium consisted of 80 ml of glacial acetic acid, 15 ml of carbon tetrachloride, 7 ml of methanol, 2 ml 6*N* sulfuric acid, and 2 ml of mercuric chloride solution (10 g of mercuric chloride in 100 ml of 95% ethanol). This method was adopted in 1955 by ASTM (12) and the results for olefins of various structures are presented in Table XIII. This method (ASTM Designation D-1158-55-T), however, has been superseded by ASTM Designation D-1159-64 (13), which is an uncatalyzed version of D-1158-55-T (11).

TABLE XIII

Analytical Data by the ASTM Electrometric Bromide–Bromate Method (11) (Catalyzed by Mercuric Chloride)

Compound	Purity, %	Extent of reaction, % of theor.
1-Hexene	99	100.5
2-Hexene	95	100.2
1-Tetradecene	99.7	101.6
1-Hexadecene	97	101.2
1-Octadecene	95	96.8
2-Methyl-1-pentene	95	94.4
4-Methyl-1-pentene	95	99.5
4-Methyl-*cis*-2-pentene	99.7	101.9
4-Methyl-*trans*-2-pentene	99.4	101.0
2,3-Dimethyl-1-butene	99.8	114.6
2,3-Dimethyl-2-butene	99.7	101.5
2-Ethyl-1-butene	95	106.3
2,3,3-Trimethyl-1-butene	98.6	117.9
Diisobutylene	—[a]	99.4
Cyclopentene	99	99.9
Cyclohexene	100	102.0
d-Limonene	—[b]	99.3
2-Methyl-1,3-butadiene	99	51.0
Styrene	—	99.1
α-Methylstyrene	—	98.3

[a]Purified by silica gel percolation.
[b]Eastman white label product.

The results by the mercuric chloride-catalyzed electrometric bromide–bromate method are high not only for the branched olefins, but also for the straight-chain olefins. These anomalous results can be explained by the observations made during the study of controlled mercury catalysis (see Section VI-A-1-a-(3)). The bromide ions added during the titration will complex mercuric chloride to form bromodichloromercurate(II) ion. This complex ion can be oxidized to the extremely reactive bromine chloride complex of mercury(II) in high local concentrations. A big advantage of the coulometric method is the minimization of high local concentrations of the reagent.

Unger (356) made a critical comparison of the visual bromide–bromate method, the catalyzed electrometric bromide–bromate method, and a third method which was the same as the latter but without mercuric chloride catalyst. He found that the first two methods gave results which showed consistently very high positive deviations even for unbranched olefins while the noncatalyzed modification gave good results except that the results for the α-olefins were somewhat low. He concluded that the third method was the best for determining olefinc unsaturation in hydrocarbon mixtures by bromine addition.

The electrometric bromide–bromate method published in 1966 with the same designation (D 1159) omits the mercuric chloride (13). A wide variety of olefins have been screened by this uncatalyzed electrometric bromide–bromate method and the results are given in Table XIV. Omission of the catalyst has reduced the magnitude of the positive deviations of the results for the branched olefins shown by Unger. The results were low for many compounds that are usually easily brominated. The α-olefins showed a consistent negative bias.

Comparison of Tables XIII and XIV emphasizes the difficulty of setting up a single method for a group of very similar compounds with only alkyl substituents. A single method for this type of compounds is desirable because many cracked petroleum fractions are complex mixtures of compounds of the types listed in these tables. The best compromise would probably be to apply controlled mercury catalysis in which either tetrabromomercurate(II) or dibromodichloromercurate(II) complexes are present in the titration medium and the bromate is added as required. To avoid localized concentration of the reagent, coulometric generation of the active agent would be preferred over the bromate method. The extrapolation of the quantity of reagent added or generated to zero time should allow correction for substitution and other slow side reactions.

The coulometric methods of generating halogens were discussed in Section VI-A-1-a-(2)-(b). Mercury catalysis was used in most of these

TABLE XIV

Analytical Data on Olefins by Electrometric Bromide–Bromate Method (ASTM Designation 1159–64T) (13) (Mercuric chloride not added)

Compound	Purity,[a] %	Bromine number		
		Theory	Found	Deviation
Straight-Chain Olefins				
1-Pentene	99.7	228.0	208	−20
trans-2-Pentene	99.91	228.0	235	+7
1-Hexene	—[b]	189.9	181	−9
cis-2-Hexene	99.80	189.9	189	−1
trans-2-Hexene	99.83	189.9	189	−1
cis-3-Hexene	99.87	189.9	193	+3
trans-3-Hexene	99.94	189.9	191.4	+1.5
1-Heptene	99.8	162.8	136	−27
trans-2-Heptene	99.85	162.8	163	0
trans-3-Heptene	99.80	162.8	163	0
1-Octene	99.7	142.4	132	−10
2-Octene	—[b]	142.4	139	−3
trans-4-Octene	99.84	142.4	149	+7
1-Decene	99.89	114.1	111.4	−2.7
1-Dodecene	99.9	95.1	82.9	−12.2
1-Tridecene	99.8	87.7	81.4	−6.3
1-Tetradecene	99.7	81.4	70.8	−10.6
1-Pentadecene	99.8	76.0	62.9	−13.1
1-Hexadecene	99.84	71.2	62.8	−8.4
Branched Chain Olefins				
2-Methyl-1-butene	99.90	228.0	231.8	+3.8
2-Methyl-2-butene	99.94	228.0	235	+7
2,3-Dimethyl-1-butene	99.86	189.9	194	+4
3,3-Dimethyl-1-butene	99.91	189.9	167	−23
2-Ethyl-1-butene	99.90	189.9	198	+8
2,3-Dimethyl-2-butene	99.90	189.9	191	+1
2-Methyl-1-pentene	99.92	189.9	182	−8
3-Methyl-1-pentene	99.70	189.9	152	−38
4-Methyl-1-pentene	99.82	189.9	176	−14
2-Methyl-2-pentene	99.91	189.9	190	0
3-Methyl-*cis*-2-pentene	99.85	189.9	193.7	+3.8
3-Methyl-*trans*-2-pentene	99.86	189.9	191	+1
4-Methyl-*cis*-2-pentene	99.92	189.9	190	0
4-Methyl-*trans*-2-pentene	99.75	189.9	190	0
2,3,3-Trimethyl-1-butene	99.94	162.8	161	−2
3-Methyl-2-ethyl-1-butene	99.8	162.8	165.4	+2.6
2,3-Dimethyl-1-pentene	99.80	162.8	158.5	−4.3
2,4-Dimethyl-1-pentene	99.87	162.8	152.8	−10.0
2,3-Dimethyl-2-pentene	99.6	162.8	162.3	−0.5

TABLE XIV (*continued*)

Compound	Purity,[a] %	Bromine number		
		Theory	Found	Deviation
4,4-Dimethyl-*cis*-2-pentene	99.79	162.8	159	−4
4,4-Dimethyl-*trans*-2-pentene	99.91	162.8	158	−5
3-Ethyl-1-pentene	99.85	162.8	173.1	+10.3
3-Ethyl-2-pentene	99.80	162.8	165	+2
2-Methyl-1-hexene	99.88	162.8	161	−2
5-Methyl-1-hexene	99.80	162.8	154	−9
3-Methyl-*cis*-2-hexene	99.8	162.8	163.6	+0.8
2-Methyl-*trans*-3-hexene	99.9	162.8	163.4	+0.6
2-Methyl-3-ethyl-1-pentene	99.81	142.4	139.8	−2.6
2,4,4-Trimethyl-1-pentene	99.91	142.4	137.0	−5.4
2,4,4-Trimethyl-2-pentene	99.92	142.4	141.2	−1.2
Diisobutene	—[c]	142.4	139.8[c]	−2.6
2-Ethyl-1-hexene	—[d]	142.4	140.2	−2.2
2,3-Dimethyl-2-hexene	99.71	142.4	143	+1
2,5-Dimethyl-2-hexene	99.8	142.4	142.8	+0.4
2,2-Dimethyl-*trans*-3-hexene	99.80	142.4	139	−3
Triisobutene	99.0	95	57.5	−37.5
Nonconjugated Cyclic Diolefins				
4-vinyl-1-cyclohexene	99.90	295.5	210[e]	(−85)
dl-1, 8(9)-*p*-Menthadiene (dipentene)	98–100	234.6	225.2	−9.4
Conjugated Diolefins				
2-Methyl-1,3-butadiene (isoprene)	99.96	470	235.7	−234
cis-1,3-Pentadiene	99.92	470	285.3	−185
trans-1,3-Pentadiene	99.92	470	234	−236
2-Methyl-1,3-pentadiene	95+[g]	389	197.3	−192
2,3-Dimethyl-1,3-butadiene	99.93	389	186.1	−203
Nonconjugated Diolefins				
1,2-Pentadiene	99.66	470	230	−240
1,4-Pentadiene	99.93	470	185	−285
2,3-Pentadiene	99.85	470	227	−243
1,5-Hexadiene	99.89	389	352	−37
Aromatics with Unsaturated Side Chains				
Phenylethylene (styrene)	—[h]	153.4	123.6	−29.8
Methylphenylethylene (α-methylstyrene)	—[h]	135.3	133.2	−2.1
Allylbenzene	97.8	135.2	0.0	−135.2
Cyclic Olefins				
Cyclopentene	99.97	234.6	237	+2
Cyclohexene	99.98	194.6	193.2	−1.4
Cyclohexene	—[c]	194.6	192.8[c]	−1.8

(*continued*)

TABLE XIV (*continued*)

Compound	Purity,[a] %	Bromine number		
		Theory	Found	Deviation
1-Methylcyclopentene	99.86	194.6	209	+14
1-Methylcyclohexene	99.82	166	162	−4
Ethenylcyclopentane (vinylcyclopentane)	99.91	166	164	−2
Ethylidenecyclopentane	99.96	166.2	167.7	+1.5
1,2-Dimethylcyclohexene	99.94	145.0	150.9	+5.9
3-Cyclopentyl-1-propene	99.87	145.0	140.9	−4.1
Ethylidenecyclohexane	99.86	145.0	147.0	+2.0
Ethenylcyclohexane (vinylcyclohexane)	99.95	145	139	−6
1-Ethylcyclohexene	99.83	145	146.6	+1.6
Indene	—[b]	137.7	134	−4

[a]API Standard Samples, unless otherwise noted.
[b]Purity not stated.
[c]Average value obtained in September 1957 Cooperative Program on purified Eastman product.
[d]Dow Research chemical.
[e]Approximate value.
[f]Hercules Powder Co., experimental sample.
[g]From Pennsylvania State University.
[h]Eastman White Label product, distilled, 50-mm pressure just prior to test.
[i]Synthesized (352°F boiling point fraction). Purity determined by gas chromatography, impurities identified as diisobutenes.

methods and the extrapolation of the cumulative electricity was not used except in a few cases (107,155,363,364).

(3) Tribromide in Chloroform

A fresh solution of bromine in chloroform undergoes a rapid initial loss in titer and the characteristic color of bromine disappears after several days. Gal'pern and Vinogradova (130) attributed this phenomenon to the formation of hydrogen tribromide:

$$5Br_2 + 2HCCl_3 \rightarrow HBr_3 + HBr_5 + 2CBrCl_3 \quad (36)$$

This reagent continues to deteriorate at a rather rapid rate and it must be standardized two or three times per day. They obtained good results on some compounds, slightly high on others, and unaccountably high for styrene.

d. Iodine

Iodine adds very slowly to double bonds, but the addition is hastened if the reaction is catalyzed by mercury (164,165). The Hübl reagent, however, which consists of an ethanol solution of iodine and mercuric chloride, is unstable and not as good as the Wijs reagent.

A special application of iodine was made by Siggia and Edsberg (323) in the determination of vinyl ethers which react with iodine in the presence of alcohol to give the iodoacetal.

$$R{-}OCH{=}CH_2 + I_2 + CH_3OH \rightarrow \underset{\displaystyle OCH_3}{R{-}\underset{|}{O}CH}{-}CH_2I + HI \qquad (37)$$

The sample is reacted in a closed flask with vigorous shaking (shaking not necessary for nonvolatile samples) for 10 min and the excess iodine is titrated with sodium thiosulfate to the disappearance of the yellow color. The results in Table XV show that the method is accurate and reproducible.

TABLE XV

Analysis of Vinyl Alkyl Ethers by Iodine–Methanol Method (323)

Sample	Analysis by hydrolysis and bisulfite addition, %	Analysis by iodine–methanol method, %
Methyl vinyl ether	96.7	96.51
		96.48
		97.02
Methyl vinyl ether	—[a]	99.61
		99.01
Ethyl vinyl ether	—[a]	99.03
		99.75
		99.51
		99.72
n-Butyl vinyl ether	98.7	98.45
		98.85
		98.81
Isobutyl vinyl ether	—[a]	99.34
		99.35
		99.25

[a]Ether was purified by washing five times with alkaline water (pH 8) to remove acetaldehyde and alcohol, cooled to −50°C and filtered from any ice which had separated, and finally distilled from sodium.

e. Iodine Chloride

Iodine chloride is considerably more reactive than iodine owing to its higher oxidizing potential and polarity. A solution of this reagent in acetic acid is referred to as the Wijs reagent. It is much more stable and reactive than the Hübl reagent. Wijs (369) suggested a 3–10 min reaction time but today a 30 min reaction time is used. Although higher results may be obtained for conjugated oils by increasing the reagent excess and reaction time, this reagent is not sufficiently reactive for this type of material. This method has been widely adopted for the determination of unsaturation in nonconjugated fatty oils. The Wijs reagent substitutes compounds with alkyl branching on both doubly bonded carbons. The details of this method are given in Section VII-C.

A solution of iodine chloride in carbon tetrachloride was proposed by Marshall (229) for samples that are insoluble in acetic acid. Reproducible results were obtained for natural rubber hydrocarbons (188) and for butadiene and related polymers (189); however, these results did not always correspond to the true unsaturation value. This anomaly was attributed by Lee, Kolthoff, and Mairs (211) to substitution which occurred after 90–95% of the addition took place in the first several minutes. Lee, Kolthoff, and Johnson (210) obtained good results on several branched olefins and polymers by the extrapolation of 5–8 results obtained by using a varied reagent excess and a 1-hr reaction time.

Hoffman and Green (158) and Hiscox (157) found that the reaction time could be shortened to 3 min by adding mercuric acetate catalyst. The results agreed with the standard Wijs method.

Iodine chloride can be prepared in water by dissolving $^1/_{60}$ mole potassium iodate and $^1/_{30}$ mole potassium iodide in 200 ml of water, adding 300 ml hydrochloric acid, and diluting to 1 liter (168). This reagent is said to be very stable. The oil samples are dissolved in ether, which extracts the iodine chloride from the two-phase system, thus optimizing the concentration of the reactants.

Graupner and Aluise (140) proposed a rapid method for determining the halogen ratio of Wijs reagent. The total halogen is determined by the usual iodometric titration with standard sodium thiosulfate. For the determination of the free iodine, the tedious Lopes method (219) is replaced by a direct titration with potassium iodate.

$$2I_2 + KIO_3 + 6HCl \rightarrow 6KCl + 5ICl + 3H_2O \qquad (42)$$

The titration is carried out with vigorous shaking until the color of the iodine disappears from the carbon tetrachloride phase. The details of this procedure are given in Section VII-C.

f. Iodine Bromide

Hanus (148a) introduced a halogenating reagent consisting of an iodine bromide solution in acetic acid. This reagent is very stable and is often preferred to the Wijs reagent but the application of the Hanus method was handicapped by the early popularity of the Wijs method.

In the standard Hanus method the sample is reacted with 0.2*N* iodine bromide solution for 30 min and the excess reagent is back-titrated iodometrically with 0.1*N* sodium thiosulfate. The results on nonconjugated fatty materials are precise but usually a few percent lower than those obtained by the Wijs procedure. The standard Hanus method is not applicable to conjugated oils.

Mikusch and Frazier (245) showed that by reacting the conjugated oils with 500–800% excess of double strength reagent (0.4*N* iodine bromide, often referred to as the Woburn reagent) for 3 hr at 0°C, quantitative addition to all three conjugated double bonds can be effected. Their results by the Woburn method is given in Table XVI. The high result for tall oil was attributed to rosin acid and sterols which react incompletely with Wijs reagent.

TABLE XVI

Iodine Values of Some Fatty Materials by the Woburn Method (245)

Sample	Normality of IBr soln	Halogen excess, %	Temp., °C	Time, hr	Iodine number: IBr soln	Iodine number: Wijs soln
Soybean oil	0.36	500–800	20	1	126.5	130.5
Soybean pentaerythritol ester	0.40	500–800	20	1	128.4	127.6
Soybean fatty acids	0.40	500–800	20	1	136.0	137.8
Linseed oil	0.36	500–800	20	1	185.5	186.8
Linseed fatty acids	0.40	500–800	20	1	196.4	194.2
Sardine oil	0.36	500–800	20	1	189.2	196.9
Tall oil	0.40	500–800	20	1	203.8	169.5
Walnut oil	0.32	500–800	20	1	156.0	155.2
Castor oil	0.32	500–800	20	1	88.9	87.4
				1	219.9	
Tung oil	0.40	545 ± 30	0	2.5	225.4	224.2[a]
				3	226.6	
				5	226.8	
β-Eleostearic acid	0.32	550–600	20	1	272.7	273.7[a]
	0.40	600–2000	0	3	273.8	
9,11-Linoleic acid	0.32	400–500	20	1	183.3	
			0	3	182.6	181.2[b]
			0	5	183.7	
Oiticica oil	0.40	800–1000	0	3	203.9	—

[a]Probable value from hydrogenation.
[b]Theoretical iodine value.

In applications of the Hanus reagents to petroleum products, it showed some promise in the determination of conjugated dienes by increasing the reagent excess and the reaction time. However, Winward and Garner (374) showed that this dependency on reagent excess is due mainly to substitution of certain aromatic types, e.g., mesitylene and isodurene, and to a lesser extent, methyl-substituted naphthalenes.

g. Bromine Chloride

The bromine chloride method is essentially a bromide–bromate method in which the bromide–bromate ratio is fixed at 2:1 instead of 5:1 and a large excess of hydrochloric acid is used. The resulting oxidation product is bromine chloride, as was shown earlier in equation (21). The bromine chloride can be generated as needed or a bromine chloride stock solution can be prepared. Burger and Schulek (69) prepared 0.1*N* bromine chlcride standard solution in the following manner: Dissolve 2.7835 g of potassium bromate and 3.967 g of potassium bromide in about 300–400 ml of water, add 365 ml of 20% hydrochloric acid, and after a few minutes, dilute the solution with water to 1000 ml. To prepare the 0.1*N* standard bromate solution, simply omit the hydrochloric acid.

The authors applied these reagents to cinnamic aldehyde, crotonic aldehyde, and acrolein. The aldehyde group was not attacked and the recoveries were very good, allowing a reaction time of 5 min. The reaction was complete almost instantaneously when only a small excess (5–10%) of bromine chloride was present. Reaction periods from 0.5–15 min and reagent excess ranging from 4 to 170% had no significant effect. However, it was interesting to note that in the case of acrolein there was an inverse relationship between recovery and bromine chloride excess; this type of phenomenon is rare in halogenation reactions and it may be peculiar to this compound or just a coincidence.

Belcher and Fleet (34) applied this reagent to the submicrodetermination of olefinic unsaturation. They reacted 40–90 μg of sample (dissolved in either 1 or 2 drops of 1*N* sodium hydroxide or 1 ml of methanol) with 0.02–0.10 ml of 0.1*N* bromine chloride solution at 9–10°C. The excess bromine chloride was back-titrated iodometrically using micro techniques. Their results are listed in Table XVII. The numbers in parentheses refer to the number of results used in obtaining the average found values.

They found that when mercury was used to catalyze the reaction, extensive substitution could occur with some samples and this side reaction was greatly enhanced by excessive amounts of reagent. They reported that acetic acid was unsatisfactory as a solvent because of some side

TABLE XVII

Analytical Data by Submicro Bromine Chloride Method (34)

Compounds	% >C=C< Theory	% >C=C< Found	Average deviation, %	Conditions
	α,β-Unsaturated Compounds			
Maleic acid	20.69	20.62 (12)	±0.32	1 hr, Hg^{2+}
Fumaric acid	20.69	20.56 (6)	±0.36	$1^1/_4$–$1^1/_2$ hr, Hg^{2+}
Cinnamic acid	16.21	16.04 (9)	±0.29	$^1/_2$ hr, MeOH
Crotonic acid	27.90	27.86 (6)	±0.40	$^1/_2$ hr, Hg^{2+}
Itaconic acid	18.46	18.41 (6)	±0.30	1–$1^1/_2$ hr, Hg^{2+}
p-Nitrocinnamic acid	12.44	12.13 (8)	±0.93	2–$2^1/_2$ hr, MeOH
β,β-Dimethylacrylic acid	25.52	25.82 (5)	±1.23	2–$2^1/_2$ hr, MeOH
Cinnamaldehyde	18.17	18.64 (6)	±0.50	$^1/_4$–$^1/_2$ hr, MeOH
Acrolein	42.84	41.38 (6)	±0.53	$^1/_4$ hr, MeOH, −9°C
	Conjugated Dienes			
trans-Stilbene	13.33	13.38 (9)	±0.36	$^1/_4$–$^7/_{12}$ hr, MeOH
Triphenylethylene	9.37	9.66 (4)	±0.20	1 hr, MeOH
trans-1,4-Diphenylbuta-1,3-diene	14.28	13.88 (5)	±0.69	$^1/_2$–$^3/_4$ hr, MeOH
trans, trans-Hexa-2,4-dienoic acid (sorbic acid)	42.84	42.92 (4)	±1.54	$1^1/_4$–2 hr, MeOH
Stilbestrol dipropionate	6.31	12.02 (5)	±0.46	$^1/_2$–$1^1/_4$ hr, MeOH
	Vinyl Compounds			
Styrene	23.06	22.75 (5)	±0.24	5 min, MeOH
N-Vinyl-2-pyrrolidone	21.61	22.56 (6)	±0.23	$^1/_4$–$^1/_2$ hr, MeOH
4-Vinylcyclohexene	44.41	46.20 (6)	±0.85	$^1/_4$–$^2/_3$ hr, MeOH

reactions and the iodine–starch endpoint was poor in this medium. Ethanol, 2-methoxyethanol, and dioxane were found to be too rapidly oxidized even at 0°C.

The use of electrogenerated bromine chloride complex of mercury in CCT was covered in Section VI-A-1-a-(2)-(c).

The initial applications of bromine chloride shows that this relatively new reagent is quite promising.

h. Chlorine

Chlorine is excessively reactive and easily leads to substitution and other side reactions. However, Čuta and Kučera (93) successfully determined the unsaturation of a number of high molecular weight fatty acids by coulometric titration with chlorine electrogenerated from hydrochloric acid solution.

Mukherjee (253,254) applied aqueous hypochlorous acid to fatty glycerides and oxidized oils.

$$>C{=}C< + HOCl \rightarrow -\underset{OH}{\underset{|}{\overset{|}{C}}}-\underset{|}{\overset{Cl}{\overset{|}{C}}}- \qquad (39)$$

Their results compare favorably with the more complex standard methods and appear to be remarkably free from secondary reactions since they were not affected by variations in sample size, reagent excess, reaction time, and temperature.

McNeill (237a) determined small amounts of unsaturation in rubber using radiochlorine. He compared this method with the iodine chloride method of Lee, Kolthoff, and Johnson (210) and concluded that the chlorination method was freer from side effects. The reaction occurs entirely by substitution and each double bond consumes 2 moles of chlorine.

$$\sim CH_2-\overset{CH_3}{\overset{|}{C}}{=}CH-CH_2\sim + 2Cl_2 \rightarrow \sim CH_2-\overset{CHCl_2}{\overset{|}{C}}{=}CH-CH_2\sim + 2HCl \qquad (40)$$

i. Pseudohalogens

Thiocyanogen behaves like the halogens in its reaction with olefinic double bonds but it is appreciably less reactive.

$$>C{=}C< + (SCN)_2 \rightarrow -\underset{SCN}{\underset{|}{\overset{|}{C}}}-\underset{|}{\overset{SCN}{\overset{|}{C}}}- \qquad (41)$$

The reagent is prepared by reacting lead thiocyanate with bromine (271).

$$Pb(SCN)_2 + Br_2 \rightarrow (SCN)_2 + PbBr_2 \qquad (42)$$

The sample is reacted with excess thiocyanogen reagent and the latter is back-titrated iodometrically in the usual manner with standard sodium thiosulfate solution.

This reagent is impractical for most samples because it is not sufficiently reactive. However, this low reactivity has been utilized for the selective determination of isolated double bonds in fatty materials (321) and the determination of certain terpene hydrocarbons, e.g., pinene which absorbs excessive amounts of bromine.

2. Hydrogenation

Hydrogenation is perhaps the most nearly universal method for the determination of olefinic unsaturation and its chief advantage is that it is free from substitution errors. However, the hydrogenation rates vary widely for different compounds and some nonolefinic organic compounds also hydrogenate under certain conditions, e.g., some carbonyl and epoxy compounds, and certain methylol benzenes and benzyl amines (271). Another disadvantage of the hydrogenation method is that it is time consuming and the sample must be run singly unless several apparatus are available. In spite of these difficulties, with experience reliable results can be obtained on many samples, so this method is often considered the "referee" method. Not only is the validity of a method often judged by the compatibility of its results with the hydrogenation results, but also, the conditions of some methods are accurately controlled to give results which are compatible with the hydrogenation results.

a. General

(1) Catalysis

Olefins, which are conjugated by another double bond, carbonyl, or aromatic ring, react with nascent hydrogen, but isolated double bonds do not react with nascent or molecular hydrogen unless certain metals are used to catalyze the reaction. The activation energy of hydrogen is too high for its spontaneous reaction with olefins. Metal catalysts lower this activation energy by permitting a series of steps, the activation energy for each of which is much lower than that required for the thermal breaking of the bonds (equation (43)) (258). As indicated in the schematic repre-

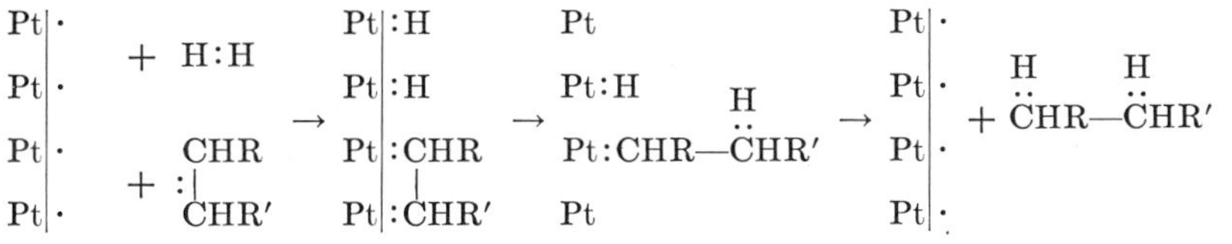

$$
\begin{array}{l} \mathrm{Pt}|\cdot \\ \mathrm{Pt}|\cdot \\ \mathrm{Pt}|\cdot \\ \mathrm{Pt}|\cdot \end{array}
\begin{array}{l} +\ \mathrm{H{:}H} \\ \\ \quad\ \mathrm{CHR} \\ +\ {:}| \\ \quad\ \mathrm{CHR'} \end{array}
\rightarrow
\begin{array}{l} \mathrm{Pt}|\mathrm{{:}H} \\ \mathrm{Pt}|\mathrm{{:}H} \\ \mathrm{Pt}|\mathrm{{:}CHR} \\ \qquad | \\ \mathrm{Pt}|\mathrm{{:}CHR'} \end{array}
\rightarrow
\begin{array}{l} \mathrm{Pt} \\ \mathrm{Pt{:}H} \qquad\quad \mathrm{H} \\ \qquad\qquad\quad\ \ddot{} \\ \mathrm{Pt{:}CHR{-}CHR'} \\ \mathrm{Pt} \end{array}
\rightarrow
\begin{array}{l} \mathrm{Pt}|\cdot \\ \mathrm{Pt}|\cdot \\ \mathrm{Pt}|\cdot \\ \mathrm{Pt}|\cdot \end{array}
\begin{array}{l} \quad \mathrm{H} \qquad \mathrm{H} \\ \quad\ \ddot{} \qquad\ \ \ddot{} \\ +\ \mathrm{CHR{-}CHR'} \end{array}
\qquad (43)
$$

sentation, the hydrogen molecule can be dissociated to give adsorbed hydrogen atoms and the olefin to give an adsorbed free biradical. Reaction of a hydrogen atom and the biradical leaves an adsorbed hydrogen atom and an adsorbed free radical. Further reaction gives the saturated hydrocarbon which is desorbed. Since the various steps in the reaction involve unpaired electrons and weak bonds, none has high activation energy.

According to the above mechanism the spatial relationship between the catalyst and the reactants is very critical. Thus, steric effects and crystal structure of the catalysts play important roles in the kinetics of hydrogenation. The facts that hydrogenation takes place predominantly by *cis* addition, that *cis* olefins hydrogenate more rapidly than *trans* olefins, and that hydrogenation is retarded by increased substitution of ethylene are compatible with the above mechanism. The choice of the optimum condition and catalyst for each compound is largely empirical. Fortunately, however, hydrogenation catalysts have been developed which show high reactivity for a variety of compounds.

The catalysts most commonly used for quantitative hydrogenation are, in their order of popularity, palladium, platinum dioxide (Adam's catalyst), platinum, and Raney nickel. Since the reaction between hydrogen and double bond occurs at or near the surface of the catalyst, it is desirable for the catalyst to possess a high surface area. Accordingly, the catalyst is usually employed as a finely divided or colloidal precipitate or mounted on the surface of a carrier such as activated carbon, silica, or zirconium oxide. It is generally true that increasing the catalyst concentration (therefore surface area) increases the rate of hydrogenation. Palladium on charcoal appears to be the best catalyst for the selective hydrogenation of ethylenic double bonds in the presence of aromatics.

Sedlak (315a) found that rhodium catalyzes the hydrogenation of aromatic compounds.

Bott and co-workers (54) prepared a platinum catalyst which was many times more active than Adam's catalyst for the hydrogenation of olefins but several times less active for the hydrogenation of benzene. The catalyst was prepared by the addition of 0.01 mmole of chloroplatinic acid in 1 ml of 95% alcohol to 0.1 mmole of tribenzylsilane in 20 ml of alcohol at 70°C. The resulting brown mixture catalyzes olefin hydrogenation at room temperature and pressure.

Brown and co-workers (63,64) reported an *in situ* preparation of fresh platinum catalyst and *in situ* generation of hydrogen by treating an acidic medium containing chloroplatinic acid and active carbon with a standard solution of sodium borohydride. The catalyst is 100% more active than Adam's catalyst. The application of the sodium borohydride solution to volumetric titration of olefins will be discussed in the "Methods" section (VI-A-2-b-(3)).

CAUTION: Raney nickel should not be allowed to go to dryness, because in the dry state, the heat produced by its oxidation by air could ignite any combustible material. Dry Adam's catalyst will glow when exposed to hydrogen because of the high heat of dissociation of the hydrogen molecule. To prevent possible explosion the catalyst must be completely wet before introducing the hydrogen (322).

(2) Solvent Effects

Acetic acid is by far the most commonly used solvent, followed by ethanol. Other solvents that do not hydrogenate may be used. In many cases the sample itself and the hydrogenation product serve as the solvent. There appears to be a definite relationship between acidity of solvent and rate of hydrogenation:

Acetic acid > Alcohol > Cyclohexane > Dioxane

Dioxane is the strongest base since it is a fairly good proton acceptor. A small amount of hydrochloric acid accelerates hydrogenation in neutral solvents (65).

(3) pH Effect

Sokol'skaya and co-workers (284,335–337) made extensive studies on the effect of pH on the rate of hydrogenation of various olefins using Raney nickel, platinum black, and palladium on barium sulfate (or calcium carbonate) catalysts. Using buffer solutions ranging in pH from 2.4 to 13.0, they plotted the kinetic curve and followed the catalyst potential during the course of the reaction. In general, the hydrogenations were faster in the acid medium than in the basic medium. However, there were numerous exceptions depending on the structure of the olefin. Thus, 1-hexene and allyl alcohol showed no pH dependence up to pH 7 using palladium catalyst. The reaction was unaffected by pH in the interval 2.4–13.0 for dimethylethynylcarbinol and phenylacetylene over platinum. In the reaction of cinnamyl alcohol and dimethylethynylcarbinol on nickel and that of allyl alcohol on platinum, the rate increased markedly with pH. These and other exceptions make the relationship between pH and the kinetics of hydrogenation of olefins empirical.

b. Methods

The apparatus for the quantitative hydrogenation of olefinic unsaturation has undergone a large number of changes (see pp. 260–262, reference 271), but the basic principles involved have not changed. The sample is reacted with hydrogen in an airtight chamber in the presence of a catalyst, and the consumption of the hydrogen is determined by the change in pressure or the change in volume. Atmospheric pressure is sufficient for the hydrogenation of most samples. However, for samples that hydrogenate with difficulty, Gould and Drake (138) have used pressures up to 5 atm, replacing the usual manometer with a metal syringe. Analytical data for the hydrogenation of various unsaturated compounds, including catalysts, solvents, time of reaction, and original references, are listed in Table XVI, pp. 266–270 of reference 271.

In the classical hydrogenation methods the hydrogen consumed during the reaction is not replaced. In two more recent methods the consumed hydrogen is replaced by either coulometric or chemical generation of hydrogen. These three methods are discussed below.

(1) Gasometric Method

A simplified quantitative hydrogenation apparatus according to Siggia

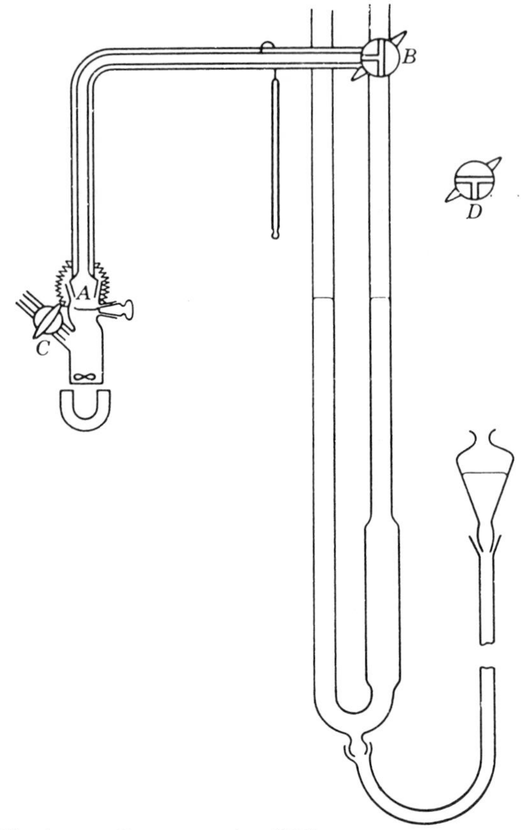

Fig. 8. Simplified hydrogenation apparatus (322).
A: Sample is suspended here during the purging step.
B: Three-way cock in the purge position with mercury level raised to the cock.
C: Two-way cock is open during the purging step and closed for the hydrogenation step.
D: Three-way cock in the hydrogenation position after buret is filled with hydrogen.

is shown in Fig. 8 (322). After adding catalyst, solvent, and sample (at position *A*) to the reaction flask, the system is flushed out with hydrogen to remove the last traces of oxygen. Then the buret arm of the manometer is filled with hydrogen and the catalyst is hydrogenated. The sample is dropped into the reaction mixture and the decrease in volume of hydrogen is noted on the buret. The reaction is complete when the volume ceases to change. (For the details of this procedure, see pp. 318–321, reference 322.)

A change in temperature during hydrogenation could cause significant errors. For accurate work, a temperature correction should be applied and low-boiling solvents should be avoided to minimize the fluctuations of solvent vapor pressure with temperature.

The precision of the method is ±5% using a 7-ml buret and ±1–2% using a 50-ml buret.

Smith (332) introduced a microhydrogenation apparatus employing a compensation flask and a differential manometer. The advantage of this method has been reviewed by Bretschneider and Burger (58). One of the apparatus (79) which evolved from the refinements contributed by many workers (see reference 271) is shown in Fig. 9. The hydrogenation apparatus consists of chambers of approximately equal volume which communicate with each other through the graduated manometer. Stopcock *d* is open only during the time when the entire system is purged with hydrogen. The solvent is added to the upper part of the graduated portion of the manometer and also to the reaction chamber and the compensation chamber. The whole apparatus is immersed in a constant-temperature bath. As hydrogenation takes place the solvent in the manometer rises into the enlarged space above the manometer. The enlarged space below the graduated region on the right arm of the manometer prevents the gas in the compensator from being carried over into the reaction chamber during a fast reaction; the capacity of this chamber is larger than the 5-ml capacity of the buret. The liquid levels in the manometer are kept constant by replacing the hydrogen consumed in the reaction chamber with mercury. The reaction is complete when the levels of the solvent in the manometer stops changing. Because of the capillary used in the manometer, the detection of the endpoint is much easier than the simpler buret type of manometer. With this apparatus, the precision is about ±1%.

A detailed procedure using this compensated hydrogenation apparatus is given in Section VII-E.

(2) Coulometric Titration

Manegold and Peters (225) were the first to use electrically generated hydrogen in place of the conventional buret. Farrington and Sawyer (121)

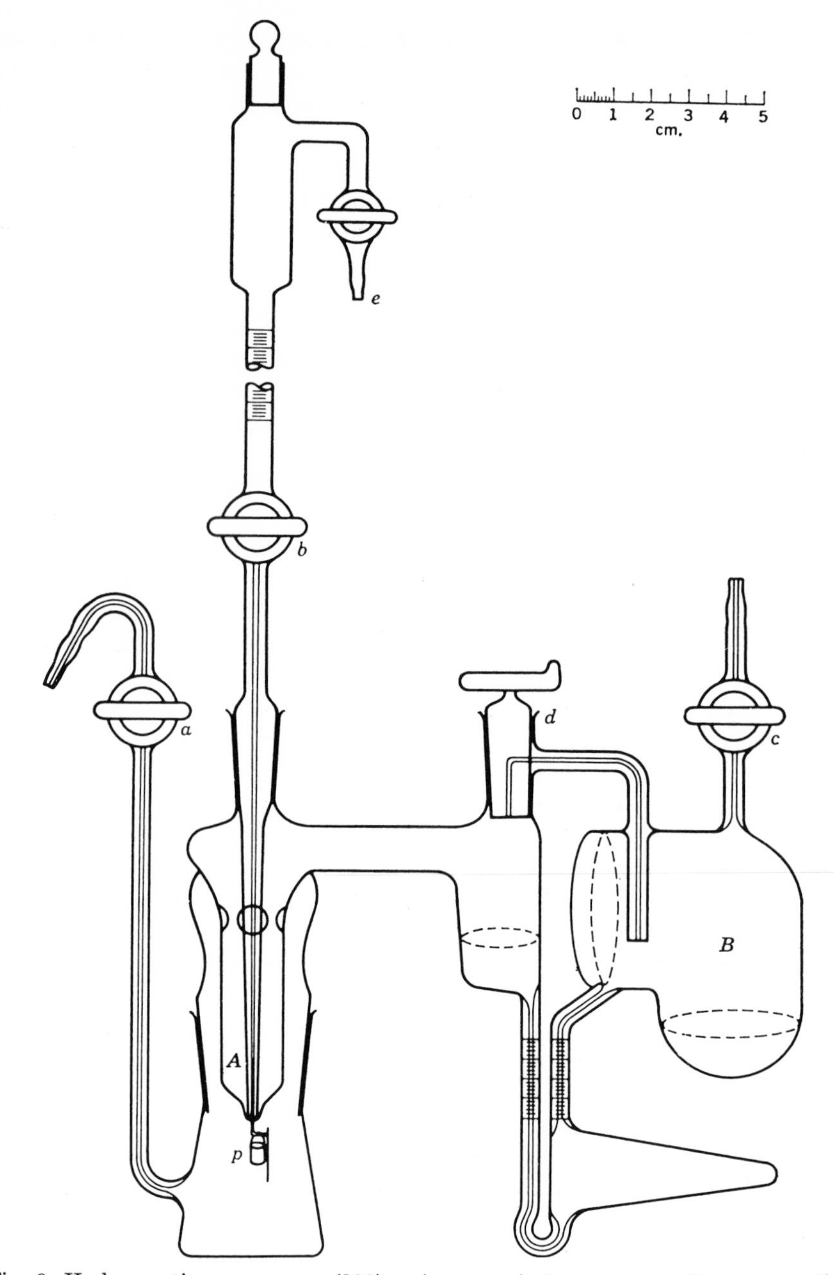

Fig. 9. Hydrogenation apparatus (322). *A*, reservoir for mercury; *B*, compensation vessel; *a*, *b*, *c*, *d*, and *e*, stopcocks; *p*, platinum beaker.

determined the relative rates of hydrogenation of olefins using electrically generated hydrogen and a platinum foil–saturated calomel electrode arrangement to follow the concentration of the hydrogen in the reaction medium.

The manually controlled apparatus of Manegold and Peters was modified by Miller and DeFord (247) to make completely automatic hydrogenation possible. A schematic diagram of their hydrogenation apparatus is shown in Fig. 10. The two most essential parts in this apparatus are in common with those of other apparatus considered previously, namely, the reaction vessel and the manometer. However, whereas the manometers in other apparatus serve as "null detectors," the manometer in this apparatus serves the function of a manostat which keeps the hydrogen pressure in the reaction chamber constant, provided there is no change in the atmospheric pressure during a titration. It also serves as the electrolysis vessel. The hydrogen is generated in the cathode arm of the manostat from a $6N$ sulfuric acid solution. The current used to generate hydrogen is integrated

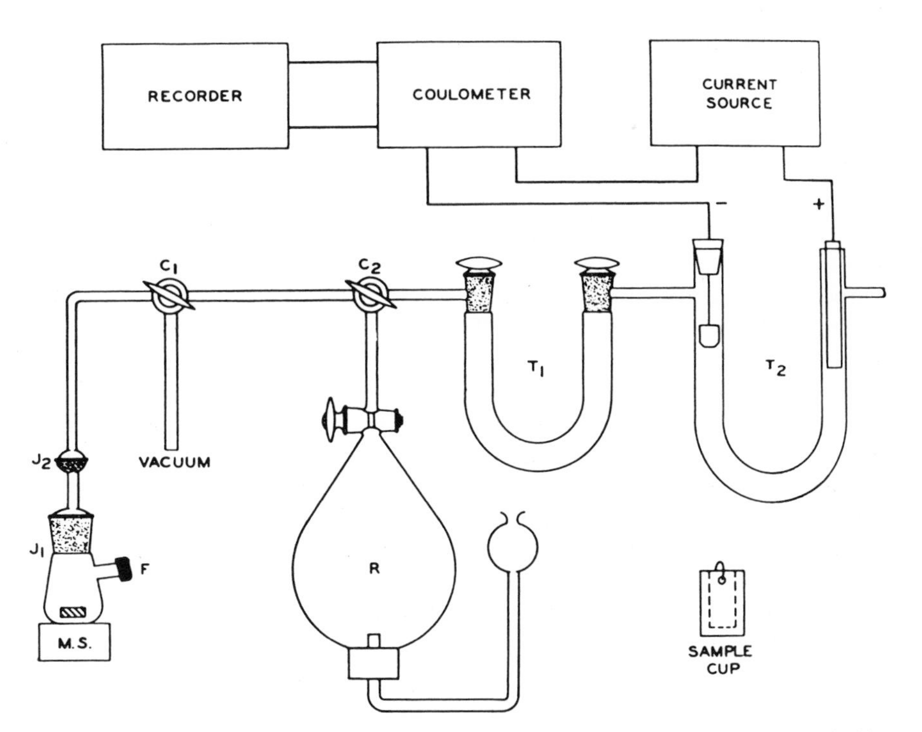

Fig. 10. Schematic diagram of the coulometric hydrogenation apparatus (247). (Courtesy of *Analytical Chemistry*.)

by the coulometer and the integral is plotted on the recorder. The titration curves are rate curves which are very similar to the cumulative coulographic titration curves described in Section VI-A-1-a-(2)-(c).

The third function of the manometer, that of automatic switching of the electrolysis circuit, can be seen best by going through a typical titration. The sample is placed in the sample cup and is suspended in the reaction vessel. The whole system is purged with hydrogen, then made airtight, and the catalyst is hydrogenated. The uptake of hydrogen by the catalyst causes the electrolyte in the cathode arm to rise until the contact of the electrolyte with the cathode closes the electrolysis circuit. When the rate of hydrogen generation exceeds the rate of hydrogen consumption by the catalyst, the increasing internal pressure pushes the electrolyte level down until the electrolyte breaks contact with the cathode, thereby opening the electrolysis circuit. Stepwise increments of hydrogen are generated in this manner until all of the catalyst is hydrogenated. The sample is dropped into the hydrogenating mixture and titration proceeds in the manner described for the hydrogenation of the catalyst.

Volatile samples are dissolved in a solvent and 1 ml of this solution is injected into the reaction vessel. Since the solution displaces an equal volume of the hydrogen from the reaction vessel, the volume, reduced to standard pressure and temperature, must be added to the coulometer reading.

Three main factors affect the accuracy of the hydrogenation: temperature, pressure, and surface tension. With an internal volume of 46.5 ml when the reaction flask is filled with 5 ml of solvent, a 1°C change in temperature during a titration is equivalent to 0.16 ml of gas. A pressure change of 1 mm is equivalent to 0.06 ml of hydrogen. Because of the surface tension effects, the liquid level in the cathode arm at the moment when contact with the cathode is broken is about 1 mm lower than the level at which contact is reestablished. With the cell used, this difference in height corresponds to a volume of about 0.13 ml. For this reason there is a dead zone corresponding to this volume of hydrogen.

In the hydrogenation of cinnamic acid, fumaric acid, benzene, naphthalene, *p*-nitrophenol, acetophenone, and phenyl propyl ketone, the accuracy was $\pm 3\%$ and the standard deviation was 2%.

In a recent refinement of the coulometric method, Burkhardt and Dirschel (70) reduced the error to less than $\pm 1\%$ for a volume of not less than 3.5 ml of hydrogen.

(3) Volumetric Titration

Brown and co-workers (63,64) introduced a novel method whereby the olefin is titrated volumetrically with a solution of sodium borohydride.

The sodium borohydride was used to generate active metal catalysts *in situ* by reducing various transition metal salts and also to generate hydrogen *in situ*.

$$NaBH_4 + HCl + 3H_2O \rightarrow NaCl + H_3BO_3 + 4H_2\uparrow \quad (44)$$

In order to make this method feasible, it was necessary to find a catalyst that met two requirements: (*1*) It must cause a liberation of the hydrogen from the sodium borohydride and (*2*) it must catalyze the addition of hydrogen to olefins sufficiently to give a reasonably short titration time. Table XVIII gives the relative rates of liberation of hydrogen from sodium borohydride in the presence of various group VIII metal catalysts and Table XIX gives the rates of hydrogenation of 1-octene in the presence of

TABLE XVIII

The Effect of the Group VIII Metals on the Time Required for the Liberation of One-Half of the Available Hydrogen from Aqueous Sodium Borohydride (63)

Metal	Compounds used	Time, min
Iron	$FeCl_2$	38
Cobalt	$CoCl_2$	9
Nickel	$NiCl_2$	18
Ruthenium	$RuCl_3$	0.3
Rhodium	$RhCl_3$	0.3
Palladium	$PdCl_2$	180
Osmium	OsO_4	18.5
Iridium	$IrCl_4$	28
Platinum	H_2PtCl_6	1

TABLE XIX

Rates of Hydrogenation of 1-Octene by Various Platinum Metal Catalysts Produced *in situ* by Borohydride Reduction (63)

		Time, min[b]	
Metal[a]	Compound used	50%	100%
Ruthenium	$RuCl_3$	70	(170)
Rhodium	$RhCl_3$	7	20
Palladium	$PdCl_2$	16	(90)
Osmium	OsO_4	45	(110)
Iridium	$IrCl_4$	32	(80)
Platinum	H_2PtCl_6	9	17
(Platinic oxide)	PtO_2	14	27

[a]0.2 mmole for 40 mmoles of 1-octene.
[b]Values in parentheses are estimated times for complete reaction.

the indicated catalysts. Rhodium and platinum are the only catalysts that met both requirements although a 17–20 minute time required for complete hydrogenation appears to be somewhat slow for a direct volumetric titration. However, rhodium is not selective for olefins, since it catalyzes the hydrogenation of aromatic compounds (315a). Thus platinum is the only catalyst applicable by this method.

Brown's apparatus for volumetric hydrogenation is shown in Fig. 11. The heart of the apparatus is the automatic valve which consists of a capillary tip of a buret dipping into a mercury well to a depth sufficient to support the column of sodium borohydride solution. As the hydrogen is utilized in the hydrogenation flask, the pressure drops 5–10 mm below atmospheric pressure, drawing a small volume of borohydride solution through the mercury seal, where it rises to the top of the mercury and runs into the flask through the small vent holes located just above the mercury interface. The acidic solution in the flask converts the borohydride into hydrogen and the resulting increase in pressure seals the valve. The

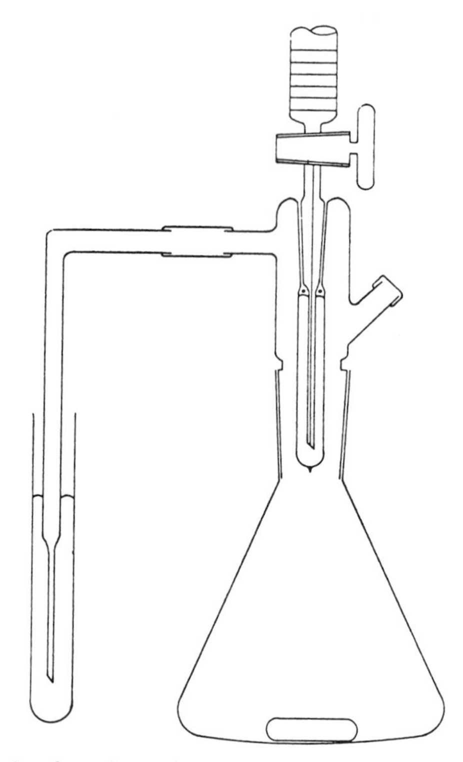

Fig. 11. Apparatus for the volumetric hydrogenation of unsaturated compounds (64). (Courtesy of *Journal of Organic Chemistry*.)

addition proceeds smoothly and automatically to the completion of the hydrogenation. The reaction vessel is isolated from the atmosphere by a mercury seal. This mercury seal also serves as a manometer because the hydrogenation takes place at pressures slightly under atmospheric. The pressure in the reaction flask is expected to drop during the titration as the height of the solution in the buret drops, but their results in Table XX did not show the effect of this dependency of the hydrogen pressure on the height of the solution in the buret.

The authors' results in Table XX compare very favorably with the classical hydrogenation methods. In fact, the results are surprisingly good, considering that each titration took only 1–2 min, whereas Table XIX shows that the catalyst system used in this method required 9 min for 50% reaction of 1-octene. Unfortunately, none of the samples was branched at the doubly bonded carbons and all of the samples precluded the *trans* configuration; hence the applicability of this method to sterically hindered olefins is not known. More recently, they have applied the method successfully to fatty oils (62b).

PROCEDURE (64)

Prepare 0.1*N* sodium hydroxide by dissolving 4.00 g of sodium hydroxide in 50.0 ml of water and diluting to 1.0 liter with absolute ethanol. Prepare 1*M* sodium borohydride by dissolving 3.95 g of sodium borohydride in 100 ml of the 0.1*N* ethanolic sodium hydroxide, filtering through glass wool if necessary. Standardize the solution by injecting 10.00 ml into aqueous acetic acid and measuring the hydrogen evolved. Prepare the 0.25 and 0.1*M* sodium borohydride solutions by diluting the 1*M* solution with the 0.1*N* sodium hydroxide.

Place 1.00 g of Darco K-B carbon, 40.0 ml of absolute ethanol, 1.00 ml of 0.02*M* chloroplatinic acid, and a Teflon-covered magnetic bar in the 125-ml flask of the apparatus shown in Fig. 11. Assemble the apparatus and immerse the flask in a beaker of water maintained at 25°C. Inject 5.00 ml of 1*M* sodium borohydride with a syringe into the vigorously stirred solution to produce the catalyst. After about 1 min, inject 2.00 ml of concentrated hydrochloric acid to destroy the excess sodium borohydride and to provide a hydrogen atmosphere. Inject a small quantity of 1-octene to bring the apparatus to equilibrium.

Carry out the analysis by injecting either the pure liquid olefins or standard solutions of the olefin in ethanol. Apply corrections for the hydrogen displaced by the sample and the reagent.

The Brown volumetric hydrogenation apparatus is commercially available. This apparatus was modified into a gasometric hydrogenator (62a). Sodium borohydride was used to synthesize the catalyst *in situ* and to purge the system initially with hydrogen. However, the hydrogen used up by the olefins was not replaced but simply displaced by adding water to a connecting chamber. The analysis was slower (30–60 min for triplicate analyses) but the accuracy was improved (0.5–1.0%).

TABLE XX

Volumetric Hydrogenation of Various Unsaturated Compounds (64)

Compound	Amount, mmoles	$NaBH_4$ soln, M	Volume of $NaBH_4$ solution	Average	Free space equivalent[c]	Olefin found, mmoles
1-Octene	20.0[a]	1.00	4.95, 4.94, 4.90, 4.96, 4.95	4.94 ± 0.02	0.32	20.08 ± 0.020
	10.0[a]	1.00	2.46, 2.44, 2.48, 2.44, 2.42	2.45 ± 0.02	1.6	9.96 ± 0.018
	5.00[a]	1.00	1.23, 1.22, 1.22, 1.24, 1.20	1.22 ± 0.02	0.08	4.96 ± 0.016
	5.00[b]	0.250	4.57, 4.59, 4.62, 4.56, 4.63	4.97 ± 0.03	0.38	4.97 ± 0.030
	1.00[b]	0.250	0.92, 0.91, 0.90, 0.92, 0.91	0.91 ± 0.01	0.08	0.99 ± 0.010
	2.00[b]	0.100	4.31, 4.39, 4.30, 4.32, 4.30	4.32 ± 0.03	0.25	1.98 ± 0.030
	1.00[b]	0.100	2.18, 2.15, 2.20, 2.15, 2.20	2.18 ± 0.03	0.13	1.00 ± 0.025
4-Methylcyclohexene	2.00[b]	0.100	4.32, 4.35, 4.38, 4.36, 4.32	4.35 ± 0.03	0.25	1.99 ± 0.025
1,5,9-Cyclododecatriene	0.67[b]	0.100	4.32, 4.33, 4.37, 4.35, 4.35	4.34 ± 0.02	0.25	1.99 ± 0.025
Ethyl oleate	2.00[b]	0.100	4.35, 4.39, 4.32, 4.35, 4.37	4.36 ± 0.02	0.25	2.00 ± 0.020
Mixture[d]	2.00[b]	0.100	4.37, 4.39, 4.40, 4.37, 4.38	4.38 ± 0.01	0.25	2.01 ± 0.012

[a]Introduced as the pure liquid.

[b]Introduced as a 1.00*M* solution in ethanol.

[c]Millimoles of hydrogen displaced by the volume of olefin or olefin solution introduced plus volume of sodium borohydride introduced (total volume in cc/25.0).

[d]A mixture of 1-octene, 4-methylcyclohexene, 1,5,9-cyclododecatriene, and ethyl oleate, prepared by mixing aliquots of the 1*N* ethanolic solutions.

3. Oxidation

Relatively few analytical methods based on the oxidation of the double bond have been developed. Among the more commonly used oxidizing agents are peracids, ozone, and permanganate. These reagents are powerful oxidizing agents and therefore lack specificity, and the standard reagents are often either unstable or difficult to prepare. However, these reagents have been very useful in the elucidation of complex olefinic structures, because they react with the double bonds to give stoichiometric amounts of end products which can be determined by other established methods.

a. Peracid

Organic peracids react with olefins to yield an epoxide which may be converted to a hydroxy ester, or in the presence of water, to the corresponding glycol:

$$\mathrm{RHC{=}CHR} + \mathrm{RCO_3H} \rightarrow \underset{\diagdown\ \diagup \atop \mathrm{O}}{\mathrm{RHC{-}CHR}} + \mathrm{RCOOH} \rightarrow \underset{\mathrm{OH}\quad\mathrm{OCOR}}{\mathrm{RHC{-}CHR}} \qquad (45)$$

$$\downarrow \mathrm{H_2O}$$

$$\underset{\mathrm{OH}\quad\mathrm{OH}}{\mathrm{RHC{-}CHR}}$$

Since the reaction involves the electrophilic attack of the π-electrons by the peroxide, electron-rich substituents, such as alkyl and phenyl groups, enhance the reactivity of the double bonds while electron-attracting substituents, such as carbonyl and carboxyl groups, decrease the rate of reaction. Thus, crotonic acid reacts slowly, while maleic and fumaric acid fail to react (271). Steric hindrance appears to have no adverse effects for alkyl-substituted olefins; the more highly substituted olefins react with greater ease.

The peroxide method cannot be used in the presence of easily oxidized compounds. Naphthalene, phenanthrene, and anthracene are readily oxidized, while benzene is not attacked. Organic sulfides, mercaptans, aldehydes, phenols, amines, and azo compounds also undergo oxidation (345). Other disadvantages of the method are long reaction times and the instability of peroxide reagents.

In the course of a search for more stable and more rapidly reacting peroxides, performic, peracetic, perpropionic, percamphoric, perfuroic, and perbenzoic acids have been investigated (271). Among these peroxides, perbenzoic acid has been the favorite choice. A method for the preparation of this reagent is given by Kolthoff, Lee, and Mairs (199). They recommend the addition of 10% benzene to the chloroform solution of perbenzoic acid to prevent the formation of oxidizing agents during storage, which leads to increase in the iodometric titer of the reagent.

Perbenzoic acid reacts more rapidly with internal double bonds than with terminal double bonds. Kolthoff and Lee (198) utilized this difference in reaction rates to determine the number of vinyl groups in a variety of rubber polymers and copolymers. The total unsaturation was determined by the iodine chloride method. The internal double bonds were determined by the perbenzoic acid method, using a calibration curve prepared from a series of oleic acid–undecylenic acid mixtures.

Saffer and Johnson (302) used an extrapolation procedure to determine the internal double bonds. They extrapolated the 17- to 24-hr portion of

TABLE XXI
Analytical Data by Perbenzoic Acid Oxidation (49)

	Unsaturation, expressed in Br No.		
Compound[a]	Theory	Found	% of Theory
1-Hexene	190	181	95.3
2-Hexene	190	193	101.6
1-Heptene	163	153	93.9
3-Heptene	163	170	104.3
Cetene	71.5	64	89.5
2-Methyl-2-butene	228	253	111.0
3,3-Dimethyl-1-butene	190	227	119.5
2,2,4-Trimethyl-1-pentene	143	143	100.0
2,2,4-Trimethyl-2-pentene	143	143	100.0
Diisobutylene	143	146	102.1
2,2,5,5-Tetramethyl-3-hexene	114	118	103.5
2-Methyl-1-undecene	95.3	96	100.7
2,3-Dimethyl-1,3-butadiene	391	448	114.6
Cyclopentene	235	264	112.3
1-Propyl-1-cyclopentene	145	146.5	101.0
Dicyclopentadiene	242	235	97.1
Cyclohexene	195	202	103.6
Limonene	235	228	97.0
Allyl alcohol	276	183	66
Methallyl alcohol	222	224	101
Allyl chloride	209	10.5	5
Methallyl chloride	177	82.5	47
Allyl acetate	160	15	9
Diallyl ether	326	80	25

[a]Purity doubtful in some cases.

the curve by analyzing three or four aliquots of the reaction mixture. The precision was about 1% but the method was not applicable to samples containing less than 70% internal double bonds.

Typical results obtained by the perbenzoic acid method are given in Table XXI (49). The results are high for some of the branched olefins and cyclic alkenes, while mildly electrophilic substituents on the doubly bonded carbons cause large negative deviations. The peracid oxidation method is not a good general method. It may be useful to differentiate between internal and end-type double bonds, but the available methods are quite tedious.

b. Ozone

The ozonization of olefins breaks both the π- and the σ-bonds to form the cyclic ozonide:

$$RCH{=}CHR' + O_3 \rightarrow RHC\underset{O-O}{\overset{O}{\langle\rangle}}CHR' \qquad (46)$$

where R′ may be an alkyl group or a hydrogen atom. In the presence of water the ozonide undergoes ozonolysis to give aldehydes under reducing conditions and acids under oxidizing conditions (equation (47)). Ketones

$$RHC\underset{O-O}{\overset{O}{\langle\rangle}}CHR' \xrightarrow{H_2O} \begin{cases} \xrightarrow{\text{reduction}} RCHO + R'CHO \\ \xrightarrow[\text{oxidation}]{} RCOOH + R'COOH \end{cases} \qquad (47)$$

result from highly substituted ethylenes (equation (48)).

$$R_2C{=}CR_2 + O_3 \rightarrow R_2C\underset{O-O}{\overset{O}{\langle\rangle}}CR_2 \xrightarrow{H_2O} 2R_2CO + H_2O_2 \qquad (48)$$

Ozone is a very powerful oxidizing agent and it reacts rapidly with isolated double bonds. The rate of ozonization is increased by increased substitution of ethylene by alkyl groups which suggests a π-complex mechanism in the formation of the ozonide (357). It reacts quantitatively with conjugated systems, but one of the double bonds is oxidized rapidly while the other is oxidized at a considerably slower rate. The reduced reactivity of the second double bond can be explained by the inductive effect of the new carbonyl group substituent formed by the ozonolysis of the first double bond. Allenic double bonds, which are not attacked by permanganate oxidation, can be ozonized and determined from the carbon dioxide formed as a cleavage product (271):

$$\rangle C{=}C{=}C\langle + 2O_3 + 2H_2O \rightarrow \rangle C{=}O + CO_2 + O{=}C\langle + 2H_2O_2 \quad (49)$$

The ozonolysis of acetylenes yields monobasic acids, while the ozonolysis of aromatic compounds, which takes place at a much lower rate, yields dibasic acids (368). Organic acids and primary and secondary amines are attacked, but alcohols, aldehydes, ketones, the C=N bond, and especially ethers are very reactive toward ozone.

Some ozonides are explosive. Such dangers can be avoided by working at low temperatures and by carrying out decomposition of the ozonide in the solution.

Two general techniques are used in the analytical application of ozone. In the titrimetric method the ozone consumed by the olefins in the formation of the ozonide is determined. The second technique involves the ozonolysis of the olefins to the oxidative or reductive cleavage products and the subsequent determination of the latter.

(1) Titrimetric (Ozonization) Methods

The development of the electrolytic ozone generator made the titrimetric method possible. A constant stream of ozone is prepared by the electrolysis of dilute sulfuric acid. The endpoint was determined colorimetrically (49), or the gas escaping from the reaction vessel was checked by a continuous iodometric titration. The results obtained by Boer and Kooyman (49) are given in Table XXII. The results are very good except for slightly high results on some of the branched olefins.

Configurations in which a cyclane ring is involved in the branching network of one of the double-bonded carbons or of an adjacent carbon are extremely prone to substitution errors by the mercury-catalyzed bromide–bromate method of Dubois and Skoog. The ozonization method showed only slight positive deviation (up to 2%) according to the work of van der Bij and Kooyman (44).

In order to overcome the problems of ozone control and endpoint detection, Guenther and co-workers (143) divided the ozone stream into two equal streams. One was passed through the olefin sample, then into a potassium iodide solution. The other stream was passed directly into a separate potassium iodide solution. The amount of ozone consumed by the olefin is calculated from the difference in the titrations of the potassium iodide solutions.

TABLE XXII

Analytical Data by the Ozonization Method of Boer and Kooyman (49)

Olefin[a]	Unsaturation, expressed in Br No.		
	Theory	Found (indicator method)	% of theory
1-Hexene	190	192	101.1
1-Heptene	163	161	98.8
3-Heptene	163	162	99.4
1-Octene	143	143	100.0
Cetene	71.5	70.5	98.6
2,3-Dimethyl-2-butene	190	192	101.1
3,3-Dimethyl-1-butene	190	189.5	99.7
Diisobutylene	143	150	104.9
Triisobutylene	95.3	95.5	100.2
2,2,5,5-Tetramethyl-3-hexene	114	119.5	104.8
1-Dodecene	95.3	95.8	100.5
2,4-Hexadiene	391	373	95.4
2,3-Dimethyl-1,3-butadiene	391	380	97.2
Cyclopentene	235	230	97.9
1-Propyl-1-cyclopentene	145	144.5	99.7
Styrene	154	156	101.3
Dimethylallyl ether	254	250	98.4
1-Chloro-3-bromo-1-propene	103	107	103.9
Cinnamic acid	108	114	105.6
Ethyl allyl phthalate	68.3	66.6	97.5
Oleic acid	56.7	51.5	90.8
Crotonaldehyde	228	226	99.1
Methacrolein	228	208	91.2
Methacrolein dimer	114	115	100.9

[a]Purity doubtful in some cases.

In spite of the promising results obtained by ozonization titration, this method has not been generally accepted because (*1*) the control of the ozone is not as straightforward as other methods, e.g., coulometric generation of bromine, (*2*) ozone escapes readily from the titration medium, and (*3*) there is no convenient endpoint detection method. The colorimetric method may be adequate for the rapidly ozonized olefins, but for the majority of unsaturated compounds that ozonize at a slower rate, the colorimetric method is expected to give premature endpoints.

(2) Ozonolysis Methods

The ozonolysis methods are based on the quantitative determination of the products of ozonolysis. These methods are particularly useful in the investigation of complex unsaturated molecules, because the final products of ozonolysis retain the structure of saturated portions of the original molecules. Doeuvre (105) used ozone for the estimation of terminal double bonds by colorimetric measurement of the formaldehyde formed by the ozonolysis. The recovery was only 90–95% in the more favorable cases. The method was improved to give 94–102% recovery by Naves (256), who used a micro technique.

Gordon and co-workers (137) used chromatographic separation and spectrophotometric (ultraviolet) determination of the 2,4-dinitrophenylhydrazones of the ozonolysis products in the C_1–C_4 range. Spasskova (338) found that ozonolysis gave the correct unsaturation values for butadiene-nitrile rubbers while an iodine bromide method gave results that were too high and the perbenzoic acid method gave results that were too low.

Giuffre and Cassani (133) applied ozone for the determinations of small amounts of unsaturation in elastomeric copolymers (e.g., ethylene-propene-cyclooctadiene). The ozonide was used to oxidize the leuco base of malachite green or an aluminum triiodide solution, and the resulting dye or free iodine was determined spectrophotometrically.

Zimmermann and co-workers (378) found that a combination of ozonization, reductive cleavage of the ozonides with lithium aluminum hydride, and paper chromatographic identification of the 3,5-dinitrobenzoyl esters of the alcohols formed, was a superior method for the determination of the distribution of double bonds in liquid olefins.

Bischoff (46) applied ozonolysis to find the bond positions in mono-olefins in chromatographic fractions from hydrocarbons obtained in the Fisher-Tropsch synthesis. Ozonolysis was carried out at high temperature to convert the fragments to the acids. The acids were converted to hydroxamic acids which were separated by paper chromatography. Since this method failed for the short-chain acids, ozonolysis was carried out in chloroform at −20 to −30°C and the resulting aldehydes and ketones were converted to dinitrophenylhydrazones, which were separated by paper chromatography.

c. Permanganate–Periodate

In neutral or alkaline medium, potassium permanganate reacts with olefins to give the glycol:

$$3RCH{=}CHR' + 2KMnO_4 + 4H_2O \rightarrow 3R\underset{\displaystyle OH}{\underset{|}{C}H}{-}\underset{\displaystyle OH}{\underset{|}{C}}HR' + 2MnO_2 + 2KOH \qquad (50)$$

Completely substituted double bonds and allenic compounds react with difficulty and rarely to completeness. The reaction is not specific for olefins since permanganate is a powerful oxidizing agent and any reducible impurity will decolorize the reagent.

Compounds having hydroxyl groups on adjacent carbon are susceptible to scission by periodic acid:

$$\underset{\text{OH}}{\text{RCH}}\text{—}\underset{\text{OH}}{\text{CHR}'} + HIO_4 \rightarrow \underset{\text{O}}{\overset{}{\text{RCH}}} + \underset{\text{O}}{\text{HCR}'} + HIO_3 + H_2O \qquad (51)$$

This reaction is specific for the vicinal hydroxyl configuration and isolated hydroxyls are not attacked.

Bricker and Roberts (60) applied reactions (50) and (51) (let R′ = H) to the determination of terminal double bonds. The formaldehyde formed was separated by steam distillation and was determined spectrophotometrically by means of chromotropic acid. This method has several limitations. The reaction is not stoichiometric and the calibration curve for one compound may not be applicable to another. Any methanol or 1,2-glycols originally present in the sample will yield formaldehyde.

Dal Nogare and co-workers (96) applied the permanganate–periodate method to determine the free methyl methacrylate in poly(methyl methacrylate). The monomer was first isolated by distilling from an acetic acid–water solution of the polymer and the monomer in the distillate was titrated with permanganate in the presence of sulfuric and periodic acids. Interference from manganese dioxide was overcome by using transmitted light or by observing the surface of the solution. The precision was 5% of the amount of monomer present. The method was also applicable to styrene, methyl acrylate, and acrylonitrile.

Lemieux and von Rudloff (213) carried out oxidative cleavage of olefins at room temperature and in neutral or slightly alkaline periodate solution containing only catalytic amounts of permanganate. The permanganate was regenerated by the oxidation of the manganate by periodate. Von Rudloff (300) obtained nearly quantitative yields of carboxylic acids (separated by liquid–liquid partition chromatography) from oleic, elaidic, eicosenoic, 10-undecenoic, linoleic, and erucic acids and from methyl linoleate. The method was also applied to terminal unsaturation (214) and to the isopropylidene group in mesityl oxide, 3-methyl-2-butenoic acid, etc. (299).

Hilditch and Lea (154) demonstrated that from the fragments formed under drastic oxidizing conditions, it is possible to determine the position of the double bond in an unsaturated ester or acid.

After investigating various methods for the determination of positions of double bonds in complex mixtures, Ucciani and co-workers (353) recommended the permanganate–periodate method. The method involved the oxidation of the mixture to mono- and dicarboxylic acids, esterification of the acids with methanol in the presence of *p*-toluenesulfonic acid, and identification of the esters by gas chromatography. They found that ozonization was not satisfactory as an oxidation procedure.

4. Specific Methods for Various Types of Unsaturation

a. Vinyl Unsaturation by Mercuration

Olefins form complexes with certain cations, e.g., platinum, palladium, copper, silver, and mercury. Mercury salts, which are most important analytically, form true addition compounds by saturating the double bond. Analogous to halogenation, the mechanism probably involves the electrophilic attack by a positively charged mercury ion followed by the nucleophilic attack of the π-complex by the solvent molecules (reactions (52)–(54)).

$$HgY_2 \rightleftharpoons HgY^+ + Y^- \qquad (52)$$

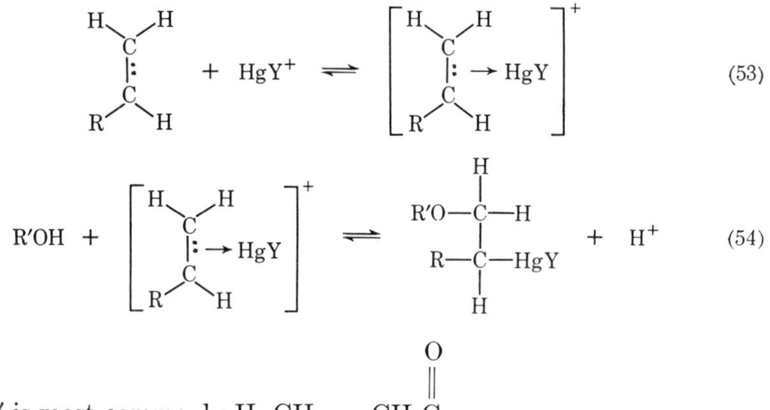

where R′ is most commonly H, CH_3, or $CH_3\overset{O}{\overset{\|}{C}}$.

The following observations support this mechanism: (*1*) The addition gives solely the *trans* product; (*2*) the solvent appears in the addition product; (*3*) the rate of mercuration is inversely proportional to the nucleophilic power of Y^-, e.g., when $Y^- = CH_3COO^-$ or NO_3^-, which does not form stable complexes with mercury(II), the reaction is rapid; when Y^- is Cl^-, the reaction is very slow; and when $Y^- = CN^-$ or CNS^-, the reaction does not occur; and (*4*) the rate of mercuration is inversely proportional to the electronegativity of the substituent, R, in equations (53) and (54); e.g., when R is a carbonyl the reaction is very slow because of the depletion of the π-electrons.

The rate of mercuration is greatly affected by stereoeffects. Thus the reaction is retarded by branching and is slow for *trans* compounds; it is selective for terminal, *cis*, and cyclic olefins.

Numerous applications have been reported where olefins in gaseous samples are determined by absorbing in mercury salt solutions and the effects of structure on mercuration of olefins have been investigated (271, pp. 303–304).

The original method of Marquardt and Luce (226) involved the hydroxymercuration of olefins in 40% aqueous dioxane. The method was applicable to styrene, ethyl vinyl benzene, divinylbenzene, α-methyl styrene, and vinyl toluene. However, the method was slow and gave low results for dichlorostyrene.

The mercuration methods most commonly used today are based on the methoxymercuration of the olefin:

$$RCH{=}CH_2 + Hg(OOCCH_3)_2 + CH_3OH \rightarrow \underset{OCH_3}{RCH}{-}\underset{HgOOCCH_3}{CH_2} + CH_3COOH \tag{55}$$

Of the four titrimetric methods which are based on equation (55), one involves the indirect titration of the acetic acid formed, two involve the direct titration of the acetic acid, and the fourth involves the titration of mercury(II) with hydrochloric acid.

(1) Indirect Titration of Acetic Acid (227,228)

Marquardt and Luce (227,228) added a measured excess of caustic and back-titrated with standard acid. The precipitation of mercuric oxide was prevented by adding acetone which formed a soluble trimercuric diacetone hydrate:

$$3Hg(OOCCH_3)_2 + CH_3COCH_3 + 6NaOH \rightarrow$$

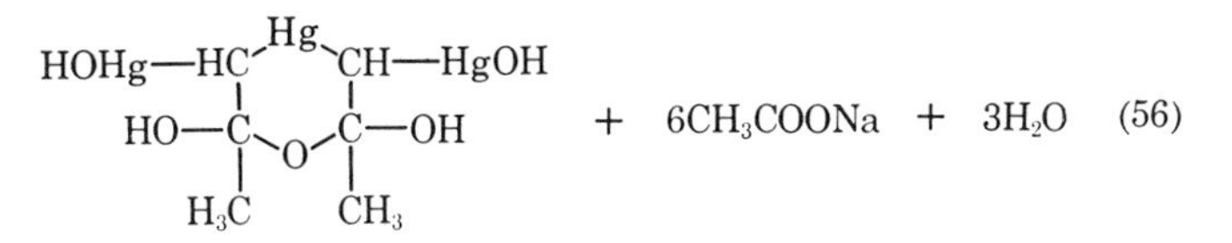

and simultaneously the reaction products of equation (55) each consume 1 equivalent of base:

$$CH_3COOH + NaOH \rightleftharpoons CH_3COONa + H_2O \tag{57}$$

$$\underset{OCH_3}{RCH}{-}\underset{HgOOCCH_3}{CH_2} + NaOH \rightleftharpoons \underset{OCH_3}{RCH}{-}\underset{HgOH}{CH_2} + CH_3COONa \tag{58}$$

The caustic consumed in equations (56) and (58) are recovered when an excess of potassium iodide is added (see reactions (59) and (60)). There-

$$\begin{array}{l} \text{HOHg—HC}\overset{\text{Hg}}{}\text{CH—HgOH} \\ \quad\;\; | \qquad\quad | \\ \text{HO—C}\underset{\text{O}}{}\text{C—OH} \\ \quad\;\; | \qquad\quad | \\ \quad \text{H}_3\text{C} \qquad \text{CH}_3 \end{array} + 12KI + 3H_2O \rightleftharpoons$$

$$6KOH + K_2HgI_4 + 2CH_3COCH_3 \quad (59)$$

$$\begin{array}{ll} \text{RCH—CH}_2 \\ \;\,| \qquad | \\ \text{OCH}_3\ \text{HgOH} \end{array} + KI \rightleftharpoons \begin{array}{ll} \text{RCH—CH}_2 \\ \;\,| \qquad | \\ \text{OCH}_3\ \text{HgI} \end{array} + KOH \quad (60)$$

fore, the net consumption of base is due to the acetic acid formed in equation (55). This net titration for acetic acid, which is equivalent to the olefin content, is obtained from the difference between the sample and blank titrations. The excess caustic is titrated to the phenolphthalein endpoint with standard hydrochloric acid.

(2) Direct Titration of Acetic Acid in the Presence of Sodium Chloride (233)

The direct titration of Martin (233) is considerably simpler than the back-titration method of Marquardt and Luce (227,228). After the methoxymercuration of the olefin [equation (55)], all of the mercury(II) including the methoxymercurial addition compound is complexed by excess chloride ions:

$$Hg(OOCCH_3)_2 + 4NaCl \rightleftharpoons Na_2HgCl_4 + 2CH_3COONa \quad (61)$$

$$\begin{array}{ll} \text{RCH——CH}_2 \\ \;\,| \qquad\quad | \\ \text{OCH}_3\ \ \text{HgOOCCH}_3 \end{array} + NaCl \rightleftharpoons \begin{array}{ll} \text{RCH——CH}_2 \\ \;\,| \qquad\quad | \\ \text{OCH}_3\ \ \text{HgCl} \end{array} + CH_3COONa \quad (62)$$

The chloro complexes are more stable than the hydroxo complexes of mercury(II), permitting the direct titration of the acetic acid. Nitrate salts catalyze the methoxymercuration although the effect is not very appreciable.

In the Martin procedure the mercuric acetate reagent is weighed out for the titration of each sample and blank. This is one of the weak points of this method because (*1*) mercuric acetate deteriorates when exposed to the air, (*2*) it is often heterogeneous and requires thorough mixing and grinding before each series of application, and (*3*) it is seldom completely soluble. The other three methoxymercuration methods use mercuric acetate solution in methanol.

(3) Direct Titration of Acetic Acid in the Presence of Sodium Bromide (173)

Johnson and Fletcher (173) used a method which is similar to the Martin method except that a methanolic mercuric acetate solution is used and sodium bromide is used to complex the mercury:

$$Hg(OOCCH_3)_2 + 4NaBr \rightleftharpoons Na_2HgBr_4 + 2CH_3COONa \qquad (63)$$

$$\underset{\substack{|\\OCH_3}}{RCH}\text{——}\underset{\substack{|\\HgOOCCH_3}}{CH_2} + NaBr \rightleftharpoons \underset{\substack{|\\OCH_3}}{RCH}\text{——}\underset{\substack{|\\HgBr}}{CH_2} + CH_3COONa \qquad (64)$$

The acetic acid formed in the mercuration step (equation (55)) is titrated with $0.1N$ potassium hydroxide to the phenolphthalein endpoint.

(4) Titration of Mercury (98)

The preceding methoxymercuration methods give high results for unsaturated esters such as vinyl acetate and vinyl benzoate, because the esters saponify quantitatively under the basic conditions of the titration as the endpoint is approached. By titrating to an acidic endpoint, Das (98) avoided the saponification error. The methoxymercuration of the sample is carried out in a 1:1 propylene glycol–chloroform medium by adding a methanolic solution of mercuric acetate. The acetatomercurate(II) complexes are titrated with standard hydrochloric acid to the thymol blue endpoint:

$$Hg(OOCCH_3)_2 + 2HCl \rightleftharpoons HgCl_2 + 2CH_3COOH \qquad (65)$$

$$\underset{\substack{|\\OCH_3}}{RCH}\text{——}\underset{\substack{|\\HgOOCCH_3}}{CH_2} + HCl \rightleftharpoons \underset{\substack{|\\OCH_3}}{RCH}\text{——}\underset{\substack{|\\HgCl}}{CH_2} + CH_3COOH \qquad (66)$$

Since each mole of excess mercuric acetate consumes 2 moles of hydrochloric acid and each mole of methoxymercuration product consumes only 1 mole of hydrochloric acid, the milliequivalence difference between the sample and blank titrations is equivalent to the millimoles of unsaturation in the sample.

In order to obtain the sharp yellow to pink endpoint, the acid titrant in equations (65) and (66) must meet two requirements: It must be a considerably stronger acid than acetic acid, and the anion must form a stable complex with mercury(II). Perchloric acid is not applicable, because although it is a stronger acid than hydrochloric acid, the perchlorate ion is one of the poorest complexing agents. Since there is a stoichiometric concentration of chloride at the endpoint, mercuric chloride tends to precipitate out, but the precipitate does not obscure the endpoint.

TABLE XXIII

Analytical Results on Olefinic Compounds by Methoxymercuration Methods (271)
(Results obtained by original authors, except where noted)

Compound	Method of Marquardt and Luce (227,228)		Method of Martin (233)		Method of Das (98)	
	Reaction conditions	Percent found	Reaction conditions	Percent found	Reaction conditions	Percent found[a]
Styrene	5 min, 25°C	99.4–99.7(5)	10–15 min, 25°C	99.7–99.9(4)	5–10 min, 30°C	99.2, 99.6
	5–10 min, 30°C	99.4, 99.7[b]				
α-Methylstyrene	—	Low[c]	—	—	—	Low[c]
Ethoxystyrene	5 min, 25°C	ca. 100[c]	—	—	—	—
Ethylvinylbenzene	5 min, 25°C	98.7–98.8(3)	10–15 min, 25°C	100.1, 100.5[d]	—	—
Divinylbenzene	5 min, 25°C	98.8–99.0(3)			—	—
Vinyltoluene	5 min, 25°C	ca. 100[c]	—	—	—	—
Vinylxylene	5 min, 25°C	ca. 100[c]	—	—	—	—
Monochlorostyrene	15 min, 25°C	ca. 100[c]	—	—	—	—
Vinylchlorostyrene	15 min, 25°C	ca. 100[c]	—	—	—	—
2,5-Dichlorostyrene	1 hr, 50°C	99.8–100.1(5)	—	—	—	—
Cyclohexene	10–15 min, 30°C	96.7, 96.3[b]	—	—	10–15 min, 30°C	96.6, 96.9

Divinyl ether	—	—	10–15 min, 25°C	96.8	—	—
Diallyl ether	—	—	10–15 min, 25°C	99.9	—	—
Allyl alcohol	30 min, 30°C	97.7[b]	10–15 min, 25°C	99.5, 98.7	30 min, 30°C	98.2, 98.2
Crotyl alcohol	—	—	10–15 min, 25°C	101.5	—	—
β-Chloroallyl alcohol	—	—	10–15 min, 25°C	15	—	—
Vinyl acetate	30 min, 30°C	88, 76[b]	10–15 min, 25°C	201	30 min, 30°C	97.4, 96.8
Allyl acetate	30 min, 30°C	33, 28[b]	—	—	30 min, 30°C	96.9–97.5(3)
Vinyl benzoate	—	—	10–15 min, 25°C	187	—	—
Diallyl phthalate	—	—	10–15 min, 25°C	94.5	—	—
Methyl acrylate	—	Low[b,c]	10–15 min, 25°C	44	—	Low[c]
Methyl methacrylate	—	Low[b,c]	10–15 min, 25°C	1.3	—	Low[c]
Diethyl maleate	—	—	10–15 min, 25°C	4.0	—	—
Diethyl itaconate	—	—	10–15 min, 25°C	Trace	—	—
Acrylonitrile	—	—	10–15 min, 25°C	3.4	—	—
2-Vinylpyridine	—	—	10–15 min, 25°C	40	—	—
N-Vinylcarbazole	—	—	10–15 min, 25°C	99.4, 99.7	—	—
Rotenone	—	—	25 min, 25°C	100.0[e]	—	—

[a]Number in parentheses indicates the number of replicate determinations.
[b]Results obtained by Das (98).
[c]Numerical data not given in the original publication.
[d]Results on a mixture, compared with the method of Marquardt and Luce (227,228).
[e]Results obtained by Hornstein (162).

The comparative analytical data for various olefins by the methoxymercuration methods of Marquardt and Luce, Martin, and Das as compiled by Polgar and Jungnickel (271) are presented in Table XXIII. The analytical data by the Johnson and Fletcher modification of the methoxymercuration method is given in Table XXIV (173). Among the four methoxymercuration methods the direct titration of acetic acid after sodium bromide addition is preferred wherever applicable. This method has been adopted tentatively by ASTM (17), and the Society of the Plastics Industry to determine small amounts of terminal unsaturation in polyoxyalkylene polyols. The details of this method are given in Section VII-F. The Das procedure is less convenient but it is the only methoxymercuration method that is applicable to unsaturated esters. None of the methoxymercuration methods is adequate for α,β-unsaturated compounds which require nucleophilic reagents.

TABLE XXIV

Analytical Results on Olefinic Compounds by the Johnson and Fletcher (Sodium Bromide) Methoxymercuration Method (173)

Compounds	Purity, wt %	Av deviation	No. of detns.	Min time, min	Temp, C°
Vinyl Ethers					
1-Butenyl methyl ether[a]	97.7	0.0	2		
Divinyl carbitol	98.4	0.1	2		
1-Propenyl ethyl ether	97.4	0.4	2		
Vinyl allyl ether	99.0	0.0	2		
Vinyl butyl Cellosolve	100.0	0.1	2		
Vinyl butyl ether[a]	98.9	0.2	3		
Vinyl (2-butylmercapto) ethyl ether	100.0	0.1	2		
Vinyl carbitol	100.8	0.1	2		
Vinyl 2-chloroethyl ether	97.3	0.0	2		
Vinyl ethyl ether[a]	98.9	0.2	14		
Vinyl hexyl carbitol	100.1	0.1	2		
Vinyl isobutyl ether[a]	98.6	0.2	2		
Vinyl propyl ether	97.0	0.3	2		
Vinyl tetradecyl ether	95.2	0.1	2		
Vinyl undecyl ether	96.2	0.1	2		
Other Unsaturated Compounds					
Allyl acetate	98.8	0.0	2	60	25
Allyl acetone	98.8	0.1	2	20	−10
Allyl alcohol	99.1	0.1	4	1	25
2-Allyl-3-methyl-2-cyclopenten-4-ol-1-one	100.1	0.0	2	10	−10
Butyl chrysanthemum monocarboxylate	95.4	0.0	3	60	25
2-Chloro-1-propenyl butyl ether	97.4	0.4	2	15	25
Cyclohexene	98.5	0.1	4	1	25
Dichlorostyrene	99.7	0.2	3	120	25

TABLE XXIV (*continued*)

Compounds	Purity, wt %	Av deviation	No. of detns.	Min time, min	Temp, °C
2,5-Dimethyl-1,5-hexadiene	94.1	0.1	3	15	25
3,4-Epoxy-1-butene	97.6	0.0	2	60	25
2-Ethoxy-3,4-dihydropyran	97.0	0.1	4	30	0
2-Ethoxy-5-methyl-3,4-dihydropyran	97.1	0.2	4	30	0
3-Ethoxy-4-propyl-5-ethyl-3, 4-dihydropyran	96.2	0.0	4	30	25
2-Formyl-3,4-dihydropyran	97.0	0.0	4	30	25
3-Hydroxy-8-nonen-2,5-dione	98.9	0.2	3	10	−10
Methallyl chloride	97.9	0.1	4	15	25
2-Methoxy-3-ethyl-3,4-dihydropyran	100.0	0.0	7	30	25
2-Methoxy-3-ethyl-4-methyl-3, 4-dihydropyran	96.1	0.0	6	30	25
2-Methoxy-3-methyl-3,4-dihydropyran	101.5	0.1	6	30	25
2-Methoxy-3-methyl-4-propyl-5-ethyl-3, 4-dihydropyran	97.2	0.1	2	120	25
4-Methyl-3,4-dihydropyran	98.7	0.0	2	15	−10
4-Methyl-1-pentene	98.2	0.0	2	30	25
α-Methylstyrene	99.2	0.1	4	5	−10
Styrene	99.6	0.1	2	10	25
Vinyl acetate[b]	99.0	0.1	4	10	25
N-Vinylpiperidone	99.1	0.0	2	10	25
N-Vinylpyrrolidone	97.0	0.3	10	10	25

[a]Use a sealed glass ampoule or an aliquot from a methanolic dilution of the sample.
[b]Each mole of vinyl acetate results in consumption of two equivalents of KOH due to ease of saponification of reaction product.

A source of error in the methoxymercuration method for vinyl and *cis* unsaturation is the possible reaction of the addition product with excess halide, which leads to low results. Marquardt and Luce (228a) developed a procedure that eliminates this interference by first carrying out the addition reaction under optimum conditions in methanol solution and then adding carbon tetrachloride and doing the titration in a two-phase system. The addition product, which is preferentially extracted into the carbon tetrachloride, is protected from the halide. This new procedure is simple and gives improved accuracy.

Another source of error in the methoxymercuration method for terminal unsaturation in polymers is the oxidation of antioxidants by mercury(II) ion (156a).

b. α,β-Unsaturated Compounds

An olefin with a strong electron-withdrawing substituent such as nitrile, carbonyl, carboxylic acid, and esters on the double bond is referred to as an

α,β-unsaturated compound. Such compounds are cationoid and require nucleophilic reagents such as bisulfite and morpholine. Mercaptan addition to α,β-unsaturated compounds has also been applied (33) but this reagent is not very stable.

(1) Bisulfite

Bisulfite ion adds to cationoid double bonds according to:

$$RCH{=}CHY + HSO_3^- \rightarrow \underset{\displaystyle SO_3^-}{RCH}{-}CH_2Y \qquad (67)$$

where Y is a strong electron-attracting group (308). The reduction in the acidity is utilized in the determination of unsaturation in such compounds.

The rate of addition of bisulfite to α,β-unsaturated compounds is greatly affected by the pH of the medium. Figure 12 shows the effect of pH

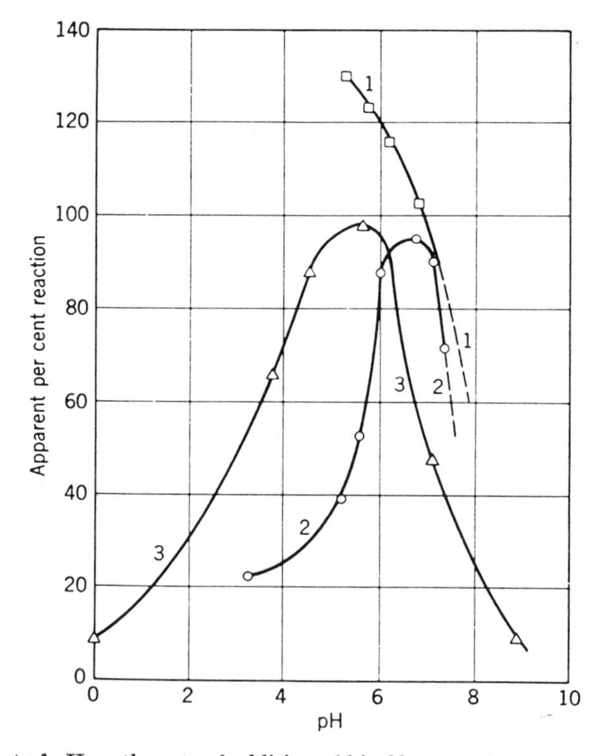

Fig. 12. Effect of pH on the rate of addition of bisulfite to α,β-unsaturated compounds. 1. Ethyl crotonate at room temperature; 2. ethyl crotonate at 100°C; 3. crotonic acid at 100°C (322).

on the rate of addition of bisulfite to ethyl crotonate and crotonic acid (322). The optimum pH range is 5.5–6.5.

Critchfield and Johnson (91) have developed the conditions necessary for the quantitative bisulfite addition to various α,β-unsaturated compounds. They use a sodium sulfite reagent which is more stable than a bisulfite reagent. The bisulfite is generated *in situ* by adding sulfuric acid just prior to the analysis. Since 25 meq of sulfuric acid is added to 50 mmoles of sodium sulfite and the sample is limited to 15 mmoles of double bonds, the pH during the addition is buffered between pK_2 of sulfurous acid to $pK_2 + 0.4$. The excess bisulfite is titrated with standard sodium hydroxide just to the disappearance of the green color of the Alizarin Yellow R–xylene cyanol mixed indicator endpoint. The analytical data obtained by the authors are given in Table XXV.

There are numerous interferences in the bisulfite method. Bisulfite adds quantitatively to aldehydes and it is oxidized readily by peroxides and other oxidizing impurities. Bisulfite also adds on to epoxides (342). Other interferences are substitution reactions involving a carbon–halogen bond and saponification of esters at elevated temperatures (322, p. 355).

The addition of morpholine to α,β-unsaturated compounds is less subject to interference than the bisulfite method.

TABLE XXV

Analytical Results on α,β-Unsaturated Compounds by the Bisulfite Addition Method (91)

Compound	Average purity, wt %		Reaction conditions	
	Sodium sulfite method[a]	Other	Temp, °C	Time, min
Acrylic acid	98.4 ± 0.2 (5)	98.7[b]	98[f]	15–50
Acrylonitrile	98.1[c]	—	25	5–30
Crotonic acid	98.7 ± 0.1 (3)	99.1[d]	98[f]	60–120
Diethyl fumarate	99.9 ± 0.1 (8)	99.9[e]	25[f,g]	15–90
Diethyl maleate	98.6 ± 0.1 (5)	98.6[e]	25[g,h]	60–90
Ethyl acrylate	99.2 ± 0.1 (6)	99.0[e]	25[g]	30–60
Maleic acid	99.2 ± 0.2 (5)	99.0[d]	98[f]	15–90
Methyl acrylate	98.4 ± 0.1 (7)	98.3[e]	25[g]	15–60

[a]Figures in parentheses represent number of determinations.
[b]Modified Kaufmann bromination of potassium salt.
[c]Standard deviation for eight degrees of freedom is 0.09.
[d]Acidity titration.
[e]Saponification.
[f]Use 15.0 ml of isopropyl alcohol as a cosolvent.
[g]Place samples and blank in a -10°C bath for 10 min before titration.
[h]Place on mechanical shaker for 15 min.

(2) Morpholine

Critchfield and co-workers (90) determined α,β-unsaturated compounds by the addition of morpholine in the presence of acetic acid catalyst:

$$\underset{\substack{\diagdown \\ C-C \\ H_2\ H_2}}{\overset{\substack{H_2\ H_2 \\ C-C \\ \diagup}}{O\quad NH}} + R-CH{=}CH-Y \xrightarrow{CH_3COOH} \underset{\substack{\diagdown \\ C-C \\ H_2\ H_2}}{\overset{\substack{H_2\ H_2 \\ C-C \\ \diagup}}{O\quad N}}-\underset{\substack{| \\ R}}{CH}-CH_2-Y \quad (68)$$

where Y is a strong electron-attracting group such as carbonyl, carboxylate, ester, or nitrile. The excess morpholine is acetylated with acetic anhydride in acetonitrile medium, and the tertiary amino addition product is titrated with methanolic hydrochloric acid to the methyl orange–xylene cyanol mixed indicator endpoint. If R in equation (68) is also a strong electron-attracting group, its inductive effect on the vicinal tertiary amino nitrogen weakens the latter to such an extent that its titration break is not sharp enough for quantitative application. For example, conductometric

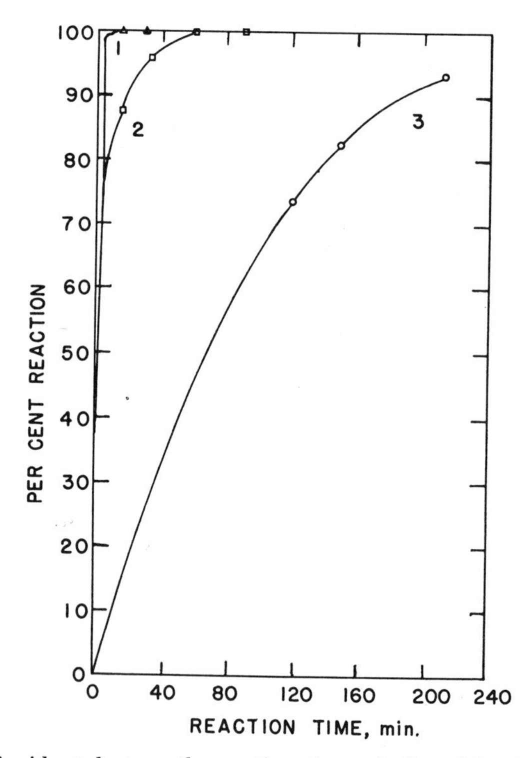

Fig. 13. Effect of acid catalysts on the reaction of morpholine with ethyl crotonate (90). (Courtesy of *Analytical Chemistry.*) (1) Acetic acid; (2) hydrochloric acid; (3) no catalyst.

titration is necessary for the determination of the morpholine addition products of diethyl fumarate and di(2-ethylhexyl) maleate.

The esters of acrylic acid react readily with morpholine without catalysis while others require catalysts. Figure 13 shows the effect of acid catalysts on the reaction of morpholine with ethyl crotonate. A fixed amount of acetic acid is used routinely for all of the α,β-unsaturated compounds.

Most epoxides react quantiatively with morpholine to form tertiary amines which are basic in acetonitrile. Organic halides interfere by reacting with morpholine to liberate halogen acids. Large quantities of aldehydes, ketones, and acid anhydrides may interfere by depleting the reagent.

The analytical results of the authors are presented in Table XXVI. The details of their methods are given in Section VII-G.

TABLE XXVI

Analytical Results on the α,β-Unsaturated Compounds by the Morpholine Addition Method (90)

Compound	Average purity[a], wt %		Reaction conditions	
	Morpholine method	Other	Time, min	Temp, °C
Acrylamide	100.0 ± 0.1 (5)	—	5–120	25
Acrylic acid	98.6 ± 0.1 (4)	98.7[b]	15–120[g]	98
Acrylonitrile	98.3[c]	—	5–60	25
Allyl cyanide	98.3 ± 0.1 (4)	98.3[d]	30–60	98
Butyl acrylate	99.8 ± 0.1 (4)	99.8[e]	5–60	25
Diethyl fumarate	99.5 ± 0.2[b] (2)	99.9[e]	30–90[h]	25
Di(2-ethylhexyl) maleate	99.7 ± 0.2[f] (2)	100.0[e]	45–90[h,i]	25
Diethyl maleate	98.4 ± 0.2[f] (2)	98.6[e]	30–90[c,i]	25
Ethyl acrylate	99.2 ± 0.2 (4)	99.0[e]	5–60	25
2-Ethylbutyl acrylate	99.6 ± 0.1 (4)	99.5[e]	5–30	25
Ethyl crotonate	100.0 ± 0.1 (3)	99.9[e]	15–60	98
2-Ethylhexyl acrylate	99.0 ± 0.1 (4)	99.0[e]	5–60	25
Methacrylonitrile	97.9 ± 0.2 (3)	—	120–240	98
Methyl acrylate	98.7 ± 0.2 (4)	98.3[e]	5–60	25
Methyl methacrylate	97.8 ± 0.1 (3)	98.8[e]	40–80	98

[a]Figures in parentheses represent number of determinations.
[b]Bromination of potassium salt.
[c]Standard deviation for five degrees of freedom is 0.11.
[d]Bromination by a modified Kaufmann procedure.
[e]Saponification.
[f]Conductometric endpoint
[g]Use 10 ml of methanol cosolvent.
[h]Use conductometric titration procedure.
[i]Use 40 ml of methanol cosolvent and 2 ml of acetic acid solution.

c. CONJUGATED DIOLEFINS—MALEIC ANHYDRIDE

The Diels-Alder (103) addition of maleic anhydride to conjugated dienes is the most important of the limited number of chemical methods available for such olefins. The reaction involves the condensation of the conjugated diene through its 1,4-carbon atoms with a dienophile to form a six-membered ring (equation (69)).

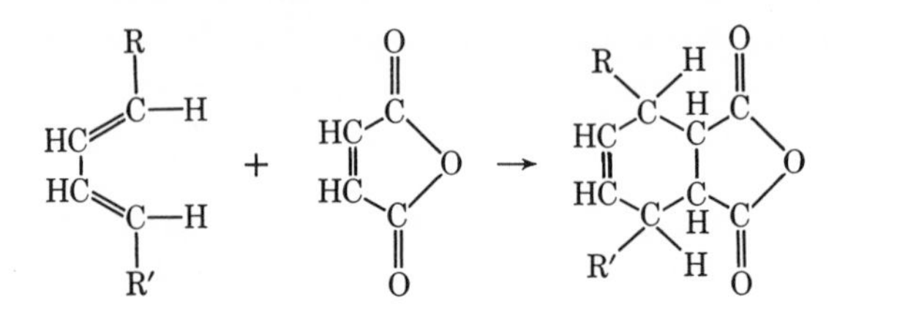

(69)

Steric effects play a very important role in the reaction. The *trans–trans* configuration shown reacts readily at room temperature while the *cis–trans* diolefin requires more drastic conditions, and the *cis–cis* diolefin is usually completely unreactive. Ellis and Jones (117) found that a small amount of iodine catalyzed the reaction. According to von Mikusch (243,244), the catalytic effect of iodine was based on the conversion of *cis* configuration (prominent in natural oils) to the reactive *trans* configuration. Five- and six-membered cyclic conjugated dienes react readily with maleic anhydride to form the bridged ring adduct, but seven-membered cyclic conjugated diolefins may require heating. When the diene system is partially in the side chain, or when the conjugation extends over two rings of a polycyclic compound, reaction may not occur to a measurable extent. Allenes and higher cumulenes do not add maleic anhydride. Excessive alkyl substitution at the ends of the diene system reduces reactivity due to steric hindrance.

Halogen atoms on the doubly bonded carbons may render the diene system completely inert toward maleic anhydride due to inductive effects which reduce the π-electron density.

Kaufmann and Baltes (184,185) applied the maleic anhydride method to the determination of conjugated dienes in fats and oils. It involves the heating of the sample with an acetone solution of maleic anhydride in a sealed tube at 100°C for 20 hr. The excess maleic anhydride is hydrolyzed; then the acid is separated from the oily layer and titrated with standard alkali.

Ellis and Jones (117) dispensed with the sealed tube by refluxing in a toluene medium, and by using a small amount of iodine catalyst they were able to reduce the reaction time to 1 hr. The 1961 revision of this method by ASTM omits the iodine catalyst and uses a 3 hr reflux time (18). The diene value is expressed as the centigrams of iodine equivalent to the maleic anhydride used per gram of sample. Cooperative precision studies indicate that this method gives diene values within 5 for different operators and within 2 for the same operator.

The application of the maleic anhydride method to petroleum products has been quite limited due to the development of spectrophotometric methods, especially for the C_4 and C_5 dienes. Tropsch and Mattox (349) determined 1,3-butadiene gasometrically by absorbing the diene in molten maleic anhydride at 100°C. The interference by the polymerization of isobutylene due to traces of maleic acid in the anhydride can be inhibited by the addition of diamylamine (62,290) or by removal of most of the isobutylene by passing the gas through 50% sulfuric acid (291).

A titrimetric maleic anhydride method developed by Grosse-Oetringhaus (141) for gasoline-range hydrocarbons gave inconsistent results. The diene number tended to increase with reaction time, which was varied from 5 to 30 hr. Satisfactory precision and accuracy is obtained by the Shell method (317) which utilizes the rapid iodine-catalyzed procedure of Ellis and Jones (117) but the sample is reacted with the 3% maleic anhydride–toluene solution in a pressure bottle in a steam bath.

d. Acetylenic Compounds

Acetylenic unsaturation may be determined by halogenation or hydrogenation as discussed earlier, but these methods lack selectivity. There are two general approaches which are highly selective for acetylenic unsaturation: hydration of the triple bond and the indirect titration of the acetylenic hydrogen.

(1) Hydration of the Triple Bond

Wagner and co-workers (359) determined mono- and dialkyl acetylenes of four or five carbons by performing the reaction with methanol in the presence of mercuric oxide and boron trifluoride catalysts. The ketal formed is distilled into an aqueous hydroxylamine hydrochloride solution. The ketal hydrolyzes to the corresponding ketone, which upon oximation releases a mole of hydrochloric acid. The hydrochloric acid is titrated potentiometrically with standard caustic, and the acetylene content is calculated from the titration value. The recovery by this method was about 92% and it was not applicable to higher boiling ketals which seemed to be unstable at the higher distillation temperatures.

The above method was improved by Siggia and co-workers (327), who carried out the hydration in one step, thereby eliminating the distillation step.

$$RC{\equiv}CR' + H_2O \xrightarrow[H_2SO_4]{HgSO_4} R\overset{\overset{O}{\|}}{C}CH_2R' \tag{70}$$

The ketone formed in equation (70) may be determined by oximation followed by acidimetric determination of the released acid,

$$R\overset{\overset{O}{\|}}{C}CH_2R' + NH_2OH{\cdot}HCl \rightarrow R\overset{\overset{NOH}{\|}}{C}CH_2R' + H_2O + HCl \tag{71}$$

or by a gravimetric method following the formation of the 2,4-dinitrophenylhydrazone (equation (72)). The accuracy of the acidimetric method

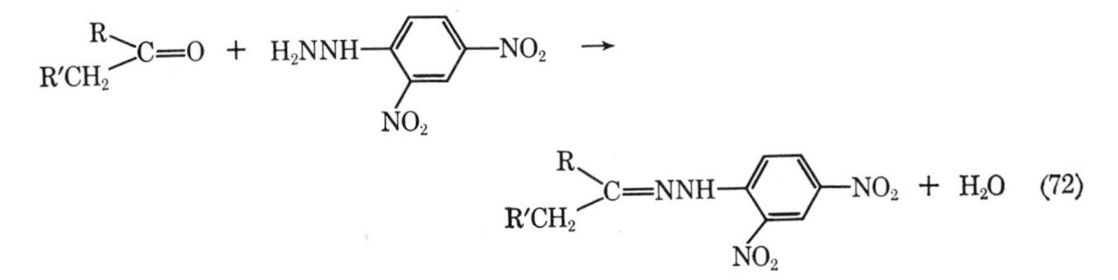

is about 100 ± 2%. The details of the procedure for these two variations are given by Siggia (322, pp. 369–371). The gravimetric method is used for ketones which cannot be oximated readily, e.g., acetophenone, formed by the hydration of phenylacetylene.

(2) Indirect Titration of the Acetylenic Hydrogen

(a) *Silver Methods*

The acetylenic hydrogen atom exhibits no acidity in itself but it reacts quite readily with certain cations such as silver, mercury(II), and copper(I). With silver nitrate, the reaction is

$$RC{\equiv}CH + AgNO_3 \rightarrow RC{\equiv}CAg + HNO_3 \tag{73}$$

The acetylide is quite insoluble and the precipitation is generally quantitative unless solubilizing groups such as hydroxyl, carboxylate, or sulfonate are present on the R group.

The acetylenic compound is determined by titrating the excess silver ions or the acid formed in reaction (73). In the argentimetric approach, a base such as ammonia is used to drive the reaction to completion. The filtrate is neutralized with nitric acid and the silver is titrated with standard $0.1N$ thiocyanate solution to the ferric alum endpoint.

Acetylenic compounds in gaseous samples may be determined by bubbling the sample through a column packed with glass beads and containing the ammoniacal silver nitrate solution.

Ammoniacal silver readily oxidizes aldehydes. This interference can be minimized by using sodium acetate as the base. Halides, sulfides, cyanides, and other substances that react with silver also interfere in both the ammonia and acetate variations of the argentimetric approach.

In the acidimetric method, most of the substances that react with the silver ions do not interfere provided that a sufficient excess of silver is used and no acids are formed.

Barnes and Molinini (28) used concentrated silver nitrate which solubilized the silver acetylide, thus allowing the titration to be carried out in a homogeneous solution:

$$AgC{\equiv}CR + AgNO_3 \rightleftharpoons \left[\begin{matrix}AgC{\equiv}CR\\ Ag\end{matrix}\right]^+ + NO_3^- \tag{74}$$

The solubilization of the acetylide is probably due to the formation of the π-complex. A $0.3M$ silver nitrate solution prevents the precipitation of 1.5 meq of acetylides in 100 ml of solution if the alkynes contain hydroxyl groups, while alkynes without any hydroxyl groups require 2–$3.5N$ silver nitrate solution. A homogeneous medium is especially desirable in the analysis of gaseous samples where precipitates are likely to obstruct volumetric apparatus, clog stopcocks, etc.

Analytical data for alkynes by the Barnes and Molinini method are presented in Table XXVII. Some of the compounds tested have very limited solubility in water but reacted quantitatively with the concentrated silver solution to produce a clear, single-phase system. One exception was 3-methyl-1-nonyn-3-ol, which had to be dissolved in 5–10 ml of 95% ethanol, and the silver reagent was added to it. This prevented the precipitation of acetylides and the analytical results were reproducible and quantitative.

Halides up to twice the concentration of the alkyne did not interfere. The endpoint was observed after the silver halide settled or the precipitate was filtered before the acidimetric titration. Although the metathesis of silver acetylides by the halides have been reported (289), no such interference is expected here, since the large excess of silver ions keeps the free halide ions at a very low concentration. Aldehydes up to about one-third of the concentration of the alkyne did not interfere. The noninterference of aldehydes is due to the fact that the pH of the solution is never main-

TABLE XXVII

Analytical Results for Alkynes by the Argento-Acidimetric Method (28)

Compound[a]	Purity, %	
	$AgNO_3$ reagent	$AgClO_4$ reagent
Alcohols		
1-Phenyl-2-propyn-1-ol	99.4	99.2
2-Propyn-1-ol[b]	98.5	98.8
2-Methyl-3-butyn-2-ol	99.0, 99.2	—
3-Methyl-1-pentyn-3-ol	99.3, 99.5	99.5
3,4-Dimethyl-1-pentyn-3-ol	99.1	99.5
3-Ethyl-1-pentyn-3-ol	98.5, 98.5	98.5
1-Ethynylcyclopentanol	97.4, 97.7	97.9
3-Methyl-1-hexyn-3-ol	98.2, 98.7	98.8
1-Ethynylcyclohexanol	97.5, 97.9	97.8
2,6-Dimethyl-4-ethynyl-heptan-4-ol	98.0, 98.0	—
3-Methyl-1-nonyn-3-ol[c]	96.3, 96.3	96.5, 96.6
Hydrocarbons		
Acetylene[d]	99.82 ± 0.32	99.85 ± 0.15
1-Propyne	99.3, 99.4	
1-Butyne[e]	98.9, 98.4	
2-Methylbuten-3-yne	99.8, 99.6	—
1-Hexyne[f]	96.1, 96.3	—
Carboxylic acid		
Propiolic acid[g]	98.0, 98.4	—
Amines		
3-Aminopropyne[h]	99.5	—
Di(2-propynyl)amine[i]	99.4	—
Miscellaneous		
Sodium acetylide[j]	97.4, 97.2	—

[a]Technical grade samples of reasonable purity; no attempts made at further purification.

[b]Boiling range 113–115°C.

[c]3.2% ketone present by hydroxylamine hydrochloride analysis.

[d]Average analysis of 14 samples ($AgNO_3$) and 4 samples ($AgClO_4$) from same cylinder of acetylene on acetone-free basis. Hempel absorbtion analysis of 24 samples from same cylinder, using bromine solution as absorbent, indicates 0.38 ± 0.08% nonacetylenics, i.e., 99.62% acetylene.

[e]1.3% nonacetylenics present by Orsat analysis.

[f]95.8% assay by potassium mercuric iodide (148).

[g]98.5% assay by titration of carboxylic acid group.

[h]98.6% assay by nonaqueous titration of amino group (268).

[i]99.3% assay by nonaqueous titration of amino group.

[j]2.1% nonacetylenics present.

tained at a high value sufficiently long to reduce silver ions. A $0.5N$ caustic solution should not be used because local excess concentration of base may cause interference. A 0.05–$0.2N$ caustic is satisfactory although $0.1N$ seems to be ideal for most compounds.

(*b*) *Copper Method* (322)

Cuprous chloride displaces acetylenic hydrogen according to:

$$2HC{\equiv}CR + Cu_2Cl_2 + C_5H_5N \rightarrow 2CuC{\equiv}CR + 2C_5H_5NHCl \qquad (75)$$

where the weak base, pyridine, drives the reaction to completion. The pyridinium chloride is titrated with standard caustic using a pH meter. An indicator endpoint cannot be used because of the intensity of the color of the solution and the presence of colored precipitate.

This method is not as accurate or precise as the silver methods; the reproducibility of this method is $\pm 2\%$. However, this method can be used for samples containing impurities that interfere with the silver method.

(*c*) *Mercury Method*

Hanna and Siggia (148) determined propargyl alcohol, 1-hexyne, 3-butyn-1-ol, and acetylene by treating the alkynes with excess potassium tetraiodomercurate(II) and a known amount of standard potassium hydroxide:

$$2HC{\equiv}CR + K_2HgI_4 + 2KOH \rightarrow Hg({-}C{\equiv}CR)_2 + 4KI + 2H_2O \qquad (76)$$

The base consumed by the reaction is equivalent to the alkyne present and is determined by titrating to the phenolphthalein endpoint with standard sulfuric acid.

The details for the indirect titration of acetylenic hydrogen by the silver, copper, and mercury methods are given by Siggia (322, pp. 381–398).

5. Analysis of Mixtures

Lee and Kolthoff (208) applied the kinetic approach in analyzing mixtures. The compounds being determined must be known so that the rate constants for each component can be obtained at the given temperature and calibration curves are plotted. The theoretical limit of resolution of their method was for compounds whose rates of reaction with a given reagent differed by more than a factor of 4. However, they showed no example of analysis of mixtures whose rates of reaction were closer than a factor of 17.

Siggia and Hanna (324) overcame the disadvantages of the preceding approach by applying second-order kinetics for the analysis of mixtures. There was no need for temperature control or knowledge of the rate constants of the compounds involved, and rates differing by only a factor of 2 were resolved.

In a second-order reaction, the rate is dependent on the concentration of two reactants in the system. If a and b are the initial concentrations of reactants A and B and if x is the concentration of A and B that has reacted at time t, then the rate of reaction is given by:

$$\frac{dx}{dt} = k(b - x)(a - x) \tag{77}$$

On integration, taking into account that $x = 0$ when $t = 0$, and $x = x$ when $t = t$:

$$k = \frac{2.303}{t(b - a)} \log \left[\frac{a(b - x)}{b(a - x)}\right] \tag{78}$$

Plots of $\log [(b - x)/(a - x)]$ vs. t yield straight lines for second-order reactions. When two second-order reactions are proceeding in the same system, a curve with two straight-line portions is obtained if the reaction rates of the two reactions are sufficiently different.

The bromination of unsaturated compounds follows a second-order rate process. The decrease in bromine concentration is followed colorimetrically. It is essential that the initial bromine and unsaturation concentrations be unequal, because then $a = b$ and $\log [(b - x)/(a - x)]$ will always be zero. The rate curve from Siggia and co-workers (325) for the bromination of a mixture of 5-methyl-1-hexyne and 2-butyne-1,4-diol is given in Fig. 14. The point in Fig. 14 are not experimental points; they were obtained from x vs. t plots. This technique averages the errors in the experimental data, and it provides a large number of points for a good plot of $\log [(b - x)/(a - x)]$ vs. t. In the region of slope *1* both components are reacting, while in the straight-line region of slope *2*, only the slower reacting component is reacting. The active component concentration is obtained by extrapolating slope *2* to point A, and drawing the horizontal line $\overline{AB}$ and the vertical line $\overline{BT}$. The value of x at time T on the x vs. t plot corresponds to the concentration of the active component. Since the total bromine value is given by the x vs. t plot, the percentage of the less active component can be calculated.

The percent active component may be calculated without using the x vs. t plot. Let $a = a_1 + a_2$, where a_1 is the more reactive unsaturated

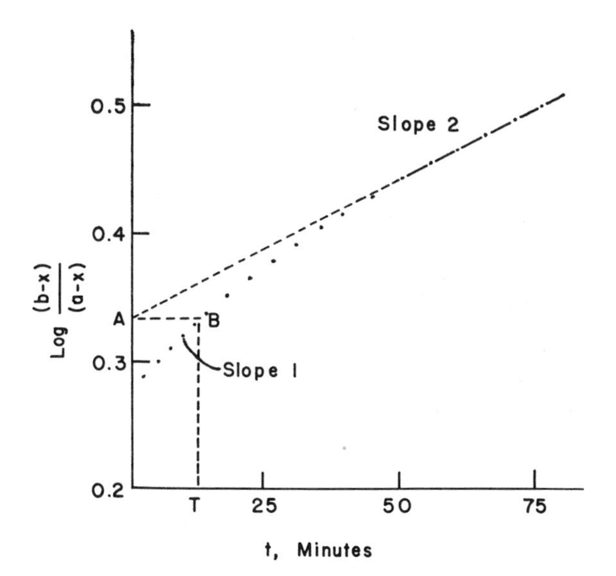

Fig. 14. Reaction rate curve for the bromination of a mixture of 5-methyl-1-hexyne and 2-butyne-1,4-diol (325). (Courtesy of *Analytical Chemistry*.)

compound. At the intercept A, $x = a_1$; therefore

$$\log [(b - x)/(a - x)] = \log [(b - a_1)/(a - a_1)] \tag{79}$$

Taking the value of $\log [(b - x)/(a - x)]$ at the intercept A,

$$\log [(b - a_1)/(a - a_1)] = A \tag{80}$$

Since b is known and a is obtained from the total bromination value, a_1 can be calculated by equation (80).

The results obtained by Siggia, Hanna, and Serencha (325) in the bromination of mixtures of unsaturated compounds are given in Table XXVIII.

In the hydrogenation of unsaturated compounds, the hydrogen pressure is kept essentially constant. Therefore, a pseudo-first-order kinetics is observed. The integrated form of a first-order reaction expression is

$$kt = 2.303 \log \left[\frac{a}{(a - x)}\right] \tag{81}$$

where a is the original concentration of unsaturation and x is the concentration reacting in time t. A plot of $\log (a - x)$ against t results in a straight line with a slope of $-(k/2.303)$. For two unsaturated compounds

TABLE XXVIII

Bromination of Mixtures of Unsaturated Compounds (325)

Mixture	Percent A Found	Percent A Present
1. A. Crotonic acid	34.7	34.5
B. Fumaric acid	27.1	27.4
2. A. Sorbic acid	67.6	70.1
B. Fumaric acid		
3. A. Sorbic acid	85.9	85.7
B. Maleic acid		
4. A. Methyl Oleate	53.1	53.8
B. Ethyl oleate	12.9	13.0
	32.4	30.8
5. A. Methyl oleate	44.1	44.9
B. *n*-Butyl oleate	62.8	61.3
	36.3	36.4
6. A. 2-Methyl-3-butyne-2-ol	27.4	27.1
B. 3-Methyl-1-pentyne-3-ol		
7. A. 2-Methyl-3-butyne-2-ol	52.2	54.8
B. 3-Methyl-1-hexyne-3-ol		
8. A. 3-Methyl-1-pentyne-3-ol	36.3	37.3
B. 3-Methyl-1-hexyne-3-ol		
9. A. 2-Methyl-3-butyne-2-ol	56.9	56.7
B. 2-Butene-1-ol		
10. A. 2-Methyl-3-butyne-2-ol	39.6	41.0
B. 2-Butyne-1,4-diol	79.5	79.1

TABLE XXIX

Hydrogenation of Mixtures of Unsaturated Compounds (325)

Mixture	Solvent	Percent A Found	Percent A Present
1. A. Maleic acid[a]	Water	50.5	49.6
B. Fumaric acid[a]			
2. A. Methyl oleate	Carbon tetrachloride	22.9	22.6
B. Ethyl oleate			
3. A. Methyl oleate	Carbon tetrachloride	8.4	8.8
B. Butyl oleate			
4. A. 5-Methyl-1-hexyne	Acetic acid–water	23.8[b]	23.6
B. 2-Butyne-1,4-diol			

[a]Sodium salts.

[b]The methyl hexyne was brominated totally faster than the butynediol.

reacting at different rates, two slopes are obtained as shown in Fig. 15. The second slope is extrapolated to zero time, and at the point of intersection, y, $x = a_1$, the concentration of the faster reacting component and

$$\log (a - a_1) = y \tag{82}$$

Since a is known from the total hydrogenation value and y is read at the intercept, this equation can be solved for a_1.

The hydrogenation results of mixtures of unsaturated compounds obtained by Siggia and co-workers are given in Table XXIX.

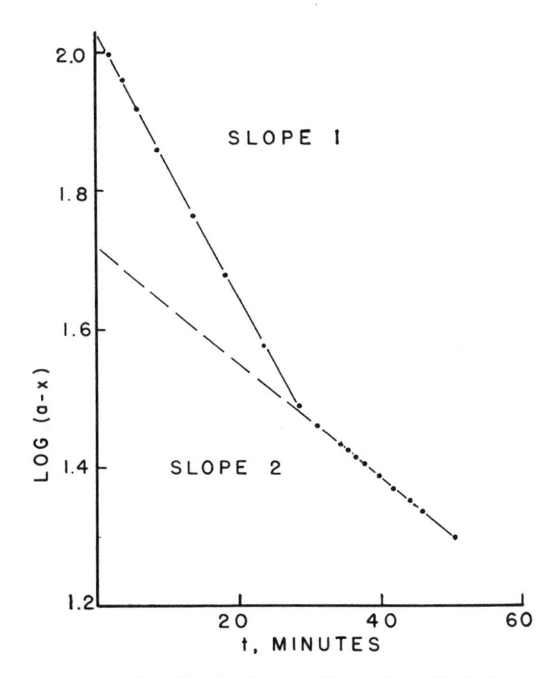

Fig. 15. Reaction rate curve for the hydrogenation of methyl oleate and ethyl oleate (325). (Courtesy of *Analytical Chemistry.*)

6. Microanalysis

Several of the standard chemical methods have been adapted for the microdetermination of unsaturation. Reid and Beddard (282) successfully applied bromine in acetic acid to the determination of low bromine absorption values ranging from 1.0 to 0.0001 g per 100 g. Less than 2% errors were obtained on allyl alcohol and crotonaldehyde solutions with bromine numbers as low as 0.001. Benson (38) adapted the Uhrig-Levin procedure (Section VI-A-1-b-(1)) for the microdetermination of *cis*-2-butene in cyclobutane. Whalley and Ellison (367) designed a micro method for oils using standard Wijs reagent and a special titrating unit with a dead-stop mechanism. Micro methods using Hanus reagent have been published by Ralls (277) and by Grunbaum and Kirk (142). Phillips and Wake (267) reported a micro method in which they used a specially designed reaction flask and iodine chloride (or bromine) in carbon tetrachloride.

A submicro method for the determination of olefinic unsaturation was reported by Belcher and Fleet (34). They used up to 0.10 ml of $0.1N$ bromine chloride and back-titrated iodometrically with up to 1.0 ml of $0.01N$ sodium thiosulfate. For samples requiring fairly long reaction times in methanolic medium, they obtained low results due to different amounts of bromine chloride in the sample and blank solutions.

Blank corrections become extremely important in microtitrations, especially in systems with high background, e.g., in the mercury-catalyzed halogenation of unsaturated compounds. The variable background for sample and blank can be avoided by applying CCT (Section VI-A-1-a-(2)-(c)). By this technique the excess halogen concentration is kept constant and the background reactions are subtracted by extrapolation of the recorded background slope. CCT has been used in our laboratory to determine quantitatively less than 1 μeq of unsaturation in chromatographic fractions of polyether polyols. The highly reactive bromine chloride complex of mercury(II) is required for the rapid titration of these low levels of high molecular weight (about 3000) unsaturated compounds, and the correction of the relatively high background is necessary.

B. SPECTRAL METHODS

The spectral methods are based on the electromagnetic wave spectra and the mass spectra of olefinic compounds. This section covers photometric methods, mass spectrometry, and nuclear magnetic resonance.

1. Photometric Methods

a. Colorimetric Method

The active methylene group of cyclopentadiene reacts with carbonyl compounds in alkaline solution to form highly colored fulvenes (equation (83)). Uhrig and co-workers (355) used benzaldehyde as the reagent in

$$\text{C}_5\text{H}_4\text{CH}_2 + \text{O}{=}\text{C}\langle^{\text{R}}_{\text{R}'} \rightarrow \text{C}_5\text{H}_4\text{C}{=}\text{C}\langle^{\text{R}}_{\text{R}'} + \text{H}_2\text{O} \qquad (83)$$

the colorimetric determination of cyclopentadiene. The yellow-orange color of the phenylfulvene formed was stable for at least 2 hr. Dicyclopentadiene, formed by spontaneous dimerization, does not react, but it can be quantitatively depolymerized and subsequently determined. Powell and co-workers (273) applied the method to mixtures of cyclopentadiene and methylcyclopentadiene using simultaneous equations.

Skoog and Dubois (331) and Roman and Smith (292) employed benzaldehyde in the colorimetric determination of indene, which undergoes a reaction similar to cyclopentadiene, but at a considerably slower rate.

A spectrophotometric method for 4-carbon and higher molecular weight olefins was developed by Altshuller and Sleva (7). The olefin was reacted with *p*-dimethylaminobenzaldehyde in concentrated sulfuric acid and the absorbance was measured at 500 mμ. The colored product is stable for several hours and the method is more sensitive than previous colorimetric methods. With gaseous olefins, 0.1 ppm or less can be determined. Formaldehyde and aromatic hydrocarbons interfere appreciably, and methods for their removal are provided. The method has been applied to the analysis of automobile exhaust gases.

Conjugated dienes can be coupled with aromatic diazo compounds. Meyer (241) obtained yellow to orange azo compounds by coupling 1,3-butadiene, 1,3-pentadiene, isoprene, 2,3-dimethyl-1,3-butadiene, and 2-methyl-1,3-butadiene with *p*-nitrobenzenediazonium sulfate and 2,4-dinitrodiazonium sulfate. These colored azo compounds were utilized in the spectrophotometric determination of conjugated diolefins by Altshuller and Cohen (9). They coupled the conjugated diolefins with *p*-nitrobenzenediazonium fluoborate in a 2-methoxyethanol–phosphoric acid medium. Isoprene-type diolefins formed azo products which absorbed strongly at 490 mμ while butadiene formed an azo compound with a maximum near 405 mμ. The absorbances at these maxima are linear with concentration between 0.3 and 30 μg/ml for isoprene-type diolefins and 20 and 200 μg/ml for 1,3-butadiene. No appreciable interference occurs from paraffinic, acetylenic, simple aromatic, and most other types of olefinic hydrocarbons. Some aldehydes, ketones, and phenols interfere moderately.

An indirect colorimetric method based on the formation of azo compounds was developed by Danish and Lidov (97) for the determination of aldrin, a chlorinated insecticide with the structure illustrated. The phenyl-

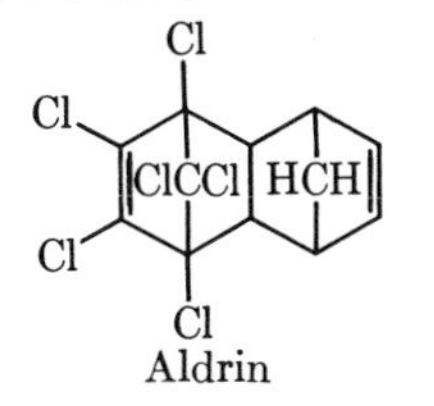

Aldrin

dihydrotriazole formed by the addition of phenyl azide to the unchlorinated bicycloheptene ring is treated with acid to form an aniline derivative by loss of nitrogen. The aniline derivative is then coupled with diazotized 2,4-dinitroaniline to give an azo compound that gives a strong red color on treatment with sulfuric acid. This procedure has been improved by O'Donnell and co-workers (262) so that as little as 0.1 ppm of aldrin residue in agricultural crop materials can be determined.

Bauer (30) differentiated the mono-, di-, tri-, and tetrasubstituted unconjugated olefins by the spectra of their π-complexes with tetracyanoethylene. In these spectra a band appears in or near the visible, with the position strongly dependent on the degree of substitution at the double bond. The wavelength variation with substitution is about 200 mμ, in contrast to about 65 mμ for the iodine complex and only about 20 mμ for the olefins themselves. Conjugated olefins and aromatic compounds interfere, not only by forming complexes absorbing at similar wavelengths, but also by reacting irreversibly with the tetracyanoethylene via a Diels-Alder reaction.

b. Ultraviolet

The simple monoolefins absorb in the far ultraviolet (below 210 mμ). Because of the high absorption by oxygen and other gases in this region, special spectrophotometers are necessary which can be evacuated or filled with a transparent gas such as hydrogen or helium.

Jones and Taylor (177), using a recording vacuum spectrophotometer with a photoelectric detector, have measured the 170–230 mμ spectra of 69 pure hydrocarbons, including all the aliphatic monoolefins through C_6 as well as cyclopentene, cyclohexene, and several diolefins, allenes, and acetylenes, mainly in the vapor phase. Except for ethylene, the monoolefins had remarkably similar spectra, particularly within groups arranged according to the number of alkyl groups attached to the double-bonded carbon atoms as shown in Table XXX. Because of the strong similarities

TABLE XXX

Far Ultraviolet Absorption Bands of Gaseous Alkenes (177)

Olefin type	Av. position of max, Å	Molar absorptivity, liters/mole cm.
1-Alkenes	1750 ± 25	11,800 ± 1200
cis-2-Alkenes	1765 ± 15	12,300 ± 200
trans-2-Alkenes	1790 ± 10	11,700 ± 800
2-Alkyl-1-alkenes	1875 ± 15	8,900 ± 1200

of the monoolefin spectra in the far ultraviolet, a total olefin determination may be possible even in complex mixtures. However, for all the olefins studied in the vapor state, the average molar absorptivity at 179.5 mμ was 9800 liters/mole-cm with a standard deviation of 2680. Solution spectra appear more favorable. Thus in *n*-heptane solution, the average molar absorptivity of six hexenes at 186.0 mμ was 9200 with a standard deviation of 930 liters/mole-cm. Diolefins with double bonds separated by two or more methylene groups give spectra resembling the monoolefins except for a twofold gain in intensity. A single methylene group between the double bonds only partially shields the groups from interaction. Allenes also were found to have a strong absorption band near 180 mμ.

Long and Neuzil (218) developed a method for distinguishing and determining olefin types by means of the ultraviolet absorption spectra of complexes formed with iodine. The positions and relative intensities of the bands are given in Table XXXI. The method is of particular interest

TABLE XXXI

Ultraviolet Absorption Bands of Olefin–Iodine Complexes (218)

Type	λ max, mμ	ϵ'
$RCH{=}CH_2$	275	12
$R_2C{=}CH_2$	290–295	25
cis-$RCH{=}CHR$	295–300	19
trans-$RCH{=}CHR$	295–300	11
$R_2C{=}CHR$	317	27
$R_2C{=}CR_2$	337	23

because tri- and tetrasubstituted olefins, which are not readily determinable by the infrared method, give characteristic strong ultraviolet absorption bands when complexed with iodine. This method is far from ideal for quantitative work, because the spectra are obtained when only a small fraction of the olefins are complexed in order to avoid the undesirable formation of the diiodo addition compounds. This is apparent from the ratio of the apparent absorptivities in Table XXXI to the molar absorptivities of the complexes which range from 10,000 to 25,000 liters/mole-cm.

The conjugated diolefins absorb in the near ultraviolet (above 210 mμ). 1,3-Butadiene absorbs strongly in the 210–220 mμ region. It may be determined at 216 mμ using 1 mm or 0.1 mm absorption cells with pressures between 50 and 500 mm (271). Although higher wavelengths may be used, interference from vinyl acetylene becomes greater. C_4 and higher paraffins, olefins, alkyl acetylenes, hydrogen, carbon monoxide, carbon dioxide, air, and water do not interfere.

Dudenbostel and Priestly (108) were able to analyze a mixture containing *trans*-1,3-pentadiene, isoprene, *cis*-1,3-pentadiene, and cyclopentadiene using absorbances at wavelengths 220, 225, 230, and 244 mμ, respectively. The concentration of each conjugated C_5 diene was obtained by solving four simultaneous equations containing the known absorptivity of each component at each wavelength.

Altshuller and co-workers (8) determined small amounts of mono- and diolefins at 300 to 310 mμ based on the absorbances produced by their reaction with concentrated sulfuric acid. Moderate interference is caused by some high molecular weight aldehydes, nitric oxide, and sulfur dioxide.

A method which is specific for and sensitive to low concentration of acetylenic compounds was developed by Siggia and Stahl (326). It is based on the formation of mercuric acetate addition products of the acetylenic compounds and the measurement of the ultraviolet absorption of these addition compounds.

The ASTM method (15) for the determination of conjugated diene acids and esters is based on the measurement of the absorbance of isooctane solutions of samples at 233 mμ. Nonconjugated mono- and polyunsaturated acids absorb strongly in the 170–200 mμ region (301); the conjugated diene and triene acids also absorb in this region but the intensity is only about 10% of the nonconjugated polyunsaturated acids. To increase the accuracy in the determination of the latter, corrections for the conjugated polyunsaturated acids may be applied (261).

The sensitivity for the nonconjugated polyunsaturated acids and esters can be increased by conversion to conjugated polyunsaturated acids by heating in potassium hydroxide–glycol solution (150).

c. Infrared

In spite of the limitations of infrared measurement of C═C stretching, it has provided useful information regarding the residual unsaturation in olefin polymers such as natural and synthetic rubbers, polyvinyl compounds, and polyacrylates (29,347).

The out-of-plane bending vibrations of C═C—H are most useful in differentiating the five classes of olefins (see Section III-D-1) other than totally substituted ethylenes (23). Coggeshall and co-workers (80,304) were able to analyze unknown mixtures of the five olefin classes, employing the characteristic wavelengths and average molar absorptivities for each class. The amount of each component was determined fairly accurately by solving simultaneous equations.

Nonconstancy of the molar absorptivities within the classes is one of the difficulties encountered in olefin-type analysis of complex mixtures. However, reasonable assumptions can be made, and average absorptivities for each class give fairly good accuracy for petroleum distillates (23,80,304). Saier and co-workers (303) calculated the statistical absorptivity for α-olefins at 11.00 μ from absorption data of C_6–C_{16} normal α-olefins (C_{18} and C_{20} calculated) and from the C number distribution obtained by chromatographic analysis. The α-olefin concentration of an unknown was obtained by dividing the observed absorbance by the statistical absorptivity. The mean error was believed to be $\pm 1\%$.

Another difficulty in the analysis of complex mixtures results from interfering absorption by paraffins and aromatics. Corrections have been employed for low concentrations of these impurities (23), but for large interferences, preliminary chromatographic separation of the saturated hydrocarbons and aromatics has been employed (80,175,304).

Stereospecificity has a large effect on the physical properties of 1,4-polybutadienes (202). In the polymerization of butadiene, the addition of the monomer may take place in three ways, giving rise to the following configurations:

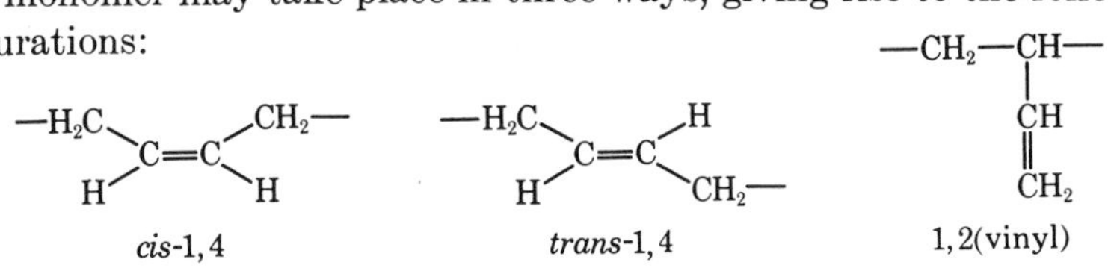

cis-1,4 *trans*-1,4 1,2(vinyl)

The *trans* and vinyl configurations are determined easily utilizing the sharp absorption bands at 10.3 and 11.0 μ (45,147,202,285,328). However, to determine the *cis* configuration, Hampton (147) used 13.8 μ, Binder (45) used 14.7 μ, and Richardson obtained the *cis* configuration by difference. Kraus and co-workers (202) found that these procedures did not give realistic material balances for total unsaturation in polybutadienes which have unsaturation distributions beyond the limits of those encountered in emulsion, sodium, potassium, or alfin polymers. The difficulty was due to the ill-defined, variable absorption of the *cis* configuration. They (202,328) overcame this difficulty by using the integrated absorbance from 12.0 to 15.75 μ for the *cis*-1,4 addition and the usual peak height method for the *trans*-1,4 and 1,2-(vinyl) additions.

The unsaturated bonds in polyethylene films may be determined from the absorbance at 910 cm^{-1} (vinyl), 885 cm^{-1} (vinylidene), and 963 cm^{-1} (*trans*-vinylene) (194). The vinylidene band is superimposed by the strong methyl group absorption. Biernacka (43) determined the absorbance of the methyl group after bromination and applied this correction to the original vinylidene absorbance. Vullo and Montaudo (358) estimated the number average molecular weight of polyethylene from infrared unsaturation data.

Szonyi and co-workers (346) utilized the absorption at 965 cm^{-1}, due to the C═C—H out-of-plane deformation vibrations, in the direct determination of *trans* unsaturation in triglycerides.

Goddu (135) used the near-infrared region for the determination of unsaturation. Bands at about 1.62 and 2.10 μ may be used for determining terminal unsaturation and a band at 2.14 μ for *cis* unsaturation. The sensitivity for terminal unsaturation can be pushed to as low as 100 ppm in many cases and the precision is about ± 1–2%. The sharp absorption at 2.14 μ by *cis* unsaturation should be very useful because of the variable and ill-defined absorption in the 12–15.75 μ region. The molar absorptivities and absorption maxima are tabulated for 16 terminal unsaturated compounds at the 1.62 μ region, 9 terminal unsaturated compounds at the 2.1 μ region, and 13 internally double-bonded compounds at the 2.14 μ region (135).

The literature cited above is only a small fraction of the papers published on the infrared analysis of unsaturated compounds. For other applications refer to page 328 of reference 271.

d. Raman Spectroscopy

The intensities of Raman lines due to C═C stretching are more independent of the effects of substituents, while the intensities of infrared absorption bands due to C═C stretching depend strongly upon the arrangement and nature of the substituents on the ethylenic carbon atoms. Raman spectroscopy is particularly useful for tetrasubstituted and centrosymmetrical unsaturated compounds, where the infrared absorption is weak or absent.

Heigl and co-workers (149) used the Raman frequency shifts in the 1640–1680 cm^{-1} region produced by C═C stretching for the determination of total olefin. They found the integrated intensity per mole of C_6–C_{10} olefins to be sufficiently constant to give satisfactory accuracy in the analysis of complex hydrocarbon mixtures. The method does not provide information regarding distribution of olefin types. However, the total olefins measured include tetrasubstituted olefins which are not determinable by infrared analysis.

McCutcheon and co-workers (234) investigated *cis–trans* isomerism about the double bonds in unsaturated fatty acid esters using Raman frequency shifts in the 1650 cm^{-1} region.

Moser and Weber (252) determined the integrated intensities of the C═C stretching vibration of 25 monoolefins, using the 459 cm^{-1} line of carbon tetrachloride as an external standard. σ_0 values, defined as $\sigma_0 = MJ/d$ (where M = molecular weight and J = relative integrated intensity), were given for the following compounds:

Compounds	σ_0
CH_2=CHR	29.5
CH_2=CRR′	33.5
cis- and *trans*-MeHC=CHR	31.5
Me_2C=CHR MeHC=CMeR RR′C=CR″R‴	Very different values; mean value = 40
3-Heptene	38

where R is a branched or unbranched alkyl group and R′ is methyl or ethyl.

2. Mass Spectrometry

Mass spectrometric analysis has been described in Part I, Volume 4, of this Treatise by Melpolder and Brown (238) and by Diebler (102). Applications of mass spectrometry to hydrocarbon mixtures were discussed by Coggeshall (80) and Brown (66).

Mass spectrometry deals with the successive processes of ion formation, ion separation according to mass, and ion abundance measurement. The characteristic mass spectra provide vital information on the chemical and structural nature of molecules and also provide a means of quantitative determination of numerous compounds.

For monoolefins, the characteristic ions belong to the series $(C_nH_{2n-1})^+$, formed by the loss of a hydrogen atom and successive methylene groups. However, this series of ions is also characteristic of cycloparaffins. Therefore, the method does not distinguish between monoolefins and cycloparaffins. Similarly, cycloolefins, diolefins, acetylenes, and bicycloparaffins are all characterized by ions of the types $(C_nH_{2n-2})^+$ and $(C_nH_{2n-3})^+$. The method is obviously too general by itself, but it can be very useful when used in conjunction with some other methods.

Thus mass spectrometry has been successfully applied to the determination of olefins by measuring the difference in characteristic mass peaks before and after treating the sample with phenyl chlorosulfide (127) or benzenesulfenyl chloride (242) to convert the olefins to low vapor pressure addition products. Others have combined mass spectrometry with column chromatography (260), gas chromatography (209), gas chromatography using mercury perchlorate-impregnated firebrick for complete olefin retention (84), and hydrogenation (215). Melpolder and co-workers (239) reported a detailed analysis of catalytically cracked naphtha by combining mass spectrometry with infrared, ultraviolet, fractional distillation, adsorption chromatography, and hydrogenation.

At high ionization voltage, fragmentation of olefin molecules occurs. Brown and Gilliams (67) studied 44 monoolefinic hydrocarbons and found that the points of rupture, and hence the major ions formed, were quite predictable from the structure of the molecules. Based on these observations, they formulated some rules for ion formation in olefins (238, p. 2018).

When the electron impact energy is dropped to about 18 V (from the usual 70 V), the extent of fragmentation is decreased and the major peaks remaining are due to the parent ion and that of one hydrogen atom less (83). Thus, mass spectra of hydrocarbon mixtures at 18 V is simpler, with various groups of compounds producing different peaks. Analytical coefficients determined for each group enable calculation of the composition of mixtures.

Lumpkin (221) used a 7 V ionizing voltage which simplified the mass spectra of high molecular weight compounds and their interpretation. The simplified spectra permit the determination of monoolefins, cyclic olefins and/or diolefins, dicyclic olefins and/or cyclic diolefins, and aromatic compounds according to molecular weight groups. Frisque and coworkers (128) combined low- and high-voltage mass spectrometric techniques to allow total analysis without the necessity of separation.

3. Nuclear Magnetic Resonance

As mentioned previously (Section III-D-4), information derived from the spin–spin interaction of protons on neighboring carbons makes NMR a very powerful tool in the study of the structure of organic molecules. Ferguson (123) combined infrared and NMR in the study of the microstructures of polychloroprenes. In addition to the identification of the structural isomers, I–IV, he found NMR evidence for sequence isomerism

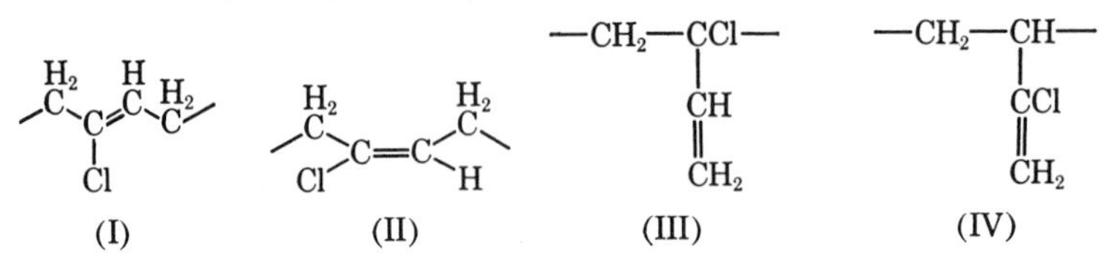

in the 1,4-polychloroprene units, as a consequence of "head-to-tail" (V), "head-to-head" (VI), and "tail-to-tail" (VII) addition of successive monomer units:

V: —CCl=CH—CH$_2$—CH$_2$CCl=CH—
VI: —CH=CCl—CH$_2$—CH$_2$—CCl=CH—
VII: —CCl=CH—CH$_2$—CH$_2$—CH=CCl—

The microstructural irregularity probably affects the elastomeric properties of neoprenes. Hence, the control and knowledge of microstructure is an important aspect of neoprene technology.

Stehling and Bartz (339b) derived comprehensive correlations of NMR chemical shifts, spin coupling constants, and characteristic spectral patterns from the spectra of 60 aliphatic monoolefins. They found that the chemical shifts of α, β, and γ CH_3's and α, β, and γ CH_2's do not overlap, which greatly facilitates the assignment of these bands in the spectra of unknown olefinic hydrocarbons. α and β CH resonances overlap those of other hydrogen types, but these can generally be distinguished by their band multiplicities. The dependence of the chemical shift of β CH's and α CH's on the presence or absence of a *cis* substituent makes it possible to distinguish between geometric isomers. The authors used these correlations to prove the structures of oligomers of monoolefins and polymers of diolefins. They also derived information concerning monomer sequence distribution and reactivity ratios for isobutylene-isoprene copolymers from the detailed interpretation of the spectra of such copolymers.

Johnson and Shoolery (174) determined unsaturation and the average molecular weight of natural fats by NMR. The proton NMR spectra of triglycerides are characterized by four sets of signals: the four glyceride methylene protons, the olefinic protons, the protons of the methylene groups attached to the doubly bonded carbon atoms, and the remaining protons on the saturated carbon atoms. The area of the signals produced by the four glyceride (C_1 and C_3) methylene protons is measured using an electronic integrator and a dc digital voltmeter. With this measurement as internal standard, the number of olefinic protons and the total number of protons can be measured accurately. These data were used to calculate the "NMR number" and the average molecular weight which are tabulated in Tables XXXII and XXXIII. The agreement of the results is very good, with the exception of tung oil for which the Wijs result is expected to be low because of the known difficulty in the halogenation of conjugated double bonds. Recently, this method was computerized so that it may be used routinely by unskilled operators (320a).

TABLE XXXII

Iodine No. Values of Various Oils by NMR and Wijs Methods (174)

Oil	NMR No.	Wijs No.
Coconut	10.5 ± 1.3	8.0–8.7
Olive	80.8 ± 0.9	83.0–85.3
Peanut	94.5 ± 0.6	95.0–97.2
Soybean	127.1 ± 1.6	125.0–126.1
Sunflower seed	135.0 ± 0.9	136.0–137.7
Safflower seed	141.2 ± 1.0	140.0–143.5
Whale	150.2 ± 1.0	149.0–151.6
Linseed	176.2 ± 1.2	179.0–181.0
Tung	225.2 ± 1.2	146.0–163.5

TABLE XXXIII

Average Molecular Weights of Oils from Saponification Values and NMR (174)

Oil	Sap. value	Sap. value mol wt	NMR mol wt
Olive	189.3	887.1	873.7 ± 5.3
Peanut	188.8	891.5	882.3 ± 7.4
Safflower seed	191.5	879.0	874.9 ± 9.3

Durbetaki and Miles (111) determined the degree of unsaturation in polybutadiene by NMR.

Bothner-By and Naar-Colin (53) determined the chemical shifts and spin–spin coupling constants from the high resolution NMR spectra of propene, 1-butene, and 1-hexene.

C. GAS–LIQUID PARTITION CHROMATOGRAPHY

Gas chromatography (GC) is a method of separation in which the gaseous or vaporized components to be separated are distributed between two phases: a fixed stationary phase with large surface area and a moving gas phase. Gas–solid chromatography (GSC) is the specific term applied to the method in which the fixed phase is a solid adsorbant; it was discussed earlier under Adsorption Chromatography in Section IV-A. Gas–liquid chromatography (GLC) is the method in which the fixed phase is a liquid supported on a solid. This section will deal with GLC.

In theory, both frontal and displacement analysis may be used in GLC, but in practice, elution analysis is used almost exclusively. In elution analysis, the sample to be analyzed is introduced into the column and is carried through the column by a stream of inert gas such as nitrogen or helium. The components of the sample are selectively retarded by the liquid phase. Under a given set of conditions, the time required for a substance to emerge from the column is characteristic for the substance, and the height or area of the peak is proportional to the amount present. The popularity of elution analysis is due to its speed and versatility in analyzing complex mixtures. A significant advantage of the elution technique is the self-purging method of operation, i.e., the column is returned to its original condition at the end of each analysis.

Gas–liquid partition chromatography has the following advantages over liquid–liquid partition chromatography (37):

1. The low viscosity of the gas permits the use of very long columns so that high efficiencies are obtained in the separations.

2. Analysis may be completed rapidly since high gas flow rates can be used.

3. The carrier gas molecules are much less strongly adsorbed than the components to be separated so that the equilibrium between the fixed and mobile phases is largely independent of the carrier gas.

4. There are many simple, convenient, fast-responding, and sensitive devices available for measuring concentrations of components in a gas. By combining gas chromatography with identification methods such as mass spectrometry (136), infrared (366), and NMR (25), the potency of the method is enhanced greatly.

5. There is a wide choice of useful liquids for GLC which permit separations of substances with nearly the same boiling points and also of those with quite similar chemical structures.

Various supports, partitioning liquids, and carrier gases have been used in the separation and determination of unsaturated compounds. Table XXXIV lists some published combinations for different types of alkenes. The detectors—among which the most commonly used are thermal conductivity (katharometer), flame ionization, and electron capture—are not listed.

TABLE XXXIV

Combinations of Supports, Partitioning Liquids, and Carrier Gases Used in the Anlaysis of Mixtures Containing Alkenes by GLC

Type of compound	Support	Liquid phase	Carrier gas	Ref.
C_4 alkenes	Cu capillary	Squalene	Ar	379
C_2–C_5 alkenes	Chromosorb	Squalane, adiponitrile	He	94
C_2–C_7 alkenes	Glass capillary	Silicone Fluid 96	N_2	270
C_2–C_8 alkenes	30–60 mesh firebrick	$AgNO_3$ in glycerol	He	32
C_6–C_8 alkenes	30–60 mesh Chromosorb P	$AgNO_3$ in benzyl nitrile	He	25
C_6–C_{12} alkenes	Nylon capillary	Tritolyl phosphate	—	296
C_{10}–C_{22} alkenes	30–60 mesh C22 firebrick	Apiezon L	He	216
C_2–C_4 alkenes, acetylenes	—	40:60 Apiezon B–dibutyl maleate	—	250
Cracked gases	C–22 firebrick	2,5-Hexanedione	—	104
Liquid petroleum fractions	Celite 545	Hexatriacontane and a mixture of *o*- and *p*-isomers of benzyldiphenyl	N_2	100
Polyenes, O-containing	NaCl	Poly(dimethylsiloxane)	H_2	298
Styrene–polystyrene	Celite	Tritolyl phosphate	1:3 N_2–H_2	61

The number of peaks in a chromatogram gives the minimum number of components in a sample since it is possible that two or more components may have the same retention time. Incomplete resolution of components with slightly different retention times is very common in the chromatography of complex mixtures. The analysis of such mixtures can be simplified considerably by comparing the chromatograms before and after the separation of one of the homologous groups. The reactivity of the π-electrons of the olefins has been utilized in such subtractive chromatographic techniques. Thus the olefins have been removed by precolumns of silver nitrate on aluminum oxide (315); sulfuric acid on silica gel (232); maleic anhydride, selective for conjugated diolefins, on silica gel (170); mercuric perchlorate on firebrick (124) or Chromosorb P (5); and by absorbing in mercuric perchlorate (231) or mercuric sulfate–sulfuric acid (166) solution. Using a dual detector chromatograph, differential chromatograms of olefins can be obtained by absorbing (167) or catalytically hydrogenating (40) the olefins present in the gaseous mixture. Kolb and Kaiser (195) identified the olefinic materials resulting from the pyrolysis of polyethylene at 1000°C by comparing the usual chromatogram with one obtained after catalytic hydrogenation of the same pyrolysis product.

Complexation of olefins by tri-*o*-thymotide (222), 1,3,5-trinitrobenzene (95), and the *N*-dodecylsalicylaldimines and methyl-*n*-octylglyoximes of certain cations (72) in the partitioning liquid is utilized in the selective retardation of the components; for example, the retardation by trinitrobenzene increases with increasing double bond substitution.

The selective affinity of molecular sieves for normal alkanes and alkenes has been utilized in improving the resolution of mixtures containing branched and cyclic hydrocarbons (4,59). The sensitivity for C_2–C_3 olefins can be increased 10–100 times by concentrating on Molecular Sieve 5A at low temperature, followed by the chromatographic analysis of the products desorbed at 300°C (255).

The quantitative determination of double bond positions in unsaturated fatty acids (176,203) and glycerides (203,350) was made by combining oxidative cleavage and the GLC analysis of the resulting mono- and diacids (350) or of the esters formed from these acids (176,203).

Ackman (2) examined the effect of temperature, chain length, number of double bonds, and position of double bonds on the separation factors. The latter may be used to identify unknowns or to predict retention times of methyl esters of unsaturated acids.

Schneider and co-workers (308a) studied the effect of the concentration of squalane on graphited carbon black on the chromatogram. When the graphite was coated with 0.06–0.4% squalane, they found that the separation chracteristics of the graphite were retained, but with improved resolution and peak symmetry. Impregnation of the graphite with greater amounts of squalane changed the characteristics to those of the stationary liquid phase.

D. POLAROGRAPHY

As stated earlier (Section III-C), isolated double bonds are not reducible at the dropping mercury electrode, but conjugated double bonds are reducible. The half-wave potentials and diffusion coefficients of phenyl-substituted olefins and acetylenes, and the half-wave potentials and the i_d/C values (constants derived from Ilkovic equation, see pp. 63, 68, reference 200) of substituted stilbenes and conjugated polyunsaturated hydrocarbons have been tabulated by Kolthoff and Lingane (200). The half-wave potentials of most of these conjugated unsaturated hydrocarbons range from -1.8 to -2.6 V vs. S.C.E. (saturated calomel electrode). Since the alkali metals are reducible in this range, tetrabutylammonium is used as the supporting electrolyte. Dioxane (50–75%) is used to solubilize the sample.

The polycyclic aromatic compounds are reducible in the same region as the conjugated unsaturated hydrocarbons. These compounds have relatively high diffusion current constants and therefore may set serious limitations in the polarographic analysis of complex hydrocarbon mixtures.

The polarographic reduction of unsaturated acids and carbonyl compounds, including polyene aldehydes, terpene aldehydes, and steroids, has been covered thoroughly by Kolthoff and Lingane (200). Unlike the conjugated unsaturated hydrocarbons, the half-wave potentials of the unsaturated carbonyl and carboxylic compounds are pH sensitive, being shifted to negative potentials by increasing pH. The corresponding shift for the esters of the unsaturated acids is considerably smaller.

Warshowsky and co-workers (365) analyzed a mixture of the barium salts of maleic and fumaric acids. They obtained separate and distinct waves in a $1M$ ammonium hydroxide–ammonium chloride buffer at a pH of about 8.

The addition products resulting from the methoxymercuration of terminal and cyclic unbranched double bonds yield two waves, the first of which is reversible. Fleet and Jee (125a) used the a.c. polarograms of the first waves for the quantitative determination of these olefins.

Polarography has been used to determine methyl vinyl ketone (129), unsaturated monocyclic and bicyclic terpene ketones (313,314), vinyl cyanide and styrene monomers in the copolymer (92), and 1,2- or 4,4-unsaturated 3-ketosteroids such as testosterone, progesterone, corticosterone, and desoxycorticosterone (115,274), cortisone (274), isomeric cholestenones (3,274), methyltestosterone, and pregnenin-17-ol-3-one (305).

E. REFRACTIVE INDEX

The difference in the refractometric properties of saturated, olefinic, and aromatic hydrocarbons has been utilized in the column chromatographic separation of the hydrocarbons (Section IV-A-1). Refractometric properties may be used for the quantitative determination of hydrocarbons in simple mixtures such as conjugated and nonconjugated olefins (205). Refractive index has been correlated with unsaturation in soybean, linseed, and other vegetable oils (223) and for the determination of styrene in synthetic rubber (375).

VII. RECOMMENDED METHODS

After reviewing the numerous methods that have been used in the determination of unsaturated compounds, it may be concluded that there is no universal method for unsaturation. The methods recommended in this section are believed to be reliable and should take care of the majority of unsaturated compounds. A convenient table has been compiled by Polgár and Jungnickel (271, pp. 361–366) based on the applicability of various methods depending on the type of sample. Highly specialized methods, such as spectrophotometry, mass spectrometry, NMR, and gas chromatography, have been covered adequately elsewhere in this Treatise. Therefore, these specialized methods have been omitted although they are very useful for specific types of unsaturated compounds and complex mixtures.

A. BROMINE IN CARBON TETRACHLORIDE (271)

Reagent

Dissolve 17.2 g (5.5 ml) of bromine in carbon tetrachloride and dilute to 1 liter. Standardize this reagent (0.2*N*) daily as follows: Place 25 ml of carbon tetrachloride and 100 ml of water in a glass-stoppered flask and cool the mixture in an ice bath for 10 min. Add 15 ml of reagent and store the flask in the ice bath in the dark for 10 min. Add 15 ml of 20% potassium iodide and titrate with standard 0.1*N* sodium thiosulfate to the starch endpoint.

Procedure

Dissolve the sample in 25 ml of carbon tetrachloride in a glass-stoppered flask, add 100 ml of water, and cool the flask in an ice bath for 10 min. Place a light shield around the neck of the flask. Add the reagent in 65–70% excess and store the flask (kept in the ice bath) in the dark.

Exactly 10 min after the addition of the reagent, add 15 ml of 20% potassium iodide solution and titrate the liberated iodine with 0.1N sodium thiosulfate to the starch endpoint. Repeat the determination with 20 min reaction time. If these two results differ by 2% or more, make a third determination with 30 min reaction time, and extrapolate the results to zero time.

B. CUMULATIVE COULOGRAPHIC TITRATION (155)

Reagents

Reagent A: Dissolve 10 g of sodium bromide in 100 ml of water and dilute to about 1 liter with acetic acid. The tribromide ion is electrogenerated from this solution.

Reagent B: Dissolve 10 g of sodium bromide and 45 g of mercuric chloride in 100 ml of water and dilute to about 1 liter with acetic acid. The positively charged bromine chloride complex of mercury(II) is electrogenerated from this medium.

Both reagents are indefinitely stable and need not be standardized. Electrolysis of a solution consisting of a 15:35 volume ratio of Reagents A and B will liberate the zero-charged bromine complex of mercury; the latter's reactivity and substitution tendency lie between those of the negatively and positively charged oxidizing species generated from Reagents A and B, respectively.

Procedure

Polarize the platinum indicating electrodes with 200 mV and adjust the "set point" to 10 μA. Place 50 ml of Reagent A into the titration cell, turn on the magnetic stirrer, and switch "Standby" to "Titrate." When the background becomes constant, add the sample by means of a dropper or a microsyringe. Weigh the sample by difference. About 10 μmoles of unsaturation is the optimum sample size (200 sec titration time at a current equivalent to 0.1 μeq/sec) although a sample with less than 1 μmole of unsaturation can be easily titrated at a lower generating current. Use appropriate aliquots of pure compounds with high unsaturation content. When the background becomes constant, terminate the titration or add another sample to the same medium provided that the background of the preceding sample does not indicate excessive side reactions. Extrapolate the background slope to the time of sample addition.

If the reaction in Reagent A was too slow, use Reagent B. Extrapolation is essential in this medium because the background is considerably higher. A large increase in the slope of the background after the endpoint compared to the initial background is a good indication of substitution interference. A 15:35 mixture of Reagents A and B often prevents or minimizes substitution errors while retaining much of the catalytic effect of B. If the reaction is still too slow, try some other method.

Calculation

$$\text{Unsaturation value} = \frac{(t)(I)}{(w)(2000\ \mu\text{eq/mmole})} = \text{mmoles/g}$$

where t = generation time in seconds, corrected for side reactions
I = current in microequivalents per second
w = weight of sample in grams

C. IODINE CHLORIDE (WIJS)

Reagent (140)

Add 5 ml of commercial iodine chloride to 1 liter of acetic acid. Store reagent (0.2N) in a dark bottle. This reagent is stable for about 30 days although longer stability has been reported.

Halogen Ratio (140)

Measure 50 ml of 1:1 hydrochloric acid and 50 ml of carbon tetrachloride into a 500-ml iodine flask. By means of a pipet, transfer 25.00 ml of the iodine chloride reagent to the flask and shake. Titrate the free iodine in the violet carbon tetrachloride layer with 0.04N KIO_3 (2.1402 g/liter since the equivalent weight = 53.505 g according to equation (38)) solution to a colorless endpoint using vigorous shaking.

On a second 25.00-ml portion of the iodine chloride reagent, determine the total halogen by adding 150 ml of water and 15 ml of 15% potassium iodide solution and titrating with standard 0.1N sodium thiosulfate solution to a starch endpoint.

Calculate the I/Cl ratio as follows:

$$(T + F)(T - F) = \text{iodine/chlorine ratio}$$

where T = milliequivalents of thiosulfate required for total halogen and F = milliequivalents of potassium iodate required for free iodine.

It is essential that the reagent contain no free chlorine. The iodine/chlorine ratio should be about 1.1:1.

Procedure (271,322)

Dissolve 2 ± 0.4 meq of sample (grams sample = (25 ± 5)/(iodine No.)) in 15–20 ml of carbon tetrachloride in a 500-ml iodine flask. Add 25.00 ml of iodine chloride reagent, swirl, and allow to stand in the dark for 30 min at 25 ± 2°C. Add 20 ml of 15% potassium iodide solution and 100 ml of water. Titrate to the starch endpoint with 0.1*N* sodium thiosulfate solution. Toward the end of the reaction, stopper the flask and shake it vigorously, so that any iodine remaining in the carbon tetrachloride may be taken up by the potassium iodide solution. Run two blanks employing the same procedure as used for the sample. Obtain the net titer by difference and calculate the iodine number according to equation (30).

D. IODINE BROMIDE (HANUS)

Reagent (271)

Dissolve 20 g of iodine bromide in 1 liter of acetic acid and store the reagent (0.2*N*) in a dark bottle.

Alternatively, dissolve 13.2 g of C.P. iodine in 1 liter of acetic acid, warming if necessary to hasten dissolution of the iodine. Cool, then add enough bromine (about 3 ml) to double the halogen content as determined by iodometric titration.

Procedure (271,322)

Dissolve 2 meq or less of sample (g sample $\lessapprox$ 25/iodine No.) in 10 ml of carbon tetrachloride or chloroform in a 250-ml iodine flask. Run a blank simultaneously using 10 ml of the same solvent. Add 25.00 ml of iodine bromide reagent to each flask, shake the flasks, and allow them to stand in the dark for 30 min with occasional shaking. Add 10 ml of 15% potassium iodide and 75 ml of water, and titrate the liberated iodine with 0.1*N* sodium thiosulfate to the starch endpoint. Obtain the net titer by difference and calculate the iodine number according to equation (30) or the bromine number according to equation (31).

E. HYDROGENATION (322)

Apparatus

Figure 9 (Section VI-A-2-b-(1)) gives a detailed design of the apparatus. (The apparatus is manufactured by Dansk Glass apparatur v/ Angelo Jensen, Vesterbrogade 126A, Copenhagen, Denmark.) The total volume of the parts on either side of the manometer is about 70 ml and the two volumes should be the same within 2–3 ml.

The graduation of the manometer need not necessarily be very accurate inasmuch as it is only used to adjust the pressure difference to zero. A 2-cm long graduation in millimeters is very convenient.

The buret is about 40 cm long and contains 5 ml. Every milliliter is graduated in 50 parts; of course, a longer and more accurate buret may be employed equally well. The tip of the buret is so fine that 1 ml of mercury flows through it in about 1 min. Therefore, it is not necessary to adjust the flow of mercury with the cock *b*.

A 25-mm long iron rod is attached to a platinum beaker *p* so that the beaker may be hooked off and dropped into the solvent by a magnet. The rod is wrapped in a platinum sheet soldered with gold, to avoid corrosion of the iron by the solvent.

The whole apparatus is supported by a clamp just beneath the cock *b* and immersed in a bath filled with water. The water in the constant-temperature bath is stirred efficiently to assure equality of temperature throughout the bath. The liquid in the hydrogenation vessel is agitated by a Teflon-jacketed iron rod, which is rotated by a revolving magnet placed under and outside the glass constant-temperature bath.

Before use, the apparatus is cleaned, and the cocks carefully greased with Apiezon. Mercury is filled into *A*, and is sucked up into the buret by applying suction to *e*. When the mercury has reached the upper mark of the buret, *b* is closed and the suction turned off. There should now be only a small amount of mercury left in *A*, just enough to cover the tip of the buret.

Procedure

Fill the solvent into *C* (about 2 ml) and into the manometer by disconnecting the cock *d*. Place the catalyst and 2 ml of the solvent in the hydrogenation vessel. Weigh the substance to be analyzed into the platinum beaker, which eventually is suspended on the glass hook. Then connect the hydrogenation vessel with the main part of the apparatus and remove the air in the whole apparatus by leading a stream of hydrogen through *c* with *c* and *a* open (Note 1).

When all of the air has been removed, first close *a* and then, shortly after, close *c*. In this way the pressure inside the apparatus will become slightly higher than that of the atmosphere. Start stirring to hydrogenate the catalyst. At the same time the hydrogen becomes saturated with the vapors of the solvent. When the catalyst is perfectly hydrogenated, the manometer will remain on level if *d* is closed. Now relieve the excess pressure by opening *a* for a moment (*d* must be open). Register the temperature and the atmospheric pressure. Close *d* and commence hydrogenation by dropping the beaker with the substance into the solvent.

Absorption of hydrogen causes the left part of the manometer to be filled with the manometer liquid (Note 2). When a reading of the consumption of hydrogen is desired, stop stirring and let mercury into *A* until the manometer again stands on level (Note 3). Hydrogenation is continued until no decrease of pressure is observed and the manometer remains on level.

Note 1. For the usual precautions concerning the employed reagents, etc., see the monograph of Pregl on quantitative organic microanalysis (5th ed., Vienna, 1947).

2. The volumes of the different parts of the apparatus are chosen so that 5 ml of hydrogen, that is, the total content of the buret, may be consumed without causing the solvent to flow into *A*. Thus the apparatus need not be attended to during hydrogenation, if only the equivalent weight of the substance is to be determined.

3. Temperature fluctuations of the order of one degree may cause a small pressure difference due to volumetric changes of the solvent and the glass. If a very accurate determination is desired, the temperature of the water bath should be kept constant within about 0.5°C.

Calculation

The volume of hydrogen V_0 (in cc at 0°C and 760 mm Hg) consumed by the sample is given by

$$V_0 = \frac{273}{760}(p - p_s)\frac{v}{T}$$

where p is the atmospheric pressure, p_s is the vapor pressure of the solvent in mm Hg at T_0, v is the volume of mercury added in cc, and T is the absolute temperature. The moles of hydrogen consumed per mole of sample is given by

$$\frac{V_0/(22{,}400\ \text{cc/mole})}{\text{Grams sample}/(\text{grams sample/mole})} = \frac{\text{moles } H_2 \text{ consumed}}{\text{moles sample}}$$

F. MERCURATION (173)

Reagent

Dissolve 40 g of reagent grade mercuric acetate in sufficient anhydrous, reagent grade methanol to make 1 liter of solution (0.12*M*). Stabilize the reagent by the addition of 3–8 drops of glacial acetic acid. Filter the reagent before using. When used in the procedure, 50 ml of the reagent should have a titration of 1–10 ml of 0.1*N* potassium hydroxide.

The procedure given below is intended for volatile samples such as vinyl ethers.

Procedure

Pipet 50 ml of the mercuric acetate reagent into each of two 250-ml glass-stoppered Erlenmeyer flasks. If a sealed glass ampoule is specified, use heat-resistant pressure bottles containing a few pieces of 8-mm glass rod. Cool the contents of the flasks between −10 and −15°C. (A bath of chipped ice and methanol can be maintained below −10°C for more than an hour without difficulty.) Reserve one of the flasks for a blank determination. Into the other flask introduce an amount of sample containing 3.0–4.0 meq of vinyl ether. Allow both the sample and the blank to stand in the bath at −10°C or lower for 10 min. To each flask add 2–4 g of sodium bromide and swirl the contents to effect solution. Add approximately 1 ml of the phenolphthalein indicator and titrate immediately with standard 0.1*N* methanolic potassium hydroxide to a pink endpoint. Do not permit the temperature of the solution to exceed 15°C during the titration. Because the method is based upon an acidimetric titration, take the usual precautions to avoid interference from carbon dioxide.

Reaction time and temperatures for other compounds are shown in Table XXIV.

G. MORPHOLINE (90)

Reagents

Hydrochloric acid (0.5*N*). Transfer 85 ml of 6*N* hydrochloric acid to a 1-liter flask and dilute to the mark with methanol. Standardize 40 ml of this reagent daily with standard 0.5*N* sodium hydroxide using phenolphthalein indicator.

Mixed indicator. Dissolve 0.15 g of methyl orange and 0.08 g of Xylene Cyanol FF in 100 ml of distilled water.

Procedure (Indicator Method)

By means of a graduated cylinder or dispensing buret, add 10 ml of morpholine to each of two 250-ml glass-stoppered Erlenmeyer flasks. For reactions at 98 ± 2°C use heat-resistant pressure bottles. Reserve one of the flasks as a blank. Into the other flask introduce an amount of sample containing not more than 23 meq of the unsaturated compound. For substantially pure material weigh to the nearest 0.1 mg. For dilute solutions the sample may be added by means of a pipet and the sample weight calculated from the specific gravity. To each flask add 7.0 ml of acetic acid solution, unless otherwise specified, and the amount of co-solvent specified in Table XXVI. Allow both the sample and the blank to stand for the time and at the temperature specified in Table XXVI. If elevated temperatures are used, carefully cool the pressure bottles to room temperature. Add 50 ml of acetonitrile to each flask by means of a graduated cylinder. While constantly swirling the flask, add 20 ml of acetic anhydride to both the sample and the blank from a suitable graduated cylinder or dispensing buret and stopper. Allow to cool to room temperature. Add 5 or 6 drops of methyl orange–xylene cyanol mixed indicator and titrate with standard 0.5*N* methanolic hydrochloric acid to the disappearance of the green color. A Fisher titrating light or similar device greatly facilitates the selection of the endpoint.

For the details of the conductometric method, see reference 90.

REFERENCES

1. Abu-Nasr, A. M., and R. T. Holman, *J. Am. Oil Chemists' Soc.*, **31,** 41 (1954).
2. Ackman, R. G., *J. Am. Oil Chemists' Soc.*, **40,** 564 (1963).
3. Adkins, H., R. M. Elofson, A. G. Rossow, and C. C. Robinson, *J. Am. Chem. Soc.*, **71,** 3622 (1949).
4. Adlard, E. R., and B. T. Whitman, *Gas Chromatog. Intern. Symp., 1961,* **3,** 371 (1962); through *Chem. Abstr.*, **58,** 4354 (1963).
5. Albert, D. K., *Anal. Chem.*, **35,** 1918 (1963).
6. Alm, R. S., R. J. P. Williams, and A. Tiselius, *Acta Chem. Scand.*, **6,** 826 (1952).
7. Altshuller, A. P., and S. F. Sleva, *Anal. Chem.*, **33,** 1413 (1961).
8. Altshuller, A. P., S. F. Sleva, and A. F. Wartburg, *Anal. Chem.*, **32,** 946 (1960).
9. Altshuller, A. P., and I. R. Cohen, *Anal. Chem.*, **32,** 1843 (1960).
10. Altshuller, A. P., and S. F. Sleva, *Anal. Chem.*, **34,** 418 (1962).
11. American Society for Testing and Materials, *1952 Book of ASTM Standards,* Philadelphia, 1953, Part V, Designation D-1158.
12. American Society for Testing and Materials, *1966 Book of ASTM Standards,* Philadelphia, Designation D-1019-62.
13. American Society for Testing and Materials, *1966 Book of ASTM Standards,* Designation D1159-64.
14. American Society for Testing and Materials, *1966 Book of ASTM Standards,* Designation D1319-55T.
15. American Society for Testing and Materials, *1966 Book of ASTM Standards,* Designation D1358-58.
16. American Society for Testing and Materials, *1966 Book of ASTM Standards,* Designation D1541-60.
17. American Society for Testing and Materials, *1966 Book of ASTM Standards,* Designation D1638-61T.

18. American Society for Testing and Materials, *1966 Book of ASTM Standards*, Designation D1961-61.
19. American Society for Testing and Materials, *1966 Book of ASTM Standards*, Designation D2002-64, Method B.
20. American Society for Testing and Materials, *1966 Book of ASTM Standards*, Designation D2003-64.
21. American Society for Testing and Materials, *1966 Book of ASTM Standards*, Designation E234-64T.
22. Anantakrishnan, S. V., and C. K. Ingold, *J. Chem. Soc.*, **1935**, 1396.
23. Anderson, J. A., Jr., and W. D. Seyfried, *Anal. Chem.*, **20**, 998 (1948).
24. Andrew, E. R., *Nuclear Magnetic Resonance*, Cambridge University Press, New York, 1955.
25. Archer, E. D., J. H. Shively, and S. A. Francis, *Anal. Chem.*, **35**, 1369 (1963).
26. Asscher, M., and D. Vofsi, *Chem. Ind.* (*London*), **1962**, 209.
27. Austin, R. R., *ISA* (*Instr. Soc. Am.*) *Trans.*, **1**, 211 (1962).
27a. Banthorpe, D. V., *Chem. Rev.*, **70**, 295 (1970).
28. Barnes, L., Jr., and L. J. Molinini, *Anal. Chem.*, **27**, 1025 (1955).
29. Barnes, R. B., U. Liddel, and V. Z. Williams, *Ind. Eng. Chem., Anal. Ed.*, **15**, 83 (1943).
30. Bauer, R. H., *Anal. Chem.*, **35**, 107 (1963).
31. Baumann, F., and D. D. Gilbert, *Anal. Chem.*, **35**, 1133 (1963).
32. Bednas, M. E., and D. S. Russel, *Can. J. Chem.*, **36**, 1272 (1958).
33. Beesing, D. W., W. P. Tyler, D. M. Kurtz, and S. A. Harrison, *Anal. Chem.*, **21**, 1073 (1949).
34. Belcher, R., and B. Fleet, *J. Chem. Soc.*, **1965**, 1740.
35. Bellamy, L. J., *The Infra-red Spectra of Complex Molecules*, Wiley, New York, 1958, pp. 34–63.
36. Benham, G. H., and L. Klee, *J. Am. Oil Chemists' Soc.*, **27**, 127 (1950).
37. Bennett, C. E., S. Dal Nogare, and L. W. Safranski, "Chromatography: Gas," in *Treatise on Analytical Chemistry*, Part I, Vol. 3, I. M. Kolthoff and P. J. Elving, Eds., Interscience, New York, 1961, pp. 1657–1723.
38. Benson, S. W., *Ind. Eng. Chem., Anal. Ed.*, **14**, 189 (1942).
39. Beral, H., B. Wermescher, L. Murea, M. Madgearu, and E. Cuciureanu, *Rev. Chim.* (*Bucharest*), **14**, 49 (1963); through *Chem. Abstr.*, **59**, 8741 (1963).
40. Berezkin, V. G., A. E. Mysak, and L. S. Polak, *Gaz. Khromatogr., Akad. Nauk SSSR, Tr. Vtoroi Vses. Konf., Moscow 1962*, 332 (Pub. 1964); through *Chem. Abstr.*, **62**, 6313 (1965).
41. Berliner, E., *J. Chem. Ed.*, **43**, 124 (1966).
42. Bible, R. H., Jr., *Interpretation of NMR Spectra. An Empirical Approach*, Plenum Press, New York, 1965.
43. Biernacka, T., *Colloq. Spectros. Intern., 9th, Lyons, 1961*, **3**, 113 (1962); through *Chem. Abstr.*, **59**, 1757 (1963).
44. Bij, J. R. van der, and E. C. Kooyman, *Rec. Trav. Chim.*, **71**, 837 (1952).
45. Binder, J. L., *Anal. Chem.*, **26**, 1877 (1954).
46. Bischoff, C., *Monatsber. Deut. Akad. Wiss. Berlin*, **3**, 674 (1961); through *Chem. Abstr.*, **58**, 2353 (1963).
47. Bjerrum, J., G. Schwarzenbach, and L. G. Sillen, *Stability Constants*, Part II, *Inorganic Ligands*, The Chemical Society, London, 1958.
48. Bobitt, J. M., *Thin Layer Chromatography*, Reinhold, New York, 1963.
49. Boer, H., and E. C. Kooyman, *Anal. Chim. Acta*, **5**, 550 (1951).

50. Boldingh, J., *Rec. Trav. Chim.*, **69**, 247 (1950).
51. Bolton, E. R., and K. A. Williams, *Analyst.*, **55**, 5 (1930).
52. Bond, G. R., Jr., *Ind. Eng. Chem., Anal. Ed.*, **18**, 692 (1946).
53. Bothner-By, A. A., and C. Naar-Colin, *Am. Chem. Soc.*, **83**, 231 (1961).
54. Bott, R. W., C. Eaborn, E. R. A. Peeling, and D. E. Webster, *Proc. Chem. Soc.*, **1962**, 337.
55. Boyd, H. M., and J. R. Roach, *Anal. Chem.*, **19**, 158 (1947).
56. Braae, B., *Anal. Chem.*, **21**, 1461 (1949).
56a. Brand, M. J. D., B. Fleet, and M. R. H. Weaver, *Analyst*, **95**, 387 (1970).
57. Bratzler, K., and H. Kleemann, *Erdöl u. Kohle*, **7**, 559 (1954).
58. Bretschneider, H., and G. Burger, *Chem. Fabrik*, **10**, 124 (1937).
59. Brenner, N., E. Cieplinski, L. S. Ettre, and V. J. Coates, *J. Chromatog.*, **3**, 230 (1960).
59a. Brezina, M., and P. Zuman, *Polarography in Medicine, Biochemistry, and Pharmacy*, Interscience, New York, 1958.
60. Bricker, C. E., and K. H. Roberts, *Anal. Chem.*, **21**, 1331 (1949).
61. Bright, K., B. J. Farmer, B. W. Malpass, and P. Snell, *Chem. Ind.* (*London*), **1965**, 610.
62. Brooks, F. R., L. Lykken, W. B. Milligan, H. R. Nebeker, and V. Zahn, *Anal. Chem.*, **21**, 1105 (1949).
62a. Brown, C. A., *Anal. Chem.*, **39**, 1883 (1967).
62b. Brown, C. A., S. C. Sethi, and H. C. Brown, *Anal. Chem.*, **39**, 823 (1967).
63. Brown, H. C., and C. A. Brown, *J. Am. Chem. Soc.*, **84**, 1493, 1494, 1495 (1962).
64. Brown, H. C., K. Sivasankaran, and C. A. Brown, *J. Org. Chem.*, **28**, 214 (1963).
65. Brown, J. H., H. W. Durand, and C. S. Marvel, *J. Am. Chem. Soc.*, **58**, 1594 (1936).
66. Brown, R. A., *Anal. Chem.*, **23**, 430 (1951).
67. Brown, R. A., and E. Gilliams, "Mass Spectra of Monoolefins," paper presented at the Meeting of ASTM Committee E-14 on Mass Spectrometry, New Orleans, La., May, 1954; through reference 238.
68. Buckwalter, H. M., and E. C. Wagner, *J. Am. Chem. Soc.*, **52**, 5241 (1930).
69. Burger, K., and E. Schulek, *Talanta*, **7**, 46 (1960).
70. Burkhardt, F., and A. Dirschel, *Mikrochim. Ichnoanalyt. Acta*, **1965**, 353.
71. Byrne, R. E., Jr., and J. B. Johnson, *Anal. Chem.*, **28**, 126 (1956).
72. Cartoni, G. P., R. S. Lowrie, C. S. G. Phillips, and L. M. Venanzi, *Gas Chromatog., Proc. Symposium, 3rd, Edinburgh, 1960*, 273; through *Chem. Abstr.*, **56**, 10942 (1962).
73. Casimir, E., and M. Dimitriu, *Petroleum-Z.*, **31**, 1 (1935); through reference 271.
74. Cassidy, H. G., *Adsorption and Chromatography*, Interscience, New York, 1951.
75. Chamberlain, N. F., "Nuclear Magnetic Resonance," in *Treatise on Analytical Chemistry*, Part I, Vol. 4, I. M. Kolthoff and P. J. Elving, Eds., Interscience, New York, 1963, pp. 1885–1958.
76. Claesson, S., *Arkiv. Kemi, Mineral. Geol. 23A*, **1**, 1 (1946).
77. Claesson, S., *Rec. Trav. Chim.*, **65**, 571 (1946).
78. Clark, A., A. Andrews, and H. W. Fleming, *Ind. Eng. Chem.*, **41**, 1527 (1949).
79. Clausson-Kaas, N., and F. Limborg, *Acta Chem. Scand.*, **1**, 884 (1947).
80. Coggeshall, N. D., "Spectroscopic Functional Group Analysis in the Petroleum Industry," in *Organic Analysis*, Vol. 1, J. Mitchell, Jr., I. M. Kolthoff, E. S. Proskauer, and A. Weissberger, Eds., Interscience, New York, 1953, pp. 403–450.
81. Conrad, A. L., *Anal. Chem.*, **20**, 725 (1948).

82. Cook, D., Y. Lupien, and W. G. Schneider, *Can. J. Chem.*, **34,** 957 (1956).
83. Cornu, A., *Publ. Group Avan. Methodes Spectrog.*, **1962,** 7; through *Chem. Abstr.*, **58,** 1482 (1963).
84. Coulson, D. M., *Anal. Chem.*, **31,** 906 (1959).
85. Cremer, E., and R. Müller, *Mikrochem. Mikrochim. Acta*, **36/37,** 553 (1951).
86. Cremer, E., and R. Müller, *Z. Elektrochem.*, **55,** 217 (1951).
87. Cremer, E., and F. Prior, *Z. Elektrochem.*, **55,** 66 (1951).
88. Criddle, D. W., and R. L. LeTourneau, *Anal. Chem.*, **23,** 1620 (1951).
89. Critchfield, F. E., *Anal. Chem.*, **31,** 1406 (1959).
90. Critchfield, F. E., G. L. Funk, and J. B. Johnson, *Anal. Chem.*, **28,** 76 (1956).
91. Critchfield, F. E., and J. B. Johnson, *Anal. Chem.*, **28,** 73 (1956).
92. Crompton, T. R., and D. Buckley, *Analyst*, **90,** 76 (1965).
93. Čuta, F., and Z. Kučera, *Chem. Listy*, **47,** 1166 (1953); through reference 271.
94. Cvejanovich, G. J., *Anal. Chem.*, **34,** 654 (1962).
95. Cvetanovic, R. J., F. J. Duncan, and W. E. Falconer, *Can. J. Chem.*, **42,** 2410 (1964).
96. Dal Nogare, S., L. R. Perkins, and A. H. Hale, *Anal. Chem.*, **24,** 512 (1952).
97. Danish, A. A., and R. E. Lidov, *Anal. Chem.*, **22,** 702 (1950).
98. Das, M. N., *Anal. Chem.*, **26,** 1086 (1954).
99. den Boer, F. C., *Z. Anal. Chem.*, **205,** 308 (1964).
100. Desty, D. H., and B. H. F. Whyman, *Anal. Chem.*, **29,** 320 (1957).
101. Dewar, M. J. S., and R. C. Fahey, *J. Am. Chem. Soc.*, **85,** 2245 (1963).
102. Diebler, V. H., "Analytical Mass Spectrometry," in *Organic Analysis*, Vol. 3, J. Mitchell, Jr., I. M. Kolthoff, E. S. Proskauer, and A. Weissberger, Eds., Interscience, New York, 1956, pp. 387–441.
103. Diels, O., and K. Alder, *Ann.*, **460,** 98 (1928) *et seq.;* through reference 271.
104. Dietz, W. A., and B. F. Dudenbostel, Jr., *Am. Chem. Soc., Div. Petrol. Chem. Preprints*, **2** (4), D 171 (1957); through *Chem. Abstr.*, **54,** 21722 (1960).
105. Doeuvre, J., *Bull. Soc. Chim. France*, **3,** 612 (1936).
106. DuBois, H. D., and D. A. Skoog, *Anal. Chem.*, **20,** 624 (1948).
107. Dubois, J. E., and Mouvier, G., *Compt. Rend.*, **255,** 1104 (1962).
108. Dudenbostel, B. F., and W. Priestly, Jr., in *The Chemistry of Petroleum Hydrocarbons*, Vol. I, B. T. Brooks, S. S. Kurtz, Jr., C. E. Bourd, and L. Schmerling, Eds., Reinhold, New York, 1954, pp. 337–349; through reference 271.
109. Duffield, J. J., and L. B. Rogers, *Anal. Chem.*, **34,** 1193 (1962).
110. Duke, J. A., and J. A. Maselli, *J. Am. Oil Chemists' Soc.*, **29,** 126 (1952).
111. Durbetaki, A. J., and C. M. Miles, *Anal. Chem.*, **37,** 1231 (1965).
112. Dutton, H. J., *J. Phys. Chem.*, **48,** 179 (1944).
113. Dutton, H. J., and C. L. Reinbold, *J. Am. Oil Chemists' Soc.*, **25,** 120 (1948).
114. Earle, F. R., and R. T. Milner, *Oil & Soap*, **16,** 69 (1939); through reference 271.
115. Edse, R., and P. Harteck, *Angew. Chem.*, **52,** 32 (1939).
116. El Khishen, S. A., *Anal. Chem.*, **20,** 1078 (1948).
117. Ellis, B. A., and R. A. Jones, *Analyst*, **61,** 812 (1936).
118. Elving, P. J., "Application of Polarography to Organic Analysis," in *Organic Analysis*, Vol. 2, J. Mitchell, Jr., I. M. Kolthoff, E. S. Proskauer, and A. Weissberger, Eds., Interscience, New York, 1954, pp. 195–236.
119. English, J., Jr., *J. Am. Chem. Soc.*, **63,** 941 (1941).
120. Ephraim, J., *Z. Angew. Chem.*, **1895,** 254.
121. Farrington, P. S., and D. T. Sawyer, *J. Am. Chem. Soc.*, **78,** 5536 (1956).

122. Fellgett, P. B., G. P. Harris, D. N. Simpson, G. B. B. M. Sutherland, H. W. Thompson, D. H. Whiffen, and H. A. Willis, *Inst. Petroleum, Hydrocarbon Res. Group Rept.* XI (1946); through reference 271.
123. Ferguson, R. C., *J. Polymer Sci., A,* **2,** 4735 (1964).
124. Ferrin, C. R., J. O. Chase, and R. W. Hurn, *Gas Chromatog., Intern. Symp., 1961,* **3,** 423 (1962); through *Chem. Abstr.,* **58,** 3869 (1963).
125. Fink, D. F., R. W. Lewis, and F. T. Weiss, *Anal. Chem.,* **22,** 858 (1950).
125a. Fleet, B., and R. D. Jee, *Talanta,* **16,** 1561 (1969).
126. Francis, A. W., *Ind. Eng. Chem.,* **18,** 821, 1095 (1926).
127. Franke, G., *Erdöl u. Kohle,* **13,** 850 (1960); *Chem. Abstr.,* **55,** 12829 (1961).
128. Frisque, A. J., H. M. Grubb, C. H. Ehrhardt, and R. W. Vander Haar, *Anal. Chem.,* **33,** 389 (1961).
128a. Fritz, J. S., and G. E. Wood, *Anal. Chem.,* **40,** 134 (1968).
129. Fulmer, E. I., J. J. Kolfenbach, and L. A. Underkofler, *Ind. Eng. Chem., Anal. Ed.,* **16,** 469 (1944).
130. Gal'pern, G. D., and J. V. Vinogradova, *Neftyanoe Khoz.,* **1,** 59 (1936); through reference 271.
131. George, J., H. Mark, and H. Wechsler, *J. Am. Chem. Soc.,* **72,** 3896 (1950).
132. George, J., H. Wechsler, and H. Mark, *J. Am. Chem. Soc.,* **72,** 3891 (1950).
133. Giuffre, L., and F. Cassani, *Chim. Ind.,* **45,** 806 (1963); through *Chem. Abstr.,* **59,** 7733 (1963).
134. Glasser, A. C., L. E. Harris, B. V. Christensen, and F. W. Bope, *J. Am. Pharm. Assoc., Sci. Ed.,* **43,** 294 (1954); through reference 271.
134a. Goddu, R. F., in *Advances in Analytical Chemistry and Instrumentation,* Vol. 1, C. N. Reilley, Ed., Interscience, New York, 1960, pp. 347–421.
135. Goddu, R. F., *Anal. Chem.,* **29,** 1790 (1957).
135a. Goddu, R. F., and D. A. Delker, *Anal. Chem.,* **30,** 2013 (1958).
136. Gohlke, R. S., *Anal. Chem.,* **31,** 535 (1959).
137. Gordon, B. E., F. Wopat, Jr., H. D. Burnham, and L. C. Jones, Jr., *Anal. Chem.,* **23,** 1754 (1951).
138. Gould, C. W., and H. J. Drake, *Anal. Chem.,* **23,** 1157 (1951).
139. Graff, M. M., and E. L. Skau, *Ind. Eng. Chem., Anal. Ed.,* **15,** 340 (1943).
140. Graupner, A. J., and V. A. Aluise, *J. Am. Oil Chemists' Soc.,* **43,** 81 (1966).
141. Grosse-Oetringhaus, H., *Petroleum Z.,* **35,** 75, 112 (1939); through reference 271.
142. Grunbaum, B. W., and P. L. Kirk, *Mikrochem. Mikrochim. Acta,* **39,** 268 (1952).
143. Guenther, K. F., G. Sosnovsky, and R. Brunier, *Anal. Chem.,* **36,** 2508 (1964).
144. Hachenberg, H., and J. Gutberlet, *Brennstoff Chem.,* **45,** 132 (1964); through *Chem. Abstr.,* **61,** 4970 (1964).
145. Hamilton, J. G., and R. T. Holman, *J. Am. Chem. Soc.,* **76,** 4107 (1954).
146. Hammar, C. G. B., *Svensk Kem. Tidskr.,* **63,** 125 (1951); through reference 271.
147. Hampton, R. R., *Anal. Chem.,* **21,** 923 (1949).
148. Hanna, J. G., and S. Siggia, *Anal. Chem.,* **21,** 1469 (1949).
148a. Hanus, J., *Z. Untersuch. Nahr.-u. Genussm.,* **4,** 913 (1901); through reference 271.
149. Heigl, J. J., J. F. Black, and B. F. Dudenbostel, Jr., *Anal. Chem.,* **21,** 554 (1949).
150. Herb, S. F., and R. W. Riemenschneider, *J. Am. Oil Chemists' Soc.,* **29,** 456 (1952).
151. Herb, S. F., L. P. Witnauer, and R. W. Riemenschneider, *J. Am. Oil Chemists' Soc.,* **28,** 505 (1951).
152. Hesse, G., H. Eilbracht, and F. Reicheneder, *Ann.,* **546,** 251 (1941).

153. Hibben, J. H., *The Raman Effect and Its Chemical Applications*, Reinhold, New York, 1939.
154. Hilditch, T. P., and C. H. Lea, *J. Chem. Soc.*, **1927**, 3106.
155. Hirozawa, S. T., presented at the Detroit Anachem Conference, Oct. 21, 1964.
156. Hirozawa, S. T., presented at the Detroit Anachem Conference, Oct. 19, 1965.
156a. Hirozawa, S. T., unpublished results.
157. Hiscox, D. J., *Anal. Chem.*, **20**, 679 (1948).
158. Hoffman, H. D., and C. E. Green, *Oil & Soap*, **16**, 236 (1939); through reference 271.
159. Holman, R. T., *J. Am. Chem. Soc.*, **73**, 5289 (1951).
160. Holman, R. T., and L. Hagdahl, *Anal. Chem.*, **23**, 794 (1951).
161. Holman, R. T., and W. T. Williams, *J. Am. Chem. Soc.*, **73**, 5285 (1951).
162. Hornstein, I., *Anal. Chem.*, **23**, 1329 (1951).
163. Huber, W., *Mikrochim. Acta*, **1960**, 44.
164. Hübl, B., *J. Soc. Chem. Ind. (London)*, **3**, 641 (1884).
165. Hübl, B., *Dinglers Polytech. J.*, **253**, 281 (1884); through reference 271.
166. Innes, W. B., and W. E. Bambrick, *J. Gas Chromatog.*, **2**, 309 (1964).
167. Innes, W. B., W. E. Bambrick, and A. J. Andreatch, *Anal. Chem.*, **35**, 1198 (1963).
168. Ito, H., and Tsukihara, K., *Yakuzaigaku*, **22**, 63 (1962); through *Chem. Abstr.*, **57**, 9972 (1962).
169. James, D. H., and C. S. G. Phillips, *J. Chem. Soc.*, **1953**, 1600.
170. Janák, J., and J. Novak, *Chem. Listy*, **51**, 1832 (1957); through *Chem. Abstr.*, **52**, 1860 (1958).
171. Janák J., and M. Rusek, *Chem. Listy*, **47**, 1190 (1953); through *Chem. Abstr.*, **48**, 3853 (1954).
172. Jasperson, H., *J. Soc. Chem. Ind. (London)*, **61**, 115 (1942).
173. Johnson, J. B., and J. P. Fletcher, *Anal. Chem.*, **31**, 1563 (1959).
174. Johnson, L. F., and J. N. Shoolery, *Anal. Chem.*, **34**, 1136 (1962).
175. Johnston, R. W. B., W. G. Appleby, and M. O. Baker, *Anal. Chem.*, **20**, 805 (1948).
176. Jones, E. P., and V. L. Davison, *J. Am. Oil Chemists' Soc.*, **42**, 121 (1965).
177. Jones, L. C., Jr., and L. W. Taylor, *Anal. Chem.*, **27**, 228, 2015 (1955).
178. Jungnickel, H. E., and W. Klinger, *Pharmazie*, **18**, 130 (1963); through *Chem. Abstr.*, **59**, 843 (1963).
179. Karasch, H., and C. M. Buess, *J. Am. Chem. Soc.*, **71**, 2724 (1949).
180. Kattwinkel, R., *Brennstoff Chem.*, **8**, 353 (1927).
181. Kaufmann, H. P., *Fette Seifen*, **46**, 268 (1939); through reference 271.
182. Kaufmann, H. P., *Studien auf dem Fettgebeit*, Verlag Chemie, Berlin, 1935; through reference 271.
183. Kaufmann, H. P., *Z. Untersuch. Lebensm.*, **51**, 3 (1926); through reference 271.
184. Kaufmann, H. P., and J. Baltes, *Fette Seifen*, **43**, 93 (1936); through reference 271.
185. Kaufmann, H. P., J. Baltes, and H. Büter, *Ber.*, **70**, 903 (1937).
186. Kaufmann, H. P., and L. Hartweg, *Ber.*, **70**, 2554 (1937).
187. Kaufmann, H. P., and W. Wolf, *Fette Seifen*, **50**, 519 (1943).
187a. Kaye, W. I., *Spectrochim. Acta*, **6**, 257 (1954).
187b. Kaye, W. I., in *The Encyclopedia of Spectroscopy*, G. L. Clark, Ed., Reinhold, New York, 1960, pp. 494–505.
188. Kemp, A. R., and G. S. Mueller, *Ind. Eng. Chem., Anal. Ed.*, **6**, 52 (1934).
189. Kemp, A. R., and H. Peters, *Ind. Eng. Chem., Anal. Ed.*, **15**, 453 (1943).

190. Kirchner, J. G., J. M. Miller, and G. J. Keller, *Anal. Chem.*, **23,** 420 (1951).
191. Kitson, R. E., *Anal. Chem.*, **25,** 1470 (1953).
192. Klee, L., and G. H. Benham, *J. Am. Oil Chemists' Soc.*, **27,** 130 (1950).
193. Knowles, G., and G. F. Lowden, *Analyst,* **78,** 159 (1953).
194. Kock, R. J. de, P. A. H. M. Hol, and H. Bos, *Z. Anal. Chem.*, **205,** 371 (1964).
195. Kolb, B., and Kaiser, K. H., *J. Gas Chromatog.*, **2,** 233 (1964).
196. Kolthoff, I. M., and F. A. Bovey, *Anal. Chem.*, **19,** 498 (1947).
197. Kolthoff, I. M., and P. J. Elving, *Treatise on Analytical Chemistry,* Part I, Vol. 3, Interscience, New York, 1961.
198. Kolthoff, I. M., and T. S. Lee, *J. Polymer Sci.*, **2,** 206 (1947).
199. Kolthoff, I. M., T. S. Lee, and M. A. Mairs, *J. Polymer Sci.*, **2,** 199 (1947).
200. Kolthoff, I. M., and J. J. Lingane, *Polarography,* Vol. II, Interscience, New York, 1952, pp. 634–646.
201. Kolthoff, I. M., and E. B. Sandell, *Textbook of Quantitative Inorganic Analysis,* 3rd ed., Macmillan, New York, 1952.
202. Kraus, G., J. N. Short, V. Thornton, *Rubber Plastics Age,* **38,** 880 (1957).
203. Kuemmel, D. F., *Anal. Chem.*, **36,** 426 (1964).
204. Kurtz, F. E., *J. Am. Chem. Soc.*, **74,** 1902 (1952).
205. Kurtz, S. S., Jr., and C. E. Headington, *Ind. Eng. Chem., Anal. Ed.*, **9,** 21 (1937).
206. Latimer, W. M., *The Oxidation States of the Elements and Their Potentials In Aqueous Solutions,* 2nd ed., Prentice-Hall, Englewood Cliffs, N. J., 1952.
207. Lederer, E., and M. Lederer, *Chromatography,* Elsevier, New York, 1953.
208. Lee, T. S., and I. M. Kolthoff, *Ann. N. Y. Acad. Sci.*, **53,** 1093 (1951).
209. Lawrey, D. M. G., and J. F. Paulson, *Anal. Chem.*, **34,** 538 (1962).
210. Lee, T. S., I. M. Kolthoff, and E. Johnson, *Anal. Chem.*, **22,** 995 (1950).
211. Lee, T. S., I. M. Kolthoff, and M. A. Mairs, *J. Polymer Sci.*, **3,** 66 (1948).
212. Leisey, F. A., and J. F. Grutsch, *Anal. Chem.*, **28,** 1553 (1956).
213. Lemieux, R. U., and E. von Rudloff, *Can. J. Chem.*, **33,** 1701 (1955).
214. Lemieux, R. U., and E. von Rudloff, *Can. J. Chem.*, **33,** 1710 (1955).
215. Lindeman, L. P., *Am. Chem. Soc., Div. Petrol. Chem., Preprints,* **7,** 15 (1962); through *Chem. Abstr.*, **61,** 498 (1964).
216. Link, W. E., H. M. Hickman, and R. A. Morrissette, *J. Am. Oil Chemists' Soc.*, **36,** 20 (1959).
217. Lips, H. J., *J. Am. Oil Chemists' Soc.*, **30,** 399 (1953).
218. Long, D. R., and R. W. Neuzil, *Anal. Chem.*, **27,** 1110 (1955).
219. Lopes, H. D. S., *J. Am. Oil Chemists' Soc.*, **28,** 390 (1951).
220. Lucas, H. J., and D. Pressman, *Ind. Eng. Chem., Anal. Ed.*, **10,** 140 (1938).
221. Lumpkin, H. E., *Anal. Chem.*, **30,** 321 (1958).
222. Maczek, A. O. S., and C. S. G. Phillips, *Gas Chromatog., Proc. Symp., 3rd, Edinburgh, 1960,* 284; through *Chem. Abstr.*, **56,** 10944 (1962).
223. Majors, K. R., and R. T. Milner, *Oil & Soap,* **16,** 228 (1938); through reference 271.
224. Manchot, W., and F. Oberhauser, *Z. Untersuch. Nahr.-u. Genussm.*, **47,** 261 (1924); through reference 271.
225. Manegold, E., and F. Peters, *Kolloid-Z.*, **85,** 310 (1938).
226. Marquardt, R. P., and E. N. Luce, *Anal. Chem.*, **20,** 751 (1948).
227. Marquardt, R. P., and E. N. Luce, *Anal. Chem.*, **21,** 1194 (1949).
228. Marquardt, R. P., and E. N. Luce, *Anal. Chem.*, **22,** 363 (1950).
228a. Marquardt, R. P., and E. N. Luce, *Anal. Chem.*, **39,** 1655 (1967).

229. Marshall, A., *J. Soc. Chem. Ind. (London)*, **19**, 213 (1900).
230. Martin, A. E., and J. Smart, *Nature*, **175**, 422 (1955).
231. Martin, R., *Anal. Chem.*, **34**, 896 (1962).
232. Martin, R. L., *Anal. Chem.*, **32**, 336 (1960).
233. Martin, R. W., *Anal. Chem.*, **21**, 921 (1949).
234. McCutcheon, J. W., M. F. Crawford, and H. L. Welsh, *Oil & Soap*, **18**, 9 (1941); through reference 271.
235. McHale, D., J. Green, and S. Marcinkiewicz, *Chem. Ind. (London)*, **1961**, 555.
236. McIlhiney, P. C., *J. Am. Chem. Soc.*, **21**, 1084 (1899).
237. McMurray, H. L., and V. Thornton, *Anal. Chem.*, **24**, 318 (1952).
237a. McNeill, I. C., *Polymer*, **4**, 15 (1963).
238. Melpolder, F. W., and R. A. Brown, "Mass Spectrometry," in I. M. Kolthoff and P. J. Elving, *Treatise on Analytical Chemistry*, Part I, Vol. 4, Interscience, New York, 1963, pp. 1959–2074.
239. Melpolder, F. W., R. A. Brown, W. S. Young, and C. E. Headington, *Ind. Eng. Chem.*, **44**, 1142 (1952).
240. Meyer, A. S., Jr., and C. M. Boyd, *Anal. Chem.*, **31**, 215 (1959).
241. Meyer, K. H., *Ber.*, **52**, 1468 (1919).
242. Mikkelsen, L., R. L. Hopkins, and D. Y. Yee, *Anal. Chem.*, **30**, 317 (1958).
243. Mikusch, J. D. von, *Angew. Chem.*, **62**, 475 (1950).
244. Mikusch, J. D. von, *Z. Anal. Chem.*, **130**, 412 (1950).
245. Mikusch, J. D. von, and C. Frazier, *Ind. Eng. Chem., Anal. Ed.*, **13**, 782 (1941).
246. Miller, J. W., and DeFord, D. D., *Anal. Chem.*, **29**, 475 (1957).
247. Miller, J. W., and D. D. DeFord, *Anal. Chem.*, **30**, 295 (1958).
248. Miller, J. W., M. D. Grimes, and B. J. Heinrich, *Proc. Am. Petrol. Inst., Sect. III*, **38**, 343 (1958).
249. Mills, I. W., S. S. Kurtz, Jr., A. H. A. Heyn, and M. R. Lipkin, *Anal. Chem.*, **20**, 333 (1948).
249a. Mitchell, J., Jr., "Functional Groups," in this Treatise, Part II, Vol. 13, 1966, pp. 20–24.
250. Miyake, H., and M. Mitooka, *Nippon Kagaku Zasshi*, **84**, 593 (1963); through *Chem. Abstr.*, **61**, 1275 (1964).
251. Morrell, J. C., and I. M. Levine, *Ind. Eng. Chem., Anal. Ed.*, **4**, 319 (1932).
252. Moser, H., and U. Weber, *Proc. Intern. Meeting Mol. Spectry., 4th, Bologna, 1959*, **3**, 1116 (1962); through *Chem. Abstr.*, **59**, 4700 (1963).
253. Mukherjee, S., *J. Am. Oil Chemists' Soc.*, **29**, 97 (1952).
254. Mukherjee, S., *J. Am. Oil Chemists' Soc.*, **32**, 351 (1955).
255. Muzyczk, J., *Chem. Anal. (Warsaw)*, **9**, 671 (1964); through *Chem. Abstr.*, **62**, 4510 (1965).
256. Naves, Y.-R., *Helv. Chim. Acta*, **32**, 1151 (1949).
257. Nicksic, S. W., and E. Rostenbach, *J. Air Pollution Control Assoc.*, **11**, 417 (1961).
258. Noller, C. R., *Textbook of Organic Chemistry*, 2nd ed., W. B. Saunders Co., Philadelphia, 1958.
259. Norris, T. A., J. H. Shively, and C. S. Constantin, *Anal. Chem.*, **33**, 1556 (1961).
260. O'Connor, J. G., and M. S. Norris, *Anal. Chem.*, **32**, 701 (1960).
261. O'Connor, R. T., D. C. Heinzelman, F. C. Pack, and R. W. Planck, *J. Am. Oil Chemists' Soc.*, **30**, 182 (1953).
262. O'Donnell, A. E., M. M. Neal, F. T. Weiss, J. M. Bann, T. J. DeCino, and S. C. Lau, *J. Agr. Food Chem.*, **2**, 573 (1954).

263. Okuda, T., *Chem. Pharm. Bull. (Tokyo)*, **10,** 96 (1962); through *Chem. Abstr.*, **57,** 9192 (1962).
264. Page, J. E., *Quart. Revs.*, **6,** 262 (1952).
265. Patton, H. W., J. S. Lewis, and W. I. Kaye, *Anal. Chem.*, **27,** 170 (1955).
266. Phillips, C. S. G., *Discussions Faraday Soc.*, **7,** 241 (1949).
267. Phillips, W. M., and W. C. Wake, *Analyst*, **74,** 306 (1949).
268. Pifer, C. W., and E. G. Wollish, *Anal. Chem.*, **24,** 300 (1952).
268a. Pincock, J. A., and K. Yates, *J. Am. Chem. Soc.*, **90,** 5643 (1968).
269. Planck, R. W., F. C. Pack, and L. A. Goldblatt, *J. Am. Oil Chemists' Soc.*, **30,** 417 (1953).
270. Polgár, A. G., J. J. Holst, and S. Groennings, *Anal. Chem.*, **34,** 1226 (1962).
271. Polgár, A., and J. L. Jungnickel, "Determination of Olefinic Unsaturation," in *Organic Analysis*, Vol. 3, J. Mitchell, Jr., I. M. Kolthoff, E. S. Proskauer, and A. Weissberger, Eds., Interscience, New York, 1956, pp. 203–386.
272. Pople, J. A., W. G. Schneider, and H. J. Bernstein, *High-Resolution Nuclear Magnetic Resonance*, McGraw-Hill, New York, 1959.
273. Powell, J. C., K. C. Edson, and E. L. Fisher, *Anal. Chem.*, **20,** 213 (1948).
274. Prelog, V., and U. Häfliger, *Helv. Chim. Acta*, **32,** 2088 (1949).
275. Prey, V., A. Berger, and H. Berbalk, *Z. Anal. Chem.*, **185,** 113 (1962).
276. Rabilloud, G., *Bull. Soc. Chim. France*, **1965,** 293.
277. Ralls, J. O., *J. Am. Chem. Soc.*, **55,** 2083 (1933); **56,** 121 (1934).
278. Rankoff, G., and D. Rankoff, *Fette, Seifen, Anstrichmittel*, **66,** 912 (1964).
279. Rasmussen, R. S., and R. R. Brattain, *J. Chem. Phys.*, **15,** 120, 131, 135 (1947).
280. Ray, N. H., *J. Appl. Chem.*, **4,** 21 (1954).
281. Reichstein, T., and J. v. Euw, *Helv. Chim. Acta*, **21,** 1197 (1938).
282. Reid, V. W., and J. D. Beddard, *Analyst*, **79,** 456 (1954).
283. Reinbold, C. L., and H. J. Dutton, *J. Am. Oil Chemists' Soc.*, **25,** 117 (1948).
284. Reshetnikov, S. M., and A. M. Sokol'skaya, *Izv. Vysshikh Uchebn. Zavedenii, Khim. i Khim. Tekhnol.*, **7,** 217 (1964); through *Chem. Abstr.*, **61,** 10555 (1964).
285. Richardson, W. S., *J. Polymer Sci.*, **13,** 229 (1954).
286. Riemenschneider, R. W., S. F. Herb, and P. L. Nichols, Jr., *J. Am. Oil Chemists' Soc.*, **26,** 371 (1949).
287. Roberts, C. B., "Titrimetric Methods," *Proc. Symp. Cornwall, Ont., Can., 1961*, pp. 21–31; through *Chem. Abstr.*, **57,** 4037 (1962).
288. Roberts, C. B., and H. A. Brejcha, *Anal. Chem.*, **35,** 1104 (1963).
289. Robey, R. F., B. E. Hudson, Jr., and H. K. Wiese, *Anal. Chem.*, **24,** 1080 (1952).
290. Robey, R. F., and C. E. Morell, and B. M. Vanderbilt, *Oil Gas J.*, **40,** 41 (1942); through reference 271.
291. Roman, W., *Anal. Chim. Acta*, **2,** 552 (1948).
292. Roman, W., and M. Smith, *Analyst*, **78,** 679 (1953).
293. Rosenmund, K. W., and H. H. Grandjean, *Erdöl u. Kohle*, **5,** 348 (1952).
294. Rosenmund, K. W., and W. Kuhnhenn, *Pharm. Zentr.*, **66,** 81 (1925).
295. Rosenmund, K. W., W. Kuhnhenn, D. Rosenburg-Gruszynski, and H. Rosetti, *Z. Untersuch. Nahr.-u. Genussm.*, **46,** 154 (1923); through reference 271.
296. Rouayheb, G. M., O. F. Folmer, and W. C. Hamilton, *Anal. Chim. Acta*, **26,** 378 (1962).
297. Rowe, G., C. C. Furnas, and H. Bliss, *Ind. Eng. Chem., Anal. Ed.*, **16,** 371 (1944).
298. Rudenko, B. A., I. I. Nazarova, and V. F. Kucherov, *Izv. Akad. Nauk SSSR, Ser. Khim.*, **1963,** 1545; through *Chem. Abstr.*, **60,** 5 (1964).

299. Rudloff, E. von, *Can. J. Chem.*, **33**, 1714 (1955).
300. Rudloff, E. von, *J. Am. Oil Chemists' Soc.*, **33**, 126 (1956).
301. Russoff, I. I., J. R. Platt, H. B. Klevens, and G. O. Burr, *J. Am. Chem. Soc.*, **67**, 673 (1945).
302. Saffer, A., and B. L. Johnson, *Ind. Eng. Chem.*, **40**, 538 (1948).
303. Saier, E. L., L. R. Cousins, and M. R. Basila, *Anal. Chem.*, **35**, 2219 (1963).
304. Saier, E. L., A. Pozefsky, and N. D. Coggeshall, *Anal. Chem.*, **26**, 1258 (1954).
305. Sartori, G., and E. Bianchi, *Gazz. Chim. Ital.*, **74**, 8 (1940); through reference 271.
306. Sax, N. I., *Dangerous Properties of Industrial Materials*, 2nd ed., Reinhold, New York, 1963.
307. Scafe, E. T., J. Herman, and G. R. Bond, Jr., *Anal. Chem.*, **19**, 971 (1947).
308. Schenck, R. T. E., and J. Danishefsky, *J. Org. Chem.*, **16**, 1683 (1951).
308a. Schneider, W., H. Bruderreck, and I. Halász, *Anal. Chem.*, **36**, 1533 (1964).
309. Schuftan, P., *Gasanalyse in der Technik*, Hirzel, Leipzig, 1931; through reference 271.
310. Schulek, E., and K. Burger, *Talanta*, **2**, 280 (1959).
311. Schulek, E., and K. Burger, *Talanta*, **7**, 41 (1960).
312. Schulek, E., K. Burger, and J. Laszlovsky, *Talanta*, **7**, 51 (1960).
313. Schwabe, K., and H. Berg, *Z. Elektrochem.*, **56**, 961 (1952).
314. Schwabe, K., G. Ohloff, and H. Berg, *Z. Elektrochem.*, **56**, 961 (1952).
315. Scott, C. G., and C. S. G. Phillips, *Nature*, **199**, 66 (1963).
315a. Sedlak, M., *Anal. Chem.*, **38**, 1503 (1966).
316. Sharefkin, J. G., and T. Sulzberg, *Anal. Chem.*, **32**, 993 (1960).
317. Shell Method Series, Designation 247/54; through reference 271.
318. Shell Method Series, Designation 221/50; through reference 271.
319. Sheppard, N., and D. M. Simpson, *Quart. Rev. (London)*, **6**, 1 (1952).
320. Sheppard, N., and G. B. B. M. Sutherland, *Proc. Roy. Soc. (London)*, **A196**, 195 (1949).
320a. Shoolery, J. N., and L. H. Smithson, *JAOCS*, **47**, 153 (1970).
321. Sidgwick, N. V., *The Chemical Elements and Their Compounds*, Vol. II, Oxford at the Clarendon Press, 1950.
322. Siggia, S., *Quantitative Organic Analysis via Functional Groups*, 3rd ed., Wiley, New York, 1963.
323. Siggia, S., and R. L. Edsberg, *Anal. Chem.*, **20**, 762 (1948).
324. Siggia, S., and J. G. Hanna, *Anal. Chem.*, **33**, 896 (1961).
325. Siggia, S., J. G. Hanna, and N. M. Serencha, *Anal. Chem.*, **35**, 362 (1963).
326. Siggia, S., and C. R. Stahl, *Anal. Chem.*, **35**, 1740 (1963).
327. Siggia, S., *Anal. Chem.*, **28**, 1481 (1956).
328. Silas, R. S., J. Yates, and V. Thornton, *Anal. Chem.*, **31**, 529 (1959).
329. Simmons, R. O., and F. W. Quackenbush, *J. Am. Oil Chemists' Soc.*, **30**, 614 (1953).
330. Sinn, V., *Ann. Chim. Anal.*, **28**, 166 (1946); *Chem. Abstr.*, **41**, 53 (1947).
331. Skoog, D. A., and H. D. Dubois, *Anal. Chem.*, **21**, 1528 (1949).
332. Smith, J. H. C., *J. Biol. Chem.*, **96**, 35 (1932); through reference 271.
333. Snyder, L. R., *Anal. Chem.*, **34**, 771 (1962).
334. Snyder, L. R., *Anal. Chem.*, **37**, 149 (1965).
335. Sokol'skaya, A. M., and S. M. Reshetnikov, *Izv. Akad. Nauk Kaz. SSR, Ser. Khim. Nauk*, **14**, 25 (1964); through *Chem. Abstr.*, **62**, 5162 (1965).
336. Sokol'skaya, A. M., and S. M. Reshetnikov, *Vestn. Akad. Nauk Kaz. SSR*, **20**, 50 (1964); through *Chem. Abstr.*, **61**, 9373 (1964).

337. Sokol'skaya, A. M., S. M. Reshetnikov, and D. V. Sokol'skii, *Dokl. Akad. Nauk SSSR*, **152,** 1369 (1963); through *Chem. Abstr.*, **60,** 1553 (1964).
338. Spasskova, A. I., *Vestn. Leningr. Univ.*, **18** (22), *Ser. Fiz. i Khim.*, **4,** 146 (1963); through *Chem. Abstr.*, **60,** 10893 (1964).
339. Spence, J. A., and Vahrman, M., *Analyst*, **91,** 324 (1966).
339a. Stahl, E., *Thin Layer Chromatography*, Academic Press, New York, 1965.
339b. Stehling, F. C., and K. W. Bartz, *Anal. Chem.*, **38,** 1467 (1966).
340. Stone, K. G., *Determination of Organic Compounds*, McGraw-Hill, New York, 1956.
341. Strain, H. H., *Chromatographic Adsorption Analysis*, Interscience, New York, 1942.
342. Sverdlov, L. M., M. G. Borisov, Yu. V. Kloch-kovskiĭ, E. P. Krainov, V. S. Kukina, and N. V. Tarasova, *Izvest. Akad. Nauk SSSR, Ser. Fiz.*, **22,** 1023 (1958); through *Chem. Abstr.*, **53,** 859 (1959).
343. Swan, J. D., *Anal. Chem.*, **26,** 878 (1954).
344. Sweetser, P. B., and C. E. Bricker, *Anal. Chem.*, **24,** 1107 (1952).
345. Swern, D., *Chem. Rev.*, **45,** 1 (1949).
346. Szonyi, C., R. S. Tait, and J. D. Craske, *J. Am. Oil Chemists' Soc.*, **39,** 276 (1962).
347. Thompson, H. W., and P. Torkington, *Trans. Faraday Soc.*, **41,** 246 (1945); *Proc. Roy. Soc. (London)*, **A184,** 3, 21 (1945).
348. Toromanoff, E., *Bull. Soc. Chim. France*, **1962,** 1190.
349. Tropsch, H., and W. J. Mattox, *Ind. Eng. Chem., Anal. Ed.*, **6,** 104 (1934).
350. Tulloch, A. P., and B. M. Craig, *J. Am. Oil Chemists' Soc.*, **41,** 322 (1964).
351. Turner, N. C., *Oil Gas J.*, **41,** 48 (1943); through reference 271.
352. Turner, N. C., *Petroleum Refiner*, **22,** 140 (1943); through reference 271.
353. Ucciani, E., J. Pasero, and M. Naudet, *Bull. Soc. Chim. France*, **1962,** 1209.
354. Uhrig, K., and H. Levin, *Ind. Eng. Chem., Anal. Ed.*, **13,** 90 (1941).
355. Uhrig, K., E. Lynch, and H. C. Becker, *Ind. Eng. Chem., Anal. Ed.*, **18,** 550 (1946).
356. Unger, E. H., *Anal. Chem.*, **30,** 375 (1958).
357. Vrbaški, T., and R. J. Cvetanovic, *Can. J. Chem.*, **38,** 1053 (1960).
358. Vullo, A. L., and G. Montaudo, *Gazz. Chim. Ital.*, **94,** 1043 (1964); through *Chem. Abstr.*, **62,** 7865 (1965).
359. Wagner, C. D., T. Goldstein, and E. D. Peters, *Anal. Chem.*, **19,** 103 (1947).
360. Walker, F. T., *J. Oil Colour Chemists' Assoc.*, **28,** 119 (1945); through reference 271.
361. Walker, F. T., and M. R. Mills, *J. Soc. Chem. Ind.*, **61,** 125 (1942); **62,** 106 (1943).
362. Walisch, W., *Ann. Univ. Saraviensis Sci.*, **7,** 289 (1958); through *Chem. Abstr.*, **53,** 16607 (1959).
363. Walisch, W., and M. R. F. Ashworth, *Mikrochim. Acta*, **1959,** 497.
364. Walisch, W., and J. E. Dubois, *Chem. Ber.*, **92,** 1028 (1959).
365. Warshowsky, B., P. J. Elving, and J. Mandel, *Anal. Chem.*, **19,** 161 (1947).
366. Westneat, D. E., and C. E. Day, Paper presented at 135th Meeting, ACS, Boston, Mass., April, 1959; through reference 37.
367. Whalley, C., and M. Ellison, *J. Oil Colour Chemists' Assoc.*, **35,** 596 (1952).
367a. Wheeler, O. W., *Chem. Rev.*, **59,** 629 (1959).
368. Wibaut, J. P., F. L. J. Sixma, L. W. F. Kamschmidt, and H. Boer, *Rec. Trav. Chim.*, **69,** 1355 (1950).
369. Wijs, J. J. A., *Ber.*, **31,** 750 (1898).
370. Wijs, J. J. A., *Z. Anal. Chem.*, **37,** 277 (1898).
371. Williams, G., *Trans. Faraday Soc.*, **37,** 749 (1941).
372. Wilson, G. E., *J. Inst. Petroleum*, **36,** 25 (1950).

373. Winkler, L. W., *Z. Untersuch. Nahr.-Genussm.*, **43,** 201 (1922); through reference 271.
374. Winward, A., and F. H. Garner, *J. Soc. Chem. Ind. (London)*, **69,** 147 (1950).
375. Wood, L. A., *Natl. Bur. Std., Misc. Publ.*, **M 185** (1947); through reference 271.
376. Zakupra, V. A., E. V. Lebedev, and V. T. Skylar, *Neftekhim., Akad. Nauk-Turkm. SSR,* **1963,** 150; through *Chem. Abstr.*, **60,** 13071 (1964).
377. Zechmeister, L., and L. Cholnoky, *Principles and Practice of Chromatography,* Chapman and Hall, London, 1943.
378. Zimmermann, G., H. G. Hauthal, and H. Schroedl, *Z. Chem.*, **2,** 277 (1962); through *Chem. Abstr.*, **55,** 62 (1963).
379. Zlatkis, A., and J. E. Lovelock, *Anal. Chem.*, **31,** 620 (1959).
380. Zuman, P., *Organic Polarographic Analysis,* Macmillan, New York, 1964.

Part II
Section B - 2

DETERMINATION OF ACYL GROUPS

BY ADAM S. INGLIS, *Division of Protein Chemistry, Wool Research Laboratories, CSIRO, Parkville, Australia*

Contents

I. INTRODUCTION

Definition of the acyl group as $R—\overset{\overset{\displaystyle O}{\|}}{C}—$ permits a consideration of a wide range of substances in this chapter. The determination of esters has already been covered fully in a separate chapter and only passing reference will be made to them here. The methods and discussion will be directed primarily to the acetyl group attached to either oxygen or nitrogen, but other important acyl groups such as formyl and benzoyl will be given special consideration.

The methods used for the estimation of the group on a macroscale indicate that it belongs to the more difficult of the group analyses, and the problems are increased by the adaption to a microscale because of the greater sensitivity to the prevailing errors. The main factors contributing to the difficulties in general chemical procedures are (*1*) in the great variation in stability of both the parent compound and the acylated product to the hydrolyzing reagent, (*2*) the blank value of the determination which may be variable and so large that it ruins the analysis, (*3*) the lack of precision arising not only from (*2*) but also from the steps involving quantitative separation and determination of the acyl group.

The following discussion of the acyl determination is largely selective in nature. Additional data and references can be found in other texts (14,-5a,44) while *Analytical Abstracts* and the biennial reviews in *Analytical Chemistry* are valuable sources of information for current work.

II. OCCURRENCE AND IMPORTANCE

The introduction and consequent determination of the acetyl group is commonly used to establish the presence of hydroxyl and amino groups in organic compounds. Various natural products—for example oils, fats, and plasticisers—are esters of acetic acid. But the determination is especially important for carbohydrates in which the percentages of carbon and hydrogen differ only slightly, because the introduction of the acetyl group, with its large molecular weight, significantly alters the analytical figures. More recently, the finding of acetyl groups in proteins (33–35,40) has accentuated the need for more accurate and improved determinations, since there may be only one acetyl group in a molecule of molecular weight 40,000. The conventional methods are inadequate for small amounts of protein; hence the requirements in this important field of research call for far more sensitive specific methods of determining acyl groups.

III. PROPERTIES OF THE ACYL GROUP

The simple conversions of acyl groups to acids by hydrolysis and to esters by alcoholic acid hydrolysis make the physical and chemical properties of acids and esters analytically important. These will therefore be considered along with useful properties of the acyl group.

A. PHYSICAL PROPERTIES

The solubility of substances containing acyl groups is generally lower than the parent hydroxyl or amino compound. This property may be utilized in extraction procedures with simple molecules but it becomes less important with increasing complexity of the molecule.

Some physical properties of formic, acetic, propionic, and benzoic acids and their methyl and ethyl esters are given in Table I. The aliphatic acids listed there are mobile liquids with pungent odors and mix with water in all proportions, but the solubility decreases as the series is ascended. The higher acids are solids. The melting points rise as the series is ascended, but not regularly. The acid with an even number of carbon atoms melts at a higher temperature than the acid with an odd number of carbon atoms which immediately follows it, thus giving two series. The aromatic acid, benzoic acid, is a solid substance, slightly soluble in water and steam volatile.

The esters are sweet-smelling volatile liquids, the boiling point rising as either the acyl or the alkyl group increases in size.

B. CHEMICAL PROPERTIES

The introduction of an acyl group into an organic compound involves a replacement of hydrogen. In the case of phenols, this leads to a loss of acidity because the ionizable hydrogen is replaced; in the case of amines, to a decreased basicity due to induction effects.

TABLE I

Physical Properties of Acids and Esters of the Common Acyl Groups (17)

Substance	Solubility, g./100 ml., 20°C.		m.p., °C.	b.p., °C.	Acyl group response		
	H_2O	Alcohol			UV	IR	NMR
Formic acid	∞	∞	8.40	100.7	Weak	Strong	Strong
Acetic acid	∞	∞	16.6	118.1	Weak	Strong	Strong
Propionic acid	∞	∞	−22	141.1	Weak	Strong	Strong
Methyl formate	30.4	∞	−99.0	31.5	Weak	Strong	Strong
Methyl acetate	31.9	∞	−98.1	57.1	Weak	Strong	Strong
Methyl propionate	65	∞	−87.5	79.9	Weak	Strong	Strong
Ethyl formate	11.8[a]	s[b]	−80.5	54.3	Weak	Strong	Strong
Ethyl acetate	8.6	∞	−83.6	77.15	Weak	Strong	Strong
Ethyl propionate	2.4	∞	−73.9	99.10	Weak	Strong	Strong
Benzoic acid	0.27[c]	47.1[d]	122	249	Strong	Strong	Strong
Methyl benzoate	0.0157[e]	∞	−12.5	199.6	Strong	Strong	Strong
Ethyl benzoate	0.08	s	−34.6	212.6	Strong	Strong	Strong

[a] 25°C. [b] Soluble. [c] 18°. [d] 15°. [e] 30°.

Distinction between an acetyl attached to carbon and one attached to either oxygen or nitrogen may be made with the iodoform reaction:

$$CH_3COCH_3 + 3I_2 + 4NaOH \rightarrow CHI_3 + CH_3COONa + 3NaI + H_2O \quad (1)$$

Either acid or alkaline hydrolysis is satisfactory for breaking an acyl bond to either oxygen or nitrogen. Because the acylate ion may be determined readily, the hydrolysis procedure has been largely used as the first step in the determination.

$$R'COOR \xrightarrow{HOH} R'COOH + ROH \quad (2)$$

or

$$R'CONHR \xrightarrow{HOH} R'COOH + RNH_2 \quad (3)$$

When the substance is heated in an alcoholic acid solution an ester is formed.

$$R'COOR + C_2H_5OH \xrightarrow{HCl} R'COOC_2H_5 + ROH \quad (4)$$

The latter may be determined by hydrolysis with standard alkali and back-titration with standard acid; hence this is also an analytically useful reaction. The formation of the sweet-smelling ester may be used as a qualitative test for the presence of an acyl group.

In the well-known Kuhn-Roth determination (29), chromic acid oxidizes acyl groups containing C–methyl groups to acetic acid, all others to carbon dioxide and water.

Other reactions could possibly be of interest for analytical purposes. Nitrous acid produces nitrogen, as well as acetic acid, with amides.

$$CH_3CONH_2 + HNO_2 \rightarrow N_2 + CH_3COOH + H_2O \quad (5)$$

The silver salts of acids can be decarboxylated with bromine as follows:

$$RCOOAg + Br_2 \rightarrow RBr + AgBr + CO_2 \quad (6)$$

IV. SEPARATION AND CONVERSION

In general, acyl–oxygen cleavage is effected readily by either acid or alkaline reagents to yield the acylate ion and a hydroxyl or amino compound. In some cases the acylate ion can be determined directly as the acid, but in general procedures the latter is often separated (after the mixture has been acidified) by distillation. The mechanisms for acid and alkaline hydrolyses differ. The common ones are outlined below; for detailed accounts of these, and for alternative possibilities, such as those involving alkyl–oxygen bond cleavages, readers are referred to standard texts (12a,23) on the subject.

A. FISSION BY ALKALI

In alkaline solution the neutral ester molecule R′COOR undergoes reaction. Acyl–oxygen fission is second order and may be represented by the following equations:

$$HO^- + R'\text{—}\overset{\displaystyle O}{\overset{\|}{C}}\text{—}OR \xrightarrow[\text{slow}]{} HO\text{—}\overset{\displaystyle O^-}{\overset{|}{C}}(R')\text{—}OR \quad (7)$$

$$HO\text{—}\overset{\displaystyle O^-}{\overset{|}{C}}(R')\text{—}OR \xrightarrow[\text{fast}]{} HO\text{—}\overset{\displaystyle O}{\overset{\|}{C}}\text{—}R' + O^-R \quad (8)$$

The basic hydrolysis of amides follows similar steps.

The hydrolysis is retarded by electropositive, and accelerated by electronegative substituents, irrespective of whether these are located on the alkyl or acyl position of the esters. This is exemplified by the following relative saponification rates:

$\frac{k}{k_{CH_3COOMe}}$	CH_3COOMe	$CHCl_2COOMe$	CH_3COOEt
	1.0	16,000	0.60

The reaction of acetates is faster than benzoates, however, because in benzoates the electron-attracting phenyl group is conjugated with the carbonyl group and it may donate π-electron density to it:

The reaction is also sterically retarded by substituents close to the reaction center:

$\frac{k}{k_{CH_3CH_2OAc}}$	CH_3CH_2OAc	$(CH_3)_3CCH_2OAc$	$(C_2H_5)_3CCH_2OAc$
	1.0	0.18	0.031

Alkaline hydrolysis generally proceeds faster than acid hydrolysis (29) although this depends to some extent upon the reagent used (22) and is not always so for *N*-acyl links. The latter are usually more resistant than *O*-acyl groups.

B. FISSION BY ACID

The bimolecular mechanism is also the usual mechanism for acid hydrolysis although a unimolecular reaction occurs in concentrated sulfuric acid where acylium ions are stable. In either case the ionic conjugate acid of the ester or amide undergoes reaction as follows:

Unimolecular

$$R'COOR + H^+ \xrightarrow[\text{fast}]{} R'CO\overset{+}{O}HR \xrightarrow[\text{slow}]{} R'\overset{+}{C}O + HOR \tag{9}$$

$$R'\overset{+}{C}O + OH_2 \xrightarrow[\text{fast}]{} R'CO\overset{+}{O}H_2 \xrightarrow[\text{fast}]{} R'COOH + H^+ \tag{10}$$

Bimolecular

$$R'COOR + H^+ \xrightarrow[\text{fast}]{} R'CO\overset{+}{O}HR \tag{11}$$

$$R'CO\overset{+}{O}HR + OH_2 \xrightarrow[\text{slow}]{} R'CO\overset{+}{O}H_2 + HOR \tag{12}$$

$$R'\overset{+}{C}OOH_2 \xrightarrow[\text{fast}]{} R'COOH + H^+ \tag{13}$$

The bimolecular reaction in acid solution is subject to steric retardation but, in contrast with the reaction in alkaline solution, polar influences are of minor importance. In very acidic media electron-attracting groups would be expected to accelerate the reaction. The stepwise reactions shown above are actually reversible, the reverse reactions being equally applicable to acid-catalyzed esterification.

V. DETECTION AND IDENTIFICATION

A. PHYSICAL METHODS

The spectrometric techniques of mass spectrometry, nuclear magnetic resonance, and infrared and ultraviolet absorption have been shown to be a very powerful combination for the determination of structure in organic compounds (41)(52a). One instrument alone may not give unequivocal evidence for the presence of a functional group but this evidence is often strengthened by that from another technique. The mass spectrum allows an empirical formula to be calculated, as well as providing a fragmentation pattern which could, for example, suggest that cleavages of acyl and methoxyl groups had occurred during the analysis. The change in a spectrum on removal of the functional group from, or on incorporation in, the parent substance usually offers confirmatory evidence for a particular structure unit. All these spectrometric techniques can also provide much additional information about the substance as a whole. Besides the texts on spectrometric techniques mentioned above, those of Bellamy (2)(2a) and Jackman (24) on infrared and nuclear magnetic resonance spectroscopy, respectively, have been particularly useful for the following discussion.

1. Ultraviolet Absorption Spectroscopy

The low-intensity band of the carbonyl group is profoundly affected by the presence of a neighboring oxygen atom on the carbonyl carbon as in acids and esters (12). A hypsochromic shift of about 70 mμ occurs for simple carboxylic esters. Ethyl acetate has a low intensity band (ϵ_{max} 60) at 204 mμ which is within the range of modern spectrophotometric instruments. In addition, the intensity of the band is usually higher than that of a ketone (ϵ_{max} 15 to 30). However, these weak bands could easily be obscured by other more active chromophores in this region. Indeed, the aromatic acyl groups show strong benzenoid absorption bands here. Benzoic acid, for example, has one strong band at 230 mμ (ϵ_{max} 10,000) and another weaker band at 270 mμ (ϵ_{max} 800).

2. Infrared Spectroscopy

Formates and acetates may be distinguished by their infrared spectra. The formate C:O band is in the region 1730 to 1715 cm.$^{-1}$ (5.78–5.8 μ) (as are those of α,β-unsaturated and acyl esters) while saturated aliphatic esters exhibit a C:O band at 1740 cm.$^{-1}$ (5.75 μ). The strong bands due to C:O stretching vibrations are also different, formates absorbing at about 1190 cm.$^{-1}$ (8.40 μ) and acetates at 1245 cm.$^{-1}$ (8.10 μ). The frequency of the C:O stretching band seems to be stabilized by the C:O group in esters. The C:O stretching modes of benzoates are slightly different from those of the aliphatic esters. Strong and usually recognizable bands occur at 1280 cm.$^{-1}$ (7.81 μ) and 1120 cm.$^{-1}$ (8.93 μ).

These characteristic absorption bands of esters can generally be distinguished from ketones. The C:O absorption of the former is usually quite strong by comparison and the C:O frequency is often higher than that of normal ketones due to the influence of the adjoining oxygen atom.

When the carbonyl group is attached to nitrogen, as in acylated amines, the bands due to C:O occur near 1660 cm.$^{-1}$ (6.02 μ) (amide I) and other amide bands characteristic of secondary amides appear. The latter show a single NH stretching mode at about 3300 cm.$^{-1}$ (3.03 μ) while there is a strong band at about 1550 cm.$^{-1}$ (6.45 μ) (amide II) and a weaker one near 1300 cm.$^{-1}$ (7.69 μ) (amide III). The frequencies of the bands alter with changes in state of the substance. Bellamy (2a) now finds it more useful to list tables showing real frequencies for specific compounds rather than charts indicating the frequency range of a class.

3. Nuclear Magnetic Resonance Spectroscopy

Addition of a CAT (47) to a nuclear magnetic resonance spectrometer has greatly enhanced its sensitivity. Concentrations of 0.5 mg./ml. in volumes as small as 50 μl. may be suitable for analysis. Since many structural features are more distinctive in nuclear magnetic resonance than in ultraviolet or infrared spectroscopy, the technique has proved to be a very valuable research tool. The charts of proton chemical shifts provided by Bovey (3a) and catalogs of spectra (46,48) are very useful in distinguishing groups and structures affecting proton shifts. However, it should be remembered that all correlation tables must be treated with reserve since certain groups, particularly aromatic nuclei, can cause anomalous shielding effects. Integration of the peak area gives a measure of the number of protons associated with a particular resonance and hence, of the number of groups involved.

Nuclear magnetic resonance provides a most convenient method of detecting the aldehyde function because of the unique line position (0.0–0.7 τ, $\tau = 10 - \delta$ p.p.m.) of the aldehydic proton. Formates may therefore be identified readily. Acetates also give characteristic spectra. The strong signal at about 9.1 τ usually given by the protons of the methyl group in saturated hydrocarbons is shifted to lower frequencies (7.8–8.0 τ) in acetates. This is a very sharp singlet.

Acylation and benzoylation of hydroxyl groups in primary alcohols produce chemical shifts of about 0.5 p.p.m. in protons on the same carbon atom; in secondary alcohols the α-proton shifts about 1.0 to 1.15 p.p.m. to lower frequencies. Benzoylation is not always consistent in this, since the phenyl group sometimes changes the shielding of other protons in the molecule. N-Acylation also produces a paramagnetic shift and N-acyl groups give sharp lines around 7.8 τ.

B. CHEMICAL METHODS

For more detailed accounts of the following tests, or for additional information, the reader is referred to the books by Cheronis and Entrikin (5) and Feigl (8).

1. Esters and Acyl Halides

Small amounts of esters and acyl halides can be determined colorimetrically using the hydroxamic acid reaction with ferric ion (5,8). The compounds must first be converted to the hydroxamic acid by heating with a saturated alcoholic hydroxylamine hydrochloride solution made alkaline with saturated alcoholic potassium hydroxide. The solution is acidified with 0.5N hydrochloric acid and a violet-red colored complex is produced upon addition of a 1% ferric chloride solution. Limits of detection for ethyl formate and ethyl acetate are 11 and 2γ, respectively.

$$RCOOR' + NH_2OH \rightarrow RCO(NHOH) + R'OH \qquad (14)$$

$$3RCO(NHOH) + FeCl_3 \rightarrow 3HCl + Fe[RCO(NHO)]_3 \qquad (15)$$

However, lactones undergo a similar reaction.

2. Formic Acid

Formic acid (1.4 γ) can be detected by the chromotropic acid (1,8-dihydroxynaphthalene-3,6-disulphonic acid) test after reduction with magnesium and 2N hydrochloric acid to formaldehyde (8). The latter gives a pink color when heated on a water bath for 10 minutes at 60° with 12N sulfuric acid and a small amount of solid chromotropic acid.

3. Acetaldehyde

The test for acetaldehyde is often used as an indirect test for other groups such as O- and N-methyl groups after an appropriate chemical reaction. A drop of a solution containing 1 γ of acetaldehyde gives a blue color when mixed with a freshly prepared mixture of equal volumes of 20% morpholine (or piperidine) and 5% sodium nitroprusside solutions (8).

Acrolein and crotonaldehyde interfere, as does propionaldehyde at high concentration. When detecting acetaldehyde vapor, ethanol must be absent because it is oxidized by the air to acetaldehyde.

4. Acetate in Formate

The calcium salts of the acids are formed by heating to dryness with calcium carbonate. Strong heating of this mixture produces acetaldehyde which is detected by the morpholine–sodium nitroprusside test (8).

$$(HCOO)_2Ca + (CH_3COO)_2Ca \rightarrow 2CaCO_3 + 2CH_3CHO \qquad (16)$$

Less than 0.4% acetic acid in formic acid gives a strong positive test. Alkali formates and acetates give a similar reaction on heating.

$$HCOONa + CH_3COONa \rightarrow Na_2CO_3 + CH_3CHO \qquad (17)$$

Lactic, butyric, and propionic acids also give the test but it is much less sensitive for butyric and propionic acids.

5. Alkyl and Aryl Acetates

A reliable test for the detection of 10 to 50 γ *O*-acetyl compounds is given by the pyrohydrolytic test with oxalic acid dihydrate at 110° (8). This reaction replaces warming with either acids or alkalis for hydrolysis of the groups; the water of crystallization of the oxalic acid brings about the hydrolysis on heating and also helps to volatilize the products. In this case, acetic acid is produced and is detected by indicator paper. *N*-Acyl substances do not react.

$$CH_3COOR + H_2O \rightarrow CH_3COOH + ROH \qquad (18)$$

6. Acetyl Groups in Proteins

Although the identification and detection of acetyl groups in proteins may be made by a hydrazinolysis procedure (33), in practice this is not a simple procedure because of the extremely small percentage of acetyl present. It involves enzymic digestion, conversion of the peptides to hydrazides, separation of the hydrazides by paper chromatography and identification of the acetylhydrazide by its position after spraying the paper with ammoniacal silver nitrate. A longer procedure (37), but one which has been used for detection of 0.1 μmole of aliphatic acyl groups, consists in coupling the hydrazide with 1-fluoro-2,4-dinitrobenzene. This acetyl DNP hydrazide is determined spectrophotometrically in 70% yield. Formyl groups give low yields; propionyl can be measured if cystine and cysteine are absent.

C. RADIOCHEMICAL METHOD

Acetylation of organic compounds with ^{14}C acetic anhydride facilitates the acetyl determination (3). The specific activity of the pure acetylated compound is compared with the specific activity of the ^{14}C acetic anhydride. The method is applicable to milligram quantities and has the virtues of simplicity and nondestruction of the sample.

VI. DETERMINATION

Because of its large molecular weight, determination of acetyl can be very valuable in structure elucidation, especially since results with an accuracy of $\pm$ 1% are possible for tractable substances. On the other hand, because many substances have very resistant acyl bonds, while other substances are degraded very readily and produce acyl groups, the reliability of the determination for unknown substances is generally open to doubt. It is very important, therefore, to obtain as much information as possible about the substance before beginning this analysis. Factors such as the general class of compound, lability of the acyl link, and the possibility of formation of acid groups other than acetic acid all help in selecting the most appropriate procedure for the determination.

When sufficient sample is available, the use of alternative procedures—for example, one involving acid and one involving alkaline hydrolysis—often indicates whether results are reliable; if they are not, when taken in conjunction with other evidence, the value more likely to be correct may be obtained. If both procedures appear to give anomalous values, then the alternative to using another procedure is to analyze the deacetylated substance and to ascertain whether it also gives anomalous results. With the variety of reagents and procedures available, a satisfactory determination should always be possible, providing there is sufficient material available.

A. CLEAVAGE CONDITIONS

1. Alkaline Hydrolysis

Hydrolysis of the acyl group may be brought about by either aqueous or methanolic sodium hydroxide solution (13). When the latter is used the methyl alcohol is distilled prior to acidification of the solution and liberation of the free acid. It was found to be more effective than acid hydrolysis (29). Thus, for acetyl groups attached to oxygen the substance was hydrolyzed for 20 minutes at 100° with an acid reagent but only 5 minutes with an alkaline reagent. Benzoyl groups, or acetyl groups linked to nitrogen often require prolonged reaction times with acid and are accordingly hydrolyzed with alkali. Acid hydrolysis is not effective for triacetylmonobenzoylglucose which requires 30 minutes hydrolysis in methanolic sodium hydroxide; aqueous alkali also gives low results after 2.5 hours hydrolysis with this substance. Similarly, the methyl ester of triacetylcholic acid is difficult to hydrolyze and requires 2.5 hours with methanolic sodium hydroxide. Replacing ethanol with the higher boiling butanol (1b), or increasing the concentration of alkali (22) gives a more effective reagent for fission of N-acetyl groups. Ma (32) saponifies with approximately 0.3 M sodium hydroxide in n-amyl alcohol in a sealed tube at 150°.

Cold dilute alkali has been used for substances such as sugars which decompose readily to give acid products (1,30,53). Borate solutions at pH 9–10 (15) and alkaline hydroxylamine (16) have been used for determination of O-acetyl groups and are satisfactory where the *N*-acetyl group is fairly resistant. Formyl groups interfere here because they are also very readily removed with alkali and can be cleaved in the cold at pH 10. These have been determined in the presence of acetyl groups after hot alkaline hydrolysis (27) using mercuric chloride to reduce the formic acid and titrating the mercurous chloride iodometrically (18).

The big drawback to hot alkaline reagents is that they often degrade the substance as well as deacylating it. This is known for carbohydrates, catechins (9), and proteins (34), although the latter could be a special case in this regard. With γ-pyrones the ring may be broken by 1% potassium hydroxide and kellin, for example, would liberate one molecule of acetic acid.

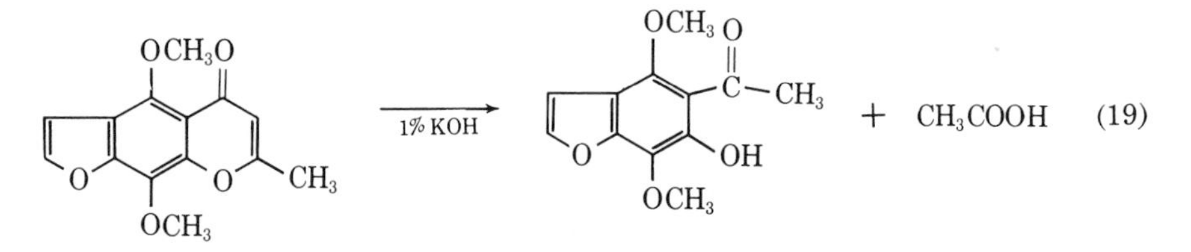

2. Acid Hydrolysis

Acid reagents are usually considered to require longer times for hydrolysis than alkaline reagents (see above) but this is not always the case. Acetanilide and phenacetin decompose readily in Wenzel's sulfuric acid but in the same period in 1*N* sodium hydroxide in 50% methanol the hydrolysis is only half complete (22). This was established by steam distillation of acetic acid and suggested that the conclusions reached by earlier workers were influenced by the mode of separation of the acetic acid liberated by the reagents. Since acetic acid was slowly distilled from a concentrated acid solution, substances could, in fact, be acid hydrolyzed during the process. Recently Tabib et. al. (45a) demonstrated this effect by analyzing a series of compounds employing alkaline hydrolysis followed by acid hydrolysis for different periods prior to distillation. In addition, their results indicated that N-acetyl groups which are attached to an alicyclic ring or to part of a heterocyclic structure were the slowest to hydrolyze in 50 per cent methanolic IN sodium hydroxide.

The fact that acetic acid is generally distilled from acid solution has meant that nonvolatile acids such as phosphoric acid (50%), sulfuric acid (33%) and *p*-toluene sulfonic acid (25%) have been employed for acid hydrolysis. *p*-Toluene sulfonic acid is the reagent that was used originally and its use has persisted over the years (7,44). It is a gentle reagent and thus not as liable to give side products which could contribute to the results. With fructose, which decomposes extremely readily to give volatile acid products, *p*-toluenesulfonic acid gives an "acetyl" value of 2.3% using the Wiesenberger procedure (51), whereas phosphoric acid produces 14% and sulfuric acid 16% for the same period of hydrolysis (21). On the other hand, it hydrolyzes more slowly and very resistant acyl links may need prolonged digestion times for quantitative fission. Quantitative removal of acetic acid from *p*-toluenesulfonic acid (21) is also slower.

3. Chromic Acid Oxidation

The oxidation mixture (29) of 5*N* chromic acid and 36*N* sulfuric acid (ratio 4:1) can be especially useful in this determination, although it has found more extensive application in the determination of C-methyl groups (13). Since any C-methyl group will give at least some acetic acid, the C-methyl value of the deacetylated substance must be known, which is a disadvantage if material is scarce, but which is also useful for elucidation of structure. In the absence of other C-methyl groups, chromic acid oxidation gives a reliable specific method for acetyl, both for sensitive substances such as sugars, and for those with resistant acyl bonds, because it oxidizes the rest of the molecule to carbon dioxide and water, thereby eliminating groups such as formyl and benzoyl.

Chromic acid oxidation has been used as an adjunct to an acetyl determination involving either acid or alkaline hydrolysis (52). The distillation unit is designed (Fig. 1) so that the volatile acids must pass through a boiling chromic acid solution before reaching the receiver. Interference from C-methyl groups in the substance is largely eliminated, while acids such as formic and salicylic are oxidized by the chromic acid solution.

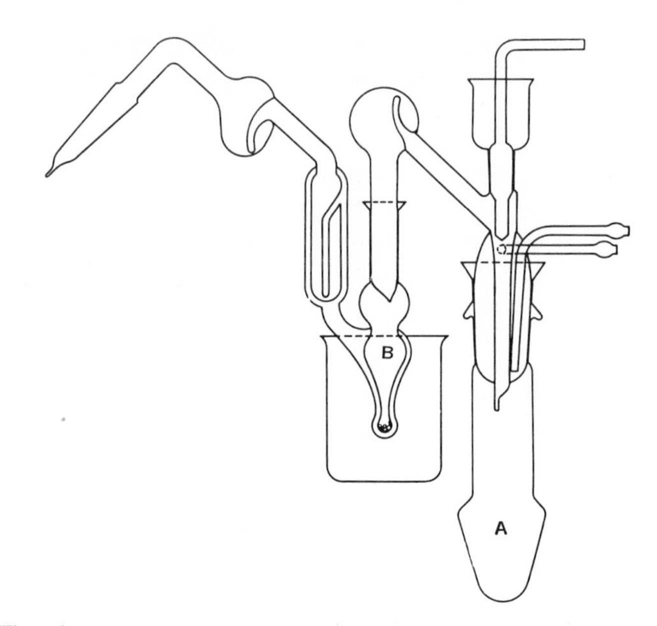

Fig. 1. Wiesenberger (52) acetyl distillation unit. Volatile vapors in A are passed through a boiling chromic acid trap B prior to entering the condenser (not shown).

4. Alcoholic Acid Hydrolysis

The transesterification procedure was developed especially for compounds that were difficult to analyze, in particular acetylated catechins and carbohydrates which may form volatile acids upon hydrolysis with aqueous acid or alkaline reagents (9). Hydrolysis by *p*-toluenesulfonic acid in ethyl alcohol was followed by quantitative distillation of the ester—the latter being effected by repeated addition and distillation of ethyl alcohol. The ester was then determined by hydrolyzing with an excess of standard alkali and back-titrating with a standard acid solution.

The reaction takes place in boiling alcohol, rather than in boiling water, so there is less danger of either decomposition of the hydrolyzed product or of formation of sulfurous acid from the reagent. On the other hand, it is a long procedure—the initial transesterification alone requires 5 to 6 hours for *N*-acetyl groups—and is not suitable for micro quantities because quantitative hydrolysis of ethyl acetate requires a relatively concentrated standard alkaline solution ($0.05N$); the back-titration would only differ by 0.2 ml. $0.05N$ solution from the initial volume of standard alkaline solution if 5 mg. of a substance containing 10% acetyl were analyzed. Wiesenberger (50) overcame this drawback by incorporating an ion-exchange step after the hydrolysis. He was then able to use a much stronger alkaline solution ($0.5N$) since the sodium ions were subsequently exchanged for hydrogen ions and the resulting solution of acetic acid was determined directly with $0.01N$ sodium hydroxide solution. Later he concentrated on more versatile and quicker procedures involving aqueous solutions (51,52).

Interest in this method of cleavage of the acyl group has been revived with the use of alternative methods for ester determination. The latter are much more sensitive and are not affected if the acid is volatile—which simplifies the procedure for quantitative distillation. Accordingly, colorimetric (31) and gas chromatographic (42) determinations of the ester have been developed after reaction with methanolic hydrochloric acid reagents. Indeed, distillation has actually been omitted in the gas chromatographic procedure, and this gives a very simple method since digestion may be carried out in a sealed tube and an aliquot of the reaction mixture injected into the gas chromatograph. Initially the method requires the determination of the areas of the methanol and methyl ester peaks given by known weights of a standard substance and the reagent in order to get a factor which is then used for unknown substances. However, the slow hydrolysis of *N*-acyl bonds is still a deficiency of the method. Although a $2N$ methanolic acid reagent is usually satisfactory, a $3.5N$ reagent is required for *S*-tribromoacetanilide (31), while acetyldiphenylamine does not give quantitative results (43).

B. ISOLATION OF THE ACYL GROUP

1. Ion Exchange

After separation, isolation of the acyl group by means of a strongly acid cation exchanger provides the most general method for determination of the group. Moreover, the apparatus and technique required are very simple and the method is rapid (5a). After hydrolysis in alcoholic alkali, which gives the sodium salt of the acid, the solution is passed down a sulfonic acid-type resin column and the effluent is titrated with standard sodium hydroxide solution to the phenolphthalein end point. For pale yellow solutions a mixture of cresol red and thymol blue, which gives a color change from yellow to violet, is better.

$$\text{R—COONa} + \text{Resin—SO}_3\text{H} \rightarrow \text{RCOOH} + \text{Resin—SO}_3\text{Na} \qquad (20)$$

All acids produced upon reaction of the substance with the alkaline reagent will be determined by this procedure.

With sulfonated phenol-formaldehyde resins, which absorb weak acids (38a), acetic acid can be determined specifically. Fructose tetraacetate and acetyl salicylic acid both gave the expected values for acetic acid after acid hydrolysis followed by distillation of the volatile acids and separation on a phenol-formaldehyde resin column (22a). Goodman et al. (12a) have recently demonstrated the determination of organic acids in mixtures by separation on Amberlite 120 and continuous monitoring of the eluate with a conductometric micro-cell.

2. Distillation

Separation of the acyl group as the acid by distillation from a nonvolatile acid solution largely overcomes difficulties due to colored substances and other acidic functions in the substance, yet at the same time is applicable to the most pertinent acyl groups such as formyl, acetyl and benzoyl. However, quantitative removal of small amounts of acid has always been a problem and various methods of distillation have been tried to achieve this end while ensuring that only the volatile acid is distilled.

a. Vacuum Distillation

The procedure originally used for acetyl determinations consisted of distillation of acetic acid at reduced pressure after hydrolysis with a solution of *p*-toluenesulfonic acid (38), and, in a modified form (44) it is still recommended for micro determinations. On the macro scale it has also been used for separating acetic acid from protein hydrolyzates (15). In this application the acetic acid was distilled from either sulfuric acid or citric acid solutions. The technique for proteins is not suitable for micro scale work and the distillate gives a variable blank value (34).

The procedure must be carried out with care, otherwise losses of acid can occur. The modifications of Elek and Harte (7) to the vacuum distillation procedure improved its reliability and their apparatus (Fig. 2) is normally used with this procedure (44). The delivery end of the condenser has a sintered glass plate sealed in to ensure efficient absorption of the distillate in the receiving flask. After the hydrolysis period the hydrolyzate is cooled and the tap on *d* opened and connected to the vacuum line. With the pressure maintained at 50 to 60 mm. pressure, the flask is raised so that the sintered disk is close to the bottom of the flask; the water surrounding reaction flask *a* is brought to boiling and the quantitative removal of acetic acid is effected by alternatively heating to dryness and carefully adding small quantities of water via *b*. Losses can occur on the glass rods in *a*, while, poor technique will result in sulfonic acid being carried into the condenser.

Bradbury (4) found that there were small compensating errors in this procedure. On the one hand there was a loss of acetic acid during the dis-

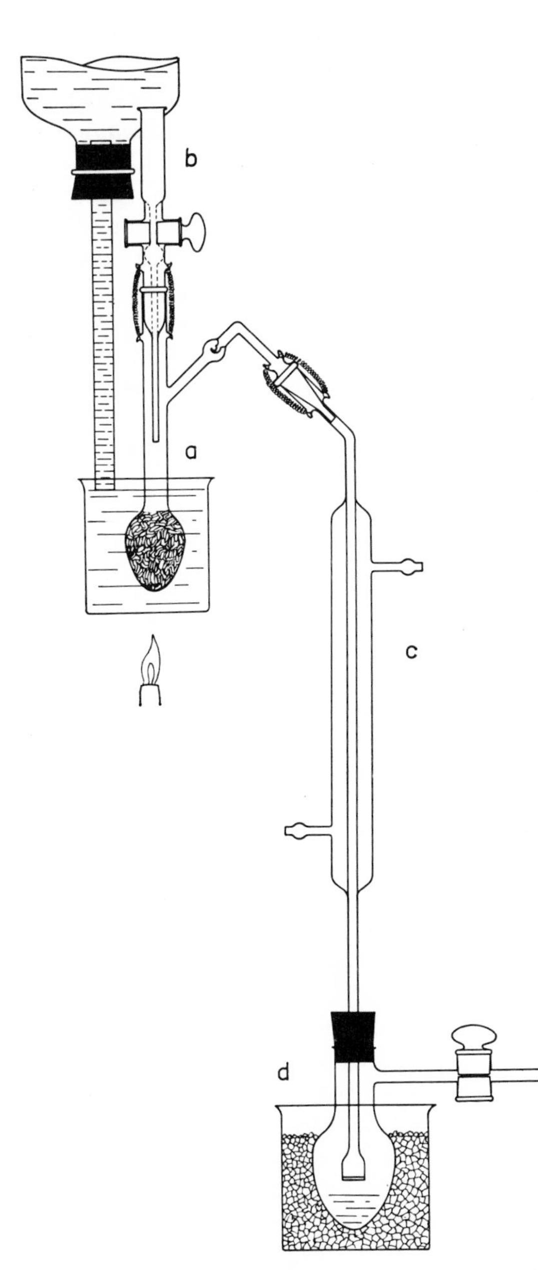

Fig. 2. Elek and Harte (7) vacuum distillation unit for acyl determinations.

tillation; on the other there was sulfate in the distillate, as well as sulfite, and only the latter was included in the blank correction. His results also showed that it was preferable to use solid *p*-toluenesulfonic acid for each determination and to titrate the acetic acid alkalimetrically instead of iodometrically.

Vacuum distillation has been recommended for a transesterification procedure with a methanolic hydrochloric acid reagent (31) but in this case the distillation conditions are not as exacting because hydrochloric acid does not interfere with the determination (colorimetric) of the ester. The apparatus simply consists of a distillation flask connected to a side-arm test tube immersed in a solid CO_2–methanol mixture.

With cryogenic vacuum distillation (40a), microgram amounts of acetic acid can be rapidly and quantitatively removed from small volumes of the hydrolyzate in hydrochloric acid. The acetic acid may be specifically determined by gas chromatography.

b. Distillation at Atmospheric Pressure

Kuhn and Roth (29) developed a versatile method of acyl determination in which the volatile acid is distilled from either sulfuric acid solution or chromic acid solution. The apparatus is simple (Fig. 3). Either acid, alkaline, or alcoholic alkaline hydrolysis conditions may be used. The reaction mixture just prior to distillation usually contains 1 ml. 12*N* sulfuric acid, 1 ml. 5*N* sodium hydroxide, and 5 ml. distilled water. The rate of distillation is maintained at about 1 ml./minute with a slow stream of oxygen passing through to prevent superheating. When the volume has been reduced to 2 to 3 ml., distilled water (5 ml.) is added through *B* by raising the stopper *S* and the process is repeated until 40 ml. (collected in portions of 20, 15, 5, and 5 ml.) has been distilled. This process requires considerable attention and it is recommended that crystals of barium chloride are added to the distillate to ensure that sulfuric acid has not been entrained in the distillate.

A much more reliable apparatus is shown in Fig. 4 and was designed specifically to prevent contamination of acetic acid with high boiling acids. The two bubbler-type traps (4 and 5) in turn condense the vapor from the boiling liquid in *DK* before it finally passes into the condenser. The distillation procedure is similar to the Kuhn-Roth procedure except that 2-ml. portions of water are added via *3* and only one titration is made on 30 ml. of distillate. When the methods were compared (21), the Wiesenberger procedure was not only faster but it gave a smaller blank value (0.14 ml. of 0.01*N*sodium hydroxide compared with 0.31 ml.).

Table II shows that both methods can give reproducible analyses. The results in column 6 were obtained without subtracting the blank values,

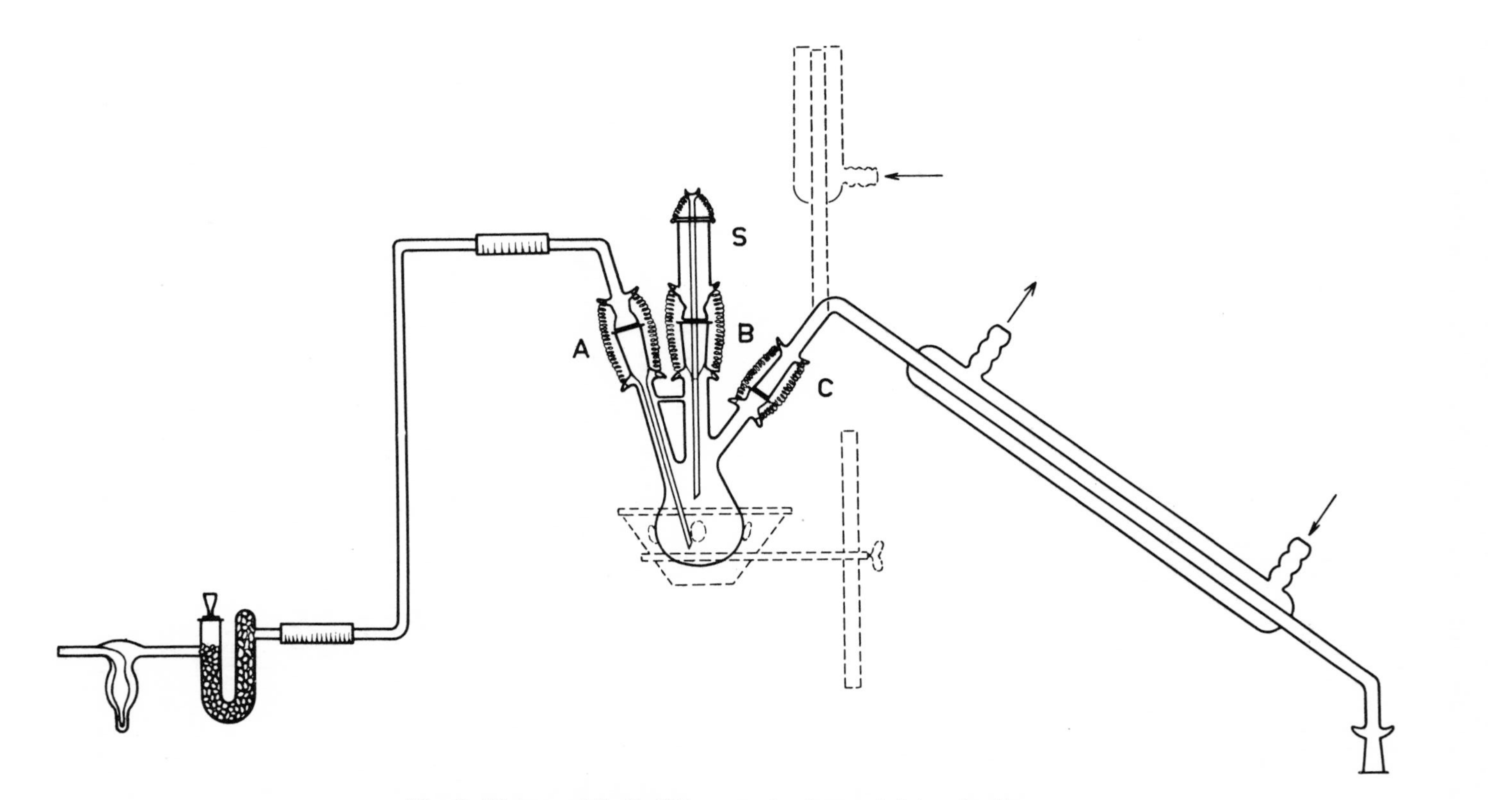

Fig. 3. Kuhn and Roth (29) apparatus for acyl determinations.

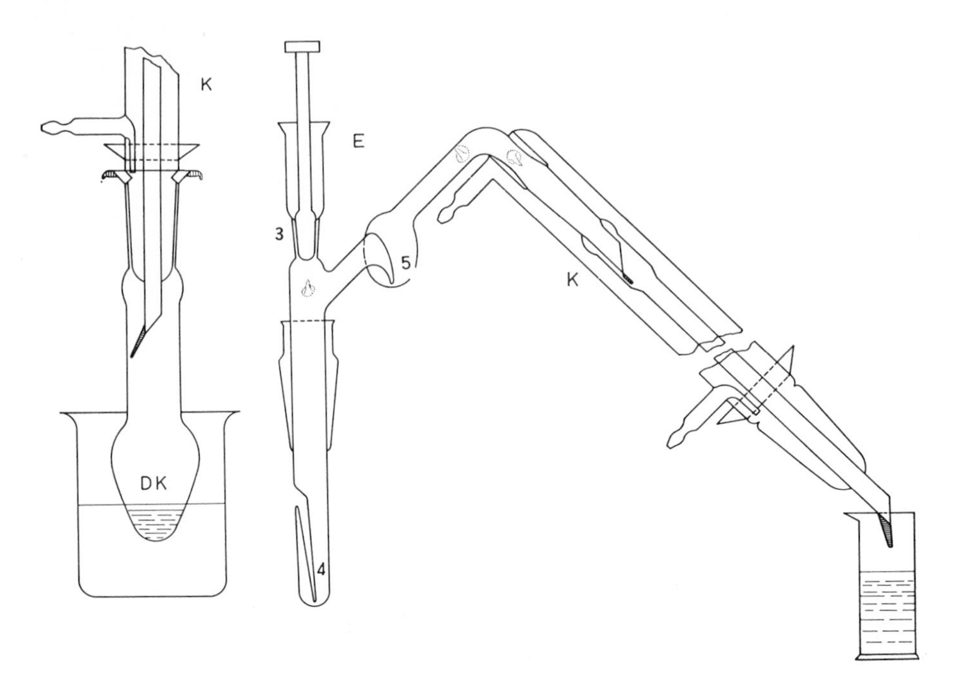

Fig. 4. Wiesenberger (51) apparatus for acyl determinations.

which is in accord with the recommended procedure for the Kuhn-Roth method (13). If blank values had been subtracted then all sets of results would have been low. This raises an elementary but critical consideration if the highest accuracy is to be obtained with the micro acetyl determination. As recommended, the procedure used for standardizing the sodium hydroxide consisted of adding 20 ml. of water (the volume of the Kuhn-Roth first distillation) to the standard acid aliquot and titrating as in an analysis. But if one titrates in this manner there is generally a blank value due to the distilled water. Similarly, a distillate gives a blank value which is dependent on the distillation volume and due at least in part to the effect of carbon dioxide on the titration. The factor for the sodium hydroxide should therefore be calculated only after subtracting this blank value due to the distilled water. Direct subtraction of the blank value of the distillate from the analysis titre can then be made without introducing errors. The results in column 7 were calculated in this way.

Attempts to reduce both the time of distillation and the dilution of the ester by only distilling a fraction of it led to a decrease in precision (42). The Wiesenberger apparatus (52) was scaled down and found suitable for distilling submicro quantities of acetic acid (1a).

TABLE II
Acetyl Results for Kuhn-Roth and Wiesenberger Methods

Method	Weight, mg.	Distil-late, ml.	Analysis titre,[a] ml., (0.01N NaOH)	Blank titre, ml.	Found, %		Calcd., %
					(1)[a]	*(2)*[b]	
KR	4.994	20	3.54	0.17	32.6	31.2	31.8
		10	0.16	0.06			
		5	0.05	0.04			
		5	0.04	0.04			
KR	5.110	20	3.44		32.1	30.6	31.8
		10	0.22				
		5	0.11				
		5	0.04				
KR	5.298	20	3.78		32.8	31.5	31.8
		10	0.19				
		5	0.04				
		5	0.04				
W	4.576	30	3.37	0.14		31.7	31.8
	5.520	30	4.14	0.14		32.3	31.8
	4.684	30	3.47	0.14		31.9	31.8

[a] Using recommended calculation (13).
[b] Results with subtraction of all blank values (see text).

c. Steam Distillation

For a semi-microdetermination Clark (6) steam distilled the acetic acid obtained after saponifying the substance in alcoholic potassium hydroxide and acidifying with sulphuric acid containing magnesium sulfate. Distillation at constant volume (20 ml) gave a recovery of 95.7 per cent acetic acid in 50 ml of distillate. When the reaction mixture was concentrated to 15 ml by additional heating during the steam distillation, recovery of acetic acid was quantitative (6a).

Later Schöniger et. al. (39) used steam distillation for the microdetermination of acetyl groups and showed that acetic acid could be removed quantitatively from a Kjeldahl distillation unit (Fig. 5). The main disadvantage of the system is that it is difficult to keep the volume of the reaction mixture down during the distillation, which militates against removal of the last traces of acetic acid; moreover, the volume of distillate is much larger than normal and leads to increased blank values. This can

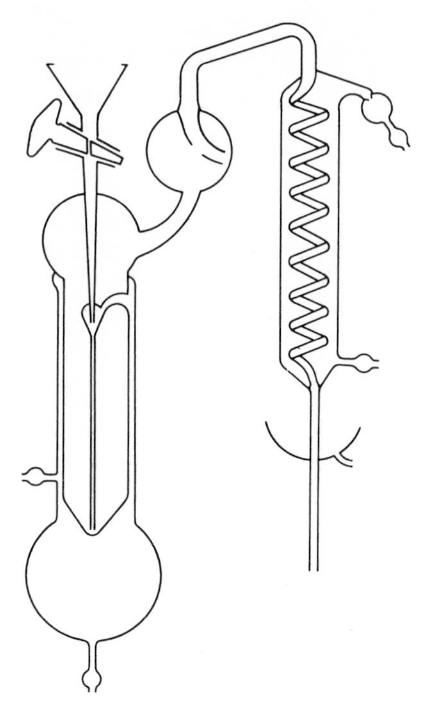

Fig. 5. Kjeldahl apparatus used by Schöniger, Lieb, and El Din Ibrahim (39) for distillation of acetic acid.

be minimized by adding saturated sodium sulfate to the mixture and distilling at a very fast rate.

Apart from adding to the usefulness of the Kjeldahl unit in the analytical laboratory, there are marked advantages of the steam distillation procedure over the distillations described in Sections VI-B-2-a and b. During the distillation, which requires only 10 to 15 minutes, no supervision is required and there is negligible risk of a blank value from the sulfuric acid in the reaction mixture; in addition, it eases the analysis of substances after hydrolysis with methanolic alkaline solutions. With other procedures (29,51) removal of methanol is often accompanied by considerable frothing, with the result that the analysis is lost or a prolonged period is required for the distillation of the methanol.

C. DETERMINATION OF THE ACYL ION

1. Alkalimetric Titration

Acetic acid has generally been determined either alkalimetrically (29) or iodometrically (10), although ethanolamine (49) and potentiometric titrations (53) have also been recommended, the latter for colored substances in particular. A comparison of these procedures was made (19) and both the alkalimetric and ethanolamine methods were recommended.

The alkalimetric determination of acetic acid presents two problems. First, as the presence of carbon dioxide will affect the titre, this must be removed from the solution without loss of acetic acid. Various methods have been used to do this. The solution was boiled for 7 to 8 seconds (10) and for 60 seconds (51). Hurka and Lieb (20) believed that a precise acetyl determination was not possible because losses of acetic acid occurred even after 5 seconds boiling but these were compensated by the incomplete removal of carbon dioxide. The technique was improved by inserting a cooling cone in the neck of the flask (26). This not only condenses the acid vapor but allows a longer period for removing carbon dioxide.

Second, the techniques for titrating weak acids vary, so the end point varies according to the technique. For example, Pregl (13) regarded the titration as complete if the pink color of phenolphthalein remained for 3 seconds after addition of a half drop of $0.01N$ alkali but Wiesenberger (51) continued the titration until the weak pink color remained for 10 to 15 seconds. Although it lengthens the procedure by a few minutes, the back-titration method of Jerie (25) gives a relatively permanent end point.

The carbon dioxide problem is accentuated if the acetic acid is separated by steam distillation because the volume of distillate is greater. The Jerie method gives a reproducible blank figure but the magnitude of the blank (22) clearly indicates the inefficiency of the boiling procedure for carbon dioxide removal. A superior procedure (54) consists in bubbling nitrogen through the solution. If the gas is passed through the distillate as it is being collected (the flask should be immersed in an ice bath to prevent loss of acetic acid) the solution can be titrated immediately after the distillation is complete. The end point is sharp and permanent as long as the gas is bubbled through the solution. This method enhances the attractiveness of the steam distillation procedure and, of course, is of general applicability.

2. Iodometric Determination

The iodometric determination of acetic and formic acids has been recommended for the acetyl and formyl determinations after hydrolysis with *p*-toluenesulfonic acid (7,44). It has been criticized on the grounds that it does not take into account any influence from either carbon dioxide or sulfuric acid, both of which may be present in the distillate (4).

Acetic acid is rather weak (Ka 1.75×10^{-5}) to be determined iodometrically and should be determined by adding excess thiosulfate to the acid-iodide-iodate solution and back-titrating with an iodine solution after 15 to 30 minutes (28). This method gives similar results to the alkalimetric method, whereas the iodometric titration method as used in the acetyl determination gives lower results for standardization of acetic acid solutions (22).

3. Colorimetric Procedure

The spot test of Feigl (8) for esters of carboxylic acids is based on their conversion to hydroxamic acids followed by the formation of a coloured complex with ferric ion (see Section V-B). This is the basis of a sensitive colorimetric determination of the esters formed by acid methanolysis of *N*-acetyl groups (31). It fills a real need in the determination, in that the usual alkaline and iodometric procedures require milligram quantities of acetyl for accurate results whereas this procedure measures decimilligram amounts. Neither hydrochloric acid nor methyl chloride interferes with the determination.

The colorimetric procedure has also been used for the determination of acetylcholine and related esters (16). The method recommended also provides, with some limitations, a rapid means of determining *O*-acetyl in the presence of *N*-acetyl groups. It depends on the rapid hydrolysis of *O*-acyl groups in alkaline hydroxylamine solution. The ferric–hydroxamic acid complex is formed after acidification and the color is read at 540 mμ. The method gives variable results for a range of esters and is not always completely specific for *O*-acetyl groups.

4. Gas Chromatography

The successful application of gas chromatography to the acyl determination (42) suggests that the older methods may soon be superseded by this modern technique. It not only fills the requirement (see Section C-3) that it is suitable for the determination of decimilligram amounts of acyl, but it also separates volatile products of the fission, thereby allowing simultaneous determination of different groups. The methyl esters of formyl, acetyl, propionyl, and trimethylacetyl groups have been determined after hydrolysis with methanolic hydrochloric acid (43). Ethanol can be used instead of methanol when the methyl esters are not resolved well. Another advantage is that the distillation step can be eliminated.

This method has the weakness that some *N*-acyl bonds are resistant to the reagent (see Section VIA-4). For analysis of biological materials there are also solubility problems. A more effective transesterification procedure, or an alternative procedure involving separation of the acids (49a) is desirable for such substances. Both solvent extraction (49a) and cryogenic distillation (40a) have been recommended for separating the acetic acid from the reaction mixture prior to gas chromatography.

D. ANALYSIS OF DIFFICULT SUBSTANCES

1. Determination of Carbohydrates

References have been made throughout the chapter to procedures that may be useful when the substance under analysis is converted to a substance which will also be determined as acyl. In some cases alkaline hydrolysis may be better, while in others, acid hydrolysis must be chosen. Acetylated carbohydrates, in particular, have been used as model compounds for workers attempting to develop general rapid methods, since sugars so readily decompose both in acid and in alkaline reagents.

Reaction in acetone solution at room temperature for 2 hours with 0.01*N* sodium hydroxide offers a satisfactory method which is largely specific for *O*-acetyl groups (1). The method is simple; the sodium hydroxide is titrated with standard acid to determine the amount consumed in the reaction. α-*N*-Methyl-L-glucosamine pentaacetate gives high values for *O*-acetyl because there is some cleavage of the *N*-acyl bond.

A great deal of work has been directed towards developing methods that are more generally applicable, and shorter reaction times at higher temperatures have been used with a variety of reagents.

Aqueous *p*-toluenesulfonic acid was suggested as the best reagent for hydrolyzing acetyl groups in carbohydrates (45), but glucose itself gave an "acetyl" value of 9% with it and the reagent was modified by half-neutralizing it with potassium hydroxide (11). The blank value of glucose was then reduced to a negligible amount while the results for glucose pentaacetate were satisfactory also. These results were not confirmed (21) using the Wiesenberger procedure with *p*-toluenesulfonic acid and aqueous 4*N* sodium hydroxide as alternative reagents. Both reagents decomposed glucose and 1.6% "acetyl" was found. Hydrolysis with 5*N* sodium hydroxide in 33% methanol followed by steam distillation is probably satisfactory for most carbohydrates (22), since fructose is easily the most susceptible of the common carbohydrates to decomposition into volatile acidic products, and the analysis of fructose tetraacetate is only 2% (relative) higher than the theoretical figure with this reagent. Results for other acetylated carbohydrates are within 1% of the calculated value.

Chromic acid oxidation is a more accurate procedure for fructose tetraacetate (22) and is the preferable reagent for carbohydrates if C-methyl groups are absent. However, care should be taken to correct for any loss of acetic acid that might occur with this reagent. Recoveries of acetic acid are slightly low when distilled from chromic acid in the Wiesenberger apparatus although quantitative from other acid reagents (21). The more recent procedures (31,42) involving methanolic hydrochloric acid have not been directed specifically to the analysis of carbohydrates. If degradation to acid products occurs only acetic acid will contribute to the gas chromatography procedure since methyl acetate is determined specifically.

2. Analysis of the Parent Substance

For many acetylated substances that do not behave ideally, it is often simpler to analyze the parent substance as well, rather than modify the method. The volatile acids derived from the parent substance can be determined and subtracted from the titre of the acetylated substance. However, one should not just make a simple subtraction of the "acetyl" value of the parent substance from the "acetyl" found for the acetylated substance. Acetylation increases the molecular weight of the compound substantially; for highly acetylated carbohydrates the weight of the parent substance produced in an analysis only corresponds to about half that of the acetylated substance, therefore the parent substance would contribute only half of its "acetyl" value to the "acetyl" value of the analysis. This is exemplified in Table III (22). The results of separate analyses for acetyl in fructose and fructose tetraacetate at first suggest that the analysis of fructose tetraacetate is unsatisfactory since the results are only 8% high, whereas fructose gives 16% "acetyl." But when the titre expected from the fructose produced in the analysis is subtracted from the analysis titre, as shown in the lower part of Table III, there is quite good agreement with the theoretical value.

If the parent substance were readily available for analysis the chromic acid oxidation procedure could be used more often. High results would arise from propionyl and other C-methyl groups rather than benzoyl and formyl groups. Information regarding the C-methyl groups in the parent substance is, of course, valuable.

3. Substances with Very Low Acyl Contents

Recently proteins have posed exacting tests for the acetyl determination. The one or two acetyl groups found per molecule demand a very sensitive specific method if they are to be detected with certainty in small samples. There is an added interest in this case because an accurate determination of acetyl assists in establishing the molecular weight of the protein.

Wool was analyzed by sulfuric acid hydrolysis followed by steam distillation only because sufficient sample was available (34). Alkaline hydrolysis gave high results. This was proved by using insulin, which has no acetyl groups, as a model compound. However, there was only a micro blank arising from a macro sample, and it is not unlikely that, in this instance, the blank is caused by incomplete removal, from the distillate, of volatile sulfur compounds produced by decomposition of cystinyl residues.

Sensitive methods of detection, such as the colorimetric and gas chromatographic procedures for esters, are more suitable for these small

TABLE III
Acetyl Results for Fructose and Fructose Tetraacetate

Reagent	Substance	Weight, mg.		NaOH, ml.	Found, %	Calc., %	Diff., %
Wenzel's sulfuric acid (f)	FTA[a]				58.0	49.4	+ 8.6
					57.4	49.4	+ 8.0
	Fructose				16.8	0.0	+16.8
					16.4	0.0	+16.4
	Correct calculation of results for FTA						
	FTA	4.252	→	5.73			
		↓					
	Fructose	2.199	→	0.85[b]			
				4.88	49.4	49.4	0.0
	FTA	6.836	→	9.12			
		↓					
	Fructose	3.536	→	1.37[b]			
				7.75	48.75	49.4	− 0.65

[a] Fructose tetraacetate.
[b] Titres due to decomposition of fructose produced by hydrolysis of 4.252 mg. and 6.836 mg. of FTA, respectively, which are then subtracted from analysis titres.

amounts of acyl, the latter also offering the possibility of simultaneous determination of different groups. The main weakness of these procedures appears to be in the alcoholysis. Methanolic hydrochloric acid is not effective in cleaving some *N*-acyl compounds (43). Gas chromatography of the acids formed by hydrolysis with 6*N* hydrochloric acid (49a) might offer advantages in such cases. Hydrolysis in 2 M HCl at 110°C for 16 hours is usually sufficient to hydrolyze acetyl groups in biological samples (40a). Excessive reaction conditions with proteins can lead to high results due to decomposition of threonine (49b).

4. Substances with Different Acyl Groups

Substances suspected of having different acyl groups in the molecule require further analysis if determined by the usual acid distillation methods. The inclusion of a trap containing the chromic acid oxidation mixture in the distillation unit (52) is effective in removing acids other than those containing C-methyl groups. Formyl and acetyl groups may be determined in the one sample after alkaline hydrolysis (27); the formyl groups are determined on the reaction mixture by the method of Hopton (18) which involves reduction of mercuric chloride by formic acid, oxidation of the resulting mercurous chloride through addition of excess iodine, and titration of unreacted iodine with sodium thiosulfate; the formyl value is subtracted from the formyl plus acetyl analysis made on the distillate. Evidence for the presence of acids other than acetic acid can be obtained if the distillate is titrated potentiometrically with sodium hydroxide, using, for example, a glass electrode as the indicator electrode in the solution which is connected via a salt bridge to a saturated calomel electrode. Formic acid (K_a 1.77 $\times$ 10^{-4}, pK_a 3.75) is stronger than acetic acid (K_a 1.75 $\times$ 10^{-5}, pK_a 4.76) and benzoic acid (K_a 6.3 $\times$ 10^{-5}, pK_a 4.20) and the pH at the inflection point is accordingly different. However, at best this is only a qualitative differentiation for mixtures of the acids since the K_a value of an acid should be 100 times larger than another for accurate potentiometric titration (30a).

Gas chromatography separates either the aliphatic esters (42) or the free acids (49a) produced on hydrolysis. Little has been done on the determination of benzoyl by this procedure. Ion exchange columns are effective for the aliphatic acids (12a).

5. Interfering Elements

The volatile hydrogen halides interfere in the acid distillation procedures. Silver sulfate is often added as a routine measure to the hydrolyzing solution (44) but this is not effective for fluorine. Cadmium sulfate has been used for preventing interference from sulfur compounds in protein analysis on the macro scale (15). Tetraacetyl-D-galactose-diethyldithioacetal gave very high results in the submicrodetermination (1a). Additions of mercuric acetate before distillation largely eliminated the interference. The colorimetric determination of the methyl esters is not affected by volatile inorganic acids (31) but the color of the hydroxamic ferric complex may not be identical from substance to substance (40).

VII. RECOMMENDED METHODS

A. ION EXCHANGE PROCEDURE

The procedure of Cheronis and Ma (5a) will be described.

Reagents

Amberlite IR-120 or an equivalent cation-exchange resin.

Hydrochloric acid 5N (approx.)

Silver nitrate 0.1*N* (approx.)

n-Amyl alcohol.

Sodium hydroxide reagent. Dissolve 3.6 g. of sodium hydroxide in 30 ml. of *n*-amyl alcohol. After standing for 2 hours at room temperature with occasional shaking, heat at 70°C until the solid dissolves.

Isopropyl alcohol–water solvent. Mix four parts of isopropyl alcohol with six parts of water.

Sodium hydroxide 0.02*N*. Standard solution.

Phenolphthalein indicator 1% in 95% ethanol.

Cresol red–thymol blue indicator. Mix 1 part of cresol red (1% aqueous solution) with 3 parts of thymol blue (1% aqueous solution).

Apparatus

An ordinary 10-ml. buret may be used for the ion exchange, that is, a column with bore 10 mm. and length 150 mm. is required. Pack the bottom of the buret with a 3-mm. layer of glass wool, prepare a slurry of 1 g. of resin with 2 ml. of hydrochloric acid and pour into the column sufficient to give a resin bed 80 mm. long. Cover the resin with a layer of glass wool, then wash with water until the washings no longer give a precipitate of silver chloride with silver nitrate reagent. The resin should always be covered with liquid and the column stoppered when not in use. Just before analysis remove the liquid above the glass wool. (Note 1).

Procedure

Accurately weigh about 0.1 meq. of the acyl group and transfer to the bottom of a thick-walled tube (10 × 75 mm.). After dissolving in 1 ml. of *n*-amyl alcohol, add 0.10 ml. of the sodium hydroxide reagent, draw out the end of the tube in the flame and seal off to give a tapered tip. Hydrolyze for 1 hour at 150°C. in a heated metal block. Cool, release the pressure in the tube by opening with a sharp flame, and then crack the top off by applying a hot glass rod to a file mark. Add 3 ml. of isopropyl alcohol–water solvent and transfer quantitatively to the ion-exchange column using three 2-ml. portions of solvent for rinsing. Open the stopcock sufficiently to give a flow rate of 1 to 2 ml./minute (Note **2**) and collect the eluate in a 125-ml. Erlenmeyer flask. When the liquid level in the column approaches the glass wool add 5 ml. of solvent and repeat until about 25 ml. of eluate is collected. Add 4 drops of phenolphthalein indicator and titrate with standardized sodium hydroxide solution until the faintly pink color which appears to persist for 30 seconds. If the eluate is slightly yellow add 4 drops of the mixed indicator and titrate to a violet end point.

Calculation for Acetyl

$$\% \text{ Acetyl} = \frac{\text{ml. } 0.02N \text{ NaOH} \times 2 \times 0.4305 \times 100}{\text{mg. sample}}$$

Notes 1. If the resin is kept under dilute hydrochloric acid between determinations the column may be used for about 20 samples.

2. Flow rates greater than 2 ml/min cause incomplete exchange.

B. VARIED HYDROLYSIS PROCEDURE

The following method is the outcome of various publications (22,29,39,-51). It is versatile, relatively simple and the apparatus may also be utilized for C-methyl and Kjeldahl nitrogen determinations.

Reagents

All reagents should be analytical grade.

5*N* *Sodium hydroxide* solution.

5*N* *Methyl alcoholic sodium hydroxide* solution. Dissolve 20 g. of sodium hydroxide in 67 ml. of distilled water and 33 ml. of methyl alcohol. The methyl alcohol should be freshly distilled after a prior 15 minute refluxing over pellets of potassium hydroxide.

Wenzel's sulfuric acid. Add 100 ml. of sulfuric acid (sp. gr. 1.84) to 200 ml. of distilled water.

Chromic acid reagent. Dissolve 17 g. of chromic anhydride in 100 ml. of distilled water and filter, if necessary. Mix four parts of this solution with 1 part of sulfuric acid (sp. gr. 1.84).

Silver sulfate.

Sodium sulfate. Saturated solution.

Sodium hydroxide. 0.01*N* Standard solution.

Phenolphthalein indicator. 1% in 95% ethanol.

Apparatus

The hydrolysis apparatus of Wiesenberger (51) can be recommended

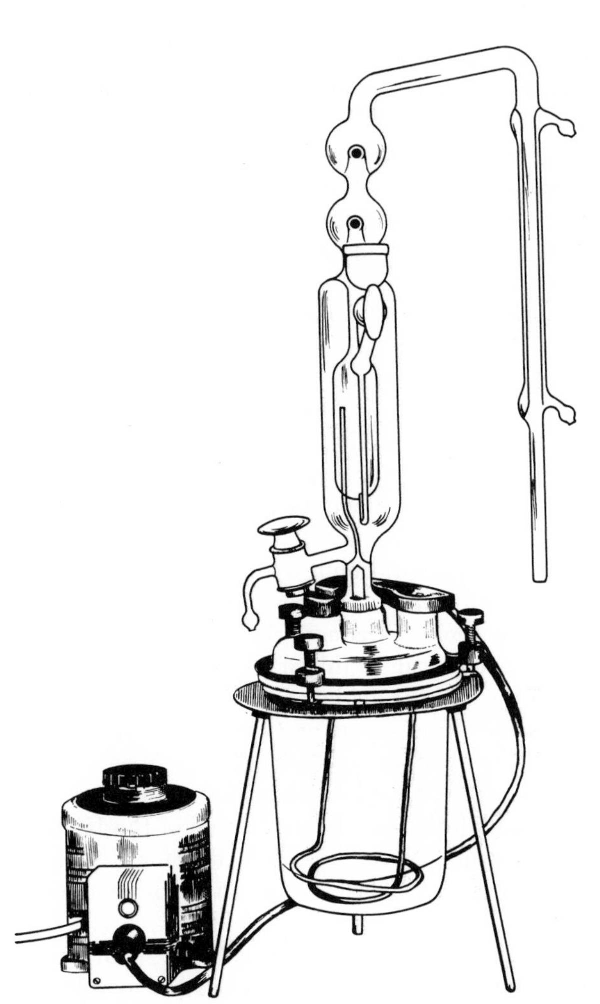

Fig. 6. One-piece Kjeldahl distillation unit (from reference 44).

(Fig. 4). The hydrolyzate may be distilled from any Kjeldahl-type unit. The apparatus in Fig. 5 can be recommended but one-piece units in which the distillation apparatus is combined with the steam generator (Fig. 6) are very convenient.

Procedure

1. Aqueous alkaline hydrolysis. Weigh 5 to 10 mg. (Note **1**) of sample and transfer to the bottom of the hydrolysis flask. Add 2 ml. of reagent and a few pieces of pumice. Lightly smear the ground joint with silicone grease, connect the water condenser and gently reflux for 30 minutes (Notes **2** and **3**).

After cooling, wash the inside of the condenser, then partially remove it from the flask and carefully wash the cone with three 0.5-ml. lots of sodium sulfate solution. Remove the condenser and wash the socket and the sides of the flask with 1 ml. of sodium sulfate solution.

Transfer quantitatively to the Kjeldahl unit, rinsing the flask with sodium sulfate solution, add 2 ml. of Wenzel's acid and wash in with 0.5 ml. distilled water (Note **4**).

Distill the acetic acid as rapidly as possible (Notes **5** and **6**) into a conical flask immersed in an ice bath. Throughout the distillation and the subsequent titration nitrogen is bubbled through the solution in the flask at a steady rate to prevent interference from carbon dioxide. After rinsing the condenser tip, carefully remove the flask from the apparatus, and titrate to the phenolphthalein end-point with $0.01N$ sodium hydroxide. The change to a faintly pink coloration is definite and permanent.

2. Alcoholic alkaline hydrolysis. Proceed as for *1.* but do not acidify with 2 ml. of Wenzel's acid. Instead, rinse the funnel with 1 ml. of sodium sulfate solution, steam distil 10 ml. of the solution to remove the methyl alcohol, then add the sulfuric acid and follow the procedure in *1.* (Note **7**).

3. Acid hydrolysis. Proceed as for *1.* but with two differences. Rather than smearing the joint of the condenser with silicone grease, moisten it with the acid. There is no need for the acidification step which is replaced by an addition of sodium sulfate to the funnel of the Kjeldahl unit.

4. Blank determination (Note **8**). Proceed as for *1*, *2*, and *3*, respectively but do not add sample to the flask. The titre obtained is subtracted from the titre of an analysis.

5. Standardization of 0.01N sodium hydroxide. Carry out the titration just as in the determination using a standard acid solution in the same conical vessel with nitrogen bubbling through the solution, but add a volume of freshly boiled distilled water equivalent to the volume obtained in an analysis. Perform a titration on the water alone, subtract the titre from the standardization titre, and calculate the factor for the sodium hydroxide.

6. Calculation. 1 ml. $0.01N$ NaOH is equivalent to 0.4305 mg. CH_3CO, 0.2902 mg. HCO, or 1.050 mg. C_6H_5CO

$$\% \, CH_3CO = \frac{(\text{Analysis titre} - \text{blank titre}) \times 0.4305 \times 100}{\text{mg. sample}}$$

The foregoing procedure should give a value of 31.84 ± 0.3% acetyl for acetanilide.

Notes 1. The quantities must be increased considerably for substances with acetyl contents less than 1%. Add 10 to 15 mg. of silver sulfate if the presence of halogen is suspected.

2. Sometimes the sample is insoluble and, if it is not wetted out by the boiling reagent, the analysis should be repeated, either with another reagent or after first dissolving in a few drops of pyridine or dioxane (36, 45*a*).

3. Electrical heating is preferable because the different heats required for different reagents can easily be reproduced.

4. Besides washing most of the sulfuric acid into the unit, the rinse with distilled water helps to keep the funnel free of sodium sulfate for the next analysis.

5. There is little danger of distilling sulfuric acid in this method.

6. The time and volume of distillate required for quantitative distillation must be determined using an acetic acid solution (or formic or benzoic acids if these are important) since it varies with the efficiency of the steam generator, the size of the apparatus, and the volume of the reagent mixture and washings. Between 100 and 150 ml. is usually distilled in 10 to 15 minutes. Once the standard conditions are determined, strictly adhere to them for best results.

7. Caution should be exercised here, otherwise siphoning can occur and the analysis will be lost. The steam distillation must be stopped to allow the acid to run into the unit but the steam must not be shut off completely and is partially diverted through the tap of the steam jacket while the acid is added. Acidification to pH3 with citric acid is better for analysis of carbohydrates.

8. The determinations of the blank and the factor for the 0.01*N* sodium hydroxide are particularly critical in this method because the large volume of distillate has a significant blank value.

C. GAS CHROMATOGRAPHY PROCEDURE

1. After pressure alcoholysis

The pressure alcoholysis procedure of Spingler and Markert (42,43) will be described.

Reagents

All chemicals are reagent grade.

2 to 2.5*N* *Methyl alcoholic hydrochloric acid.* Add 300 ml. hydrochloric acid (sp. gr. 1.18) to a 1-liter round-bottom flask fitted with a separating funnel and a tube passing into a 500-ml. Erlenmeyer flask. The tube is packed for about 10 cm. of its length with either calcium chloride or mag-

nesium perchlorate and the Erlenmeyer contains 100 ml. of methanol. Add sulfuric acid (sp. gr. 1.84) dropwise through the funnel onto the hydrochloric acid. Gently warm the latter and allow the gas to dissolve in the methanol under the pressure produced in the apparatus. After 20 to 30 minutes titrate the solution and add sufficient methanol to make the solution 2 to 2.5*N* (Note **1**). A small amount of water (not greater than 5%) can be tolerated.

Apparatus

Any gas chromatograph fitted with a column which will separate methyl formate, methyl acetate, and methanol is suitable. The recommended parameters for the Perkin-Elmer Vapor Fractometer 154B are: carrier gas, helium, 1.0 atm.; flow rate, 23 ml./minute; detector voltage, 6*V*; column K (20% Carbowax 1500 on 60 to 80 mesh Chromosorb), 2 m.; temperature, 65°; sensitivity, half max. initially, 1/128 max. for methanol.

Procedure

Weigh 1 to 5 mg. (Note **2**) of substance into a thick-walled tube (7 × 80 mm. external dimensions). Draw out the upper portion of the tube to a capillary 3 to 4 cm. long and 0.05 to 1 mm. internal diameter. Weigh the tube when cold, add 100 μl. of methanolic hydrochloric acid and weigh again. Seal the tube and place it in a boiling water bath for 30 minutes to effect hydrolysis (Note **3**). After cooling in an ice–salt mixture, crack and open the tube at the capillary, extract a sample of about 20 μl. with a syringe and inject it into the gas chromatograph (Notes **4** and **5**).

Calculation

$$\% \text{ acyl} = \frac{F \times \text{Area under ester peak} \times \text{Weight of reagent}}{\text{Area under methanol peak} \times \text{Weight of substance}}$$

where F is obtained from the chromatogram of a standard substance. The areas under the curves can be determined by multiplying the peak height by the band width at half height or by alternative methods.

Notes 1. As a reagent for formyl determinations this solution may last for only a few days, after which it is likely to contain decomposition products which are eluted with methyl formate. It is satisfactory as an acetyl reagent for at least 2 months.

2. 0.5 mg. might be better for substances with very high acetyl contents while 30 to 75 mg. and 500 μl. of reagent is desirable for very low acetyl contents (40).

3. Some *N*-acyl groups may require both longer hydrolysis times and stronger reagents. Even so, acetyldiphenylamine gives low results.

4. This method eliminates the distillation step which is common to the conventional methods. This has the virtues of simplicity, increased sensitivity, and more reproducible results but the column must be cleaned after 10 to 15 analyses, or when the methanol curve becomes asymmetric, by heating to 130 to 160°C. for 20 to 30 minutes.

5. For acyls such as propionyl, or if the acetonide (—OCH_2COCH_2O—) group is present, the parameters of the gas chromatograph may require changing to resolve the alcoholysis products from methyl acetate. The formation of the ethyl esters also provides another possible alternative.

2. After Acid Hydrolysis of Proteins and Peptides (49b)

Reagents

Chemicals are reagent grade. Tert-Butyl ethyl ether (t-BEE) should be free from acetic acid.

Apparatus

Ward et. al. (49*b*) recommend the following conditions for the gas chromatographs they employed for the analyses.

Warner-Chilcott, model 600: equipped with a direct injection all glass column and a Sr^{90} beta ionization detector, model 600-7-1; recorder, Minneapolis-Honeywell model Y143, with a disk integrator; carrier gas and detector scavenger gas, argon; chromatographic column, 0.4 × 175 cm packed with 10% LAC 2-R-446 (a polyethylene glycol adipate obtained from Warner-Chilcott Laboratories Instruments Division) on acid-washed Chromosorb W, 60/80 mesh (Johns Manville Co.); column temperature 125° with a gas flow rate of 20 ml/min through the column plus 120 ml/min scavenger through the detector; detector temperature, 140°C; detector voltage, 800; signal attenuation of 5 (usually).

F and M Model 402: equipped with a flame ionization detector, nitrogen carrier gas, flow rate 66 ml/min; attenuation, 1 × 16 up to 0.2 μg of acetic acid and 1 × 32 for 1 μg.

Procedure

Weigh sufficient protein (Note **1**) to give between 0.5 and 2 μmoles of acetic acid after hydrolysis, place in a 1 × 7.5 cm tube and add 250 μl of 6 N H_2SO_4. Seal under nitrogen and hydrolyze at 105° for 2 hours in a silicone oil bath (Note **2**). The tube should be positioned so that only the liquid-containing portion is submerged in the oil bath, thus allowing reflux from the upper walls.

After hydrolysis cool the tube in an ice bath and centrifuge to remove liquid droplets from the walls. Open the tube, add 100 mg of anhydrous Na_2SO_4 and 500 μl of t-BEE. The extraction of acetic acid into the ether is facilitated by stirring several times on a vortex stirrer. Centrifuge briefly in a clinical centrifuge, remove the t-BEE with a Pasteur pipet and transfer the ether to a small, stoppered vial containing 10 mg of anhydrous Na_2SO_4. Dry the ether for approximately ½ hour prior to gas chromatography. Four such extracts are made and analyzed individually (Note **3**).

Inject 4–5 μl into the gas chromatograph. Calculate the area of the acetic acid peak and quantitate by reference to a standard calibration curve covering the range 0.025 to 1.0 μg acetic acid (in t-BEE).

Notes 1. When acetate buffers are used in the purification of the protein, it should be dissolved in 0.01 N HCl and lyophilized prior to analysis to ensure complete removal of acetate ion. The 'blank' can be ascertained by analyzing an unhydrolyzed sample.

2. Threonine produces small amounts of acetic acid on hydrolysis. This will lead to high acetyl values, even after short hydrolysis periods, if unusually high concentrations of threonine are present.

3. Pooling the extracts helps the precision but is not advisable for low levels of acetic acid. The use of cryogenic distillation (40a) may be a suitable alternative to the extraction procedure.

REFERENCES

1. Alicino, J., *Anal. Chem.*, **20**, 590 (1948).
1a. Awashy, A. K., R. Belcher, and A. M. G. MacDonald, *Analyt. Chim. Acta*, **33**, 311 (1965).
2. Bellamy, L. J., *The Infra-Red Spectra of Complex Organic Molecules*, 2nd ed., Wiley, New York, 1958.
2a. Bellamy, L. J., *Advances in Infra Red Group Frequencies*, Methuen, London, 1968.
3. Benson, R. H., and R. B. Turner, *Anal. Chem.*, **32**, 1464 (1960).
3a. Bovey, F. A., *Nuclear Magnetic Resonance Spectroscopy*, Academic Press, New York, 1969.
4. Bradbury, R. B., *Anal. Chem.*, **21**, 1139 (1949).
5. Cheronis, N. D., and J. B. Entrikin, *Identification of Organic Compounds*, Interscience, New York, 1963.
5a. Cheronis, N. D., and T. S. Ma, *Organic Functional Group Analysis*, Interscience, New York, 1964.
6. Clark, E. P., *Ind. Eng. Chem., Anal. Ed.*, **8**, 487 (1936).
6a. Clark, E. P., *Ind. Eng. Chem., Anal. Ed.*, **9**, 539 (1937).
7. Elek, A., and R. A. Harte, *Ind. Eng. Chem., Anal. Ed.*, **8**, 267 (1936).

8. Feigl, F., *Spot Tests in Organic Analysis,* 6th Eng. ed., Elsevier, New York, 1960.
9. Freudenberg, K., and E. Weber, *Z. Angew. Chem.*, **38,** 280 (1925).
10. Friedrich, A., and S. Rapoport, *Biochem. Z.*, **251,** 432 (1932).
11. Friedrich, A., and H. Sternberg, *Biochem. Z.*, **286,** 20 (1936).
12. Gillam, A. R., and E. S. Stern, "An Introduction to Electronic Absorption Spectroscopy" in *Organic Chemistry,* 2nd ed., Edward Arnold, London, 1957.
12a. Goodman, G. W., B. C. Lewis, and A. F. Taylor, *Talanta,* **16,** 807 (1969).
12b. Gould, E. S., *Mechanism and Structure in Organic Chemistry,* Holt Rinehart and Winston, New York, 1959.
13. Grant, J., ed., *Quantitative Organic Microanalysis,* of F. Pregl, 4th Eng., ed., J. & A. Churchill, London, 1945.
14. Hall, R. T., and W. E. Shaefer, "Determination of Esters," in Mitchell, ed., *Organic Analysis,* Interscience, New York, 1954, Vol. II, pp. 19–70.
15. Herriott, R. M., *J. Gen. Physiol.*, **19,** 283 (1935).
16. Hestrin, S., *J. Biol. Chem.*, **180,** 249 (1949).
17. Hodgman, C. D., *Handbook of Chemistry and Physics,* 41st ed., Chemical Rubber Publishing Co., Cleveland, 1959.
18. Hopton, J. W., *Anal. Chim. Acta,* **8,** 429 (1953).
19. Hurka, W., *Mikrochemie,* **31,** 5 (1943).
20. Hurka, W., and H. Lieb, *Mikrochemie ver. Mikrochim. Acta,* **29,** 258 (1941).
21. Inglis, A. S., M. Sc. Thesis, 1954.
22. Inglis, A. S., *Mikrochim. Acta,* **2,** 228 (1957).
22a. Inglis, A. S., and P. W. Nicholls, unpublished work.
23. Ingold, C. K., *Structure and Mechanism in Organic Chemistry,* Bell, London, 1953.
24. Jackman, L. M., "Applications of Nuclear Magnetic Resonance Spectroscopy in Organic Chemistry," in Barton and Doering eds., *International Series of Monographs on Organic Chemistry,* Pergamon, New York, 1959, Vol. V.
25. Jerie, H., *Mikrochemie,* **40,** 189 (1952/53).
26. Kainz, G., *Mikrochemie,* **30,** 250 (1952).
27. Kan, M., F. Suzuki, and H. Kashiwagi, *Microchem. J.*, **8,** 42 (1964).
28. Kolthoff, I. M., and R. Belcher, *Volumetric Analysis III,* Interscience, New York, 1957.
29. Kuhn, R., and H. Roth, *Ber.*, **66,** 1274 (1933).
30. Kunz, A., and C. S. Hudson, *J. Am. Chem. Soc.*, **48,** 1982 (1926).
30a. Lingane, J. J., *Electroanalytical Chemistry,* Interscience, New York, 1953.
31. Ludowieg, F., and A. Dorfman, *Biochim. Biophys. Acta,* **38,** 212 (1960).
32. Ma, T. S., "Microdetermination of the Acyl Group by Ion Exchange," in Welcher, ed., *Standard Methods of Chemical Analysis,* 6th ed., Van Nostrand, New York, Vol. IIA, p. 408.
33. Narita, K., *Biochem. Biophys. Acta,* **28,** 184 (1958).
34. O'Donnell, I. J., E. O. P. Thompson, and A. S. Inglis, *Aust. J. Biol. Sci.*, **15,** 732 (1962).
35. Offer, G. W., *Biochim. Biophys. Acta,* **90,** 193 (1964).
36. Onoe, T., *Japan Analyst,* **9,** 479 (1960).
37. Phillips, D. M. P., *Biochem. J.*, **86,** 397 (1963).
38. Pregl, F., and A. Soltys, *Mikrochemie,* **7,** 1 (1929).
38a. Samuelson, O., *Ion Exchange Analytical Chemistry,* John Wiley, New York, 1953.
39. Schöniger, W., H. Lieb, and M. G. El Din Ibrahim, *Mikrochim. Acta,* 96 (1954).

40. Schroeder, W. A., J. T. Cua, G. Matsuda, and W. D. Fenninger, *Biochim. Biophys. Acta*, **63**, 532 (1962).
40a. Shepherd, G. R., and B. J. Noland, *Anal. Biochem.*, **26**, 325 (1968).
41. Silverstein, R. M., and G. C. Bassler, *Spectrometric Identification of Organic Compounds*, Wiley, New York, 1963.
42. Spingler, H., and F. Markert, *Mikrochim. Acta, 1959*, 122.
43. Spingler, H., and F. Markert, private communication.
44. Steyermark, Al., *Quantitative Organic Microanalysis*, 2nd ed., Academic, New York, 1961.
45. Suzuki, M., *J. Biochem.* (*Japan*), **27**, 367 (1938).
45a. Tabib, S. N., S. Y. Kulkarni, and V. S. Pansare, *Microchem. J.*, **13**, 98 (1968).
46. Syzmanski, H. A., and R. E. Yelin, *NMR Band Hand Book*, IFI/Plenum, New York (1968).
47. Technical Measurement Corporation, Mnemotron Division, 202 Mamaronek Ave., White Plains, N.Y.
48. Varian Associates, *High Resolution NMR Spectra Catalogue*, Palo Alto, California, 1962.
49. Viditz, F. v., *Mikrochim. Acta*, **1**, 326 (1937).
49a. Ward, D. N., and J. A. Coffey, *Biochemistry*, **3**, 1575 (1964).
49b. Ward, D. N., J. A. Coffey, D. B. Ray, and W. M. Lamkin, *Anal. Biochem.*, **14**, 243 (1966).
50. Wiesenberger, E., *Mikrochemie ver. Mikrochim. Acta*, **30**, 241 (1942).
51. Wiesenberger, E., *Mikrochemie ver. Mikrochim. Acta*, **33**, 51 (1948).
52. Wiesenberger, E., *Mikrochim. Acta, 1954*, 127.
52a. Williams, D. H., and I. Fleming, *Spectroscopic Methods in Organic Chemistry*, McGraw-Hill, London, 1966.
53. Wolfrom, M. L., M. Konigsberg, and S. Soltzberg, *J. Am. Chem. Soc.*, **58**, 490 (1936).
54. Zimmerman, W., and H. Jerie, unpublished work.

Part II
Section B-2

ORGANIC ANALYSIS: *O*-ALKYL, *N*-ALKYL, AND *S*-ALKYL

BY ADAM S. INGLIS, *Division of Protein Chemistry, Wool Research Laboratories, CSIRO, Parkville, Australia*

Contents

I. INTRODUCTION

In appraising the *O*-, *N*- and *S*-alkyl determinations some 85 years after the introduction of the methoxyl determination by Zeisel (102), one finds that the commonly used procedures today, except for the use of more elaborate apparatus and equipment, are not dissimilar to those of the early workers in the field, despite a wealth of papers describing modified Zeisel procedures. This apparent anomaly is due in part to the fact that variables in this determination are interdependent to a large extent, and this has not always been appreciated by workers suggesting changes in procedures. In Section VI, therefore, rather than discuss whole methods it is intended to analyze each part of the procedure with a view to the ramifications of its alteration on the analysis. It is hoped that this approach, when combined with a description of more recent attempts to overcome limitations of the procedure, will be of greatest value to other workers in the field and will stimulate further advances.

In order to place these and later remarks in perspective, a very brief historical account is first necessary. However, it must be stressed that this chapter is not intended as an exhaustive treatment of the subject, although some duplication of earlier work will no doubt occur in making it a unified whole. Readers are referred to other books (25,36,55,90) for additional information and to the biennial reviews in *Analytical Chemistry* for more recent work.

Zeisel (102) published his method for the determination of the alkoxy group on a macro scale in 1885. He boiled the sample (0.02 to 0.3 g.) with hydriodic acid (10 ml.; sp. gr. 1.68) to release alkyl iodide which was carried by a stream of carbon dioxide into an absorption solution of alcoholic silver nitrate. The double salt $AgI \cdot AgNO_3$ was decomposed by the addition of dilute nitric acid and the silver iodide was weighed. To avoid high results due to iodine or hydriodic acid vapors, the alkyl iodide was washed with water containing a little amorphous phosphorus; to minimize losses of alkyl iodide the flask was connected to a reversed condenser containing water at 40 to 50° and the washer was immersed in a water bath at 50 to 60°.

$$RCOOCH_3 \xrightarrow{HI} RCOOH + CH_3I \quad (1)$$

$$ROCH_3 \xrightarrow{HI} ROH + CH_3I \quad (2)$$

$$CH_3I \xrightarrow{AgNO_3} AgI \cdot AgNO_3 \xrightarrow[HNO_3]{H_2O} AgI + AgNO_3 \quad (3)$$

This method was adapted to a micro scale by Pregl (47) but he did not warm the wash solution or employ a warm water jacket. He dissolved the substance in phenol, or a mixture of phenol and acetic anhydride, prior to addition of hydriodic acid, and then boiled the solution for 20 minutes, which was sufficient time for complete demethylation and absorption. For compounds containing sulfur, cadmium sulfate was added to the wash solution to retain any hydrogen sulfide formed in the reaction. However, the method failed with sulfur compounds, not only because the hydrogen sulfide was not completely trapped, but also because mercaptans were produced with alkyl groups attached to sulfur, and the insoluble silver mercaptides could not be completely removed from the silver iodide.

$$RSH + AgNO_3 \rightarrow RSAg + HNO_3 \quad (4)$$

In 1915 Kirpal and Bühn (66) published a volumetric modification of the method in which the alkyl iodide was absorbed in pyridine; an aqueous solution of the resulting compound was mixed with potassium chromate and the iodine liberated was titrated with a silver nitrate solution. However, in 1930 Vieböck and Schwappach (96) published a better volumetric modification which was adapted to the micro scale by Vieböck and Brecher (97). The alkyl iodide was absorbed in a solution of bromine, sodium acetate, and acetic acid, oxidized to the iodate by bromine, and finally determined as iodine by addition of potassium iodide and sulfuric acid. The reaction is summarized as follows:

$$CH_3I + Br_2 \rightarrow CH_3Br + IBr \qquad (5)$$

$$I\ Br + 3H_2O + 2Br_2 \rightarrow HIO_3 + 5HBr \qquad (6)$$

Titration

$$IO_3^- + 6H^+ + 5I^- \rightarrow 3I_2 + 3H_2O \qquad (7)$$

$$2S_2O_3^{-2} + I_2 \rightarrow S_4O_6^{-2} + 2I^- \qquad (8)$$

Apart from adaption of the method to a submicro scale (15), this has been the most significant improvement in the quantitative procedure because it is rapid and convenient; it has a favorable factor—one alkoxyl group is equivalent to 6 atoms of iodine—and it is not affected by compounds containing sulfur. A vast literature has been concerned with clarifying the various contradictory reports which have arisen over the years, and while applications of modern instrumental techniques such as infrared (5) and gas chromatography (95) have specially useful applications in the determination, they do not effect any increase in speed, simblicity, or accuracy over the normal volumetric modification of the Zeisel procedure.

The determination of *S*-alkyl may be carried out by this procedure (11,12) but with a prolonged reaction time. However, the range of compounds analyzed is small and much remains to be done to establish conditions suitable for all types. The method has been particularly useful for methionine but other amino acids give low results (1).

Progress with the *N*-alkyl determination has generally followed that of the *O*-alkyl determination. The normal procedure is still essentially the method of Herzig and Meyer (51a) which was published in 1894. They formed the quaternary ammonium salt of the alkimide by heating with hydriodic acid containing ammonium iodide, then collected and determined the alkyl iodide formed by pyrolysis of the salt at 399 to 369°. The hydrolysis procedure is show and several repetitions of the process are often necessary to remove the alkyl group quantitatively. Even so, it may be inaccurate and must still be classed as a tedious, unreliable procedure.

II. OCCURRENCE AND IMPORTANCE

O-Alkyl groups are found in an abundance of natural products. Sugars, lignins, volatile oils, and alkaloids all possess these functional groups. The *N*-alkyl groups are also an important constituent of the alkaloids. The alkoxyl determination has been of great value in the elucidation of the structure of these substances, since a good determination can fix the minimum molecular weight; in the case of the methoxyl group it is equivalent to an element of atomic weight 31 and, especially when coupled with an oxygen determination, gives a clear indication of the functional nature of the oxygen groups in the molecule. It is also of value in establishing the presence of alcohol of crystallization. Of the *O*-, *N*- and *S*-alkyl determinations, the methoxyl and ethoxyl determinations are the most important, and the chapter will be concerned primarily with these.

In industry the alkoxyl determination can be used as a control procedure for esters which are not suitable for saponification values and it has been used for the determination of compounds of silicon (76), boron (2), and aluminum (29).

III. PROPERTIES OF GROUPS

The *O*-, *N*- and *S*-alkyl groups are present in a vast number of substances. The *O*-alkyl determination may be required on alcohols, esters, acetals, or ethers ranging from highly volatile liquids to high melting point solids. These substances also exhibit great differences in other physical properties, such as solubility, as well as in chemical reactivity. However, the physical properties of the alkyl iodides are of common interest, since most methods of determination involve an initial fission of the substance with hydriodic acid to produce the alkyl iodide.

A. PHYSICAL PROPERTIES

Table I gives some physical properties of alkyl iodides from methyl to amyl. The solubility in water is quite low, but this could lead to losses in micro and submicro determinations, especially for passage of methyl iodide through aqueous scrubbers. The boiling points of the iodides become higher with increasing molecular weight. This pattern is followed by the chlorides and bromides, the chlorides of an alkyl group having the lowest boiling point and the iodide the highest.

B. PHARMACOLOGICAL PROPERTIES

Substances containing the *O*-alkyl group are important in the medical field. Ethanol and esters such as ethyl acetate are used as stimulants;

TABLE I

Physical Properties of Alkyl Iodides (52) and Alkoxyl Groups

Iodide	Molecular weight	Density, g/ml.	m.p., °C.	b.p., °C.	Solubility, g./100 ml., 20°C.		O-Alkyl response		
					H_2O	Alcohol	UV	IR	NMR
Methyl	141.95	2.2799	− 66.1	42.5	1.4	∞	Weak	Strong	Strong
Ethyl	155.98	1.933	−108.5	72.2	0.4	s.	Weak	Strong	Strong
n-Propyl	170.01	1.747	−101.4	102.4	0.0867	∞	Weak	Strong	Strong
Isopropyl	170.01	1.703	− 90.8	89.5	0.14	∞	Weak	Strong	Strong
Allyl	167.98	1.848	− 99.3	103	i.	s.	Weak	Strong	Strong
n-Butyl	184.03	1.617	−103.5	131	0.0202[a]	∞	Weak	Strong	Strong
sec-Butyl	184.03	1.595	−104.0	117.5	i.	s.	Weak	Strong	Strong
tert-Butyl	184.03	1.571	− 33.65	100[a]	i.	∞	Weak	Strong	Strong
n-Amyl	198.06	1.517	− 85.6	156	i.	s.	Weak	Strong	Strong

[a] 17.5°.

diethyl ether is a common anesthetic. Alkaloids form an extremely important group of compounds of pronounced physiological activity and constitute the active principles of the common vegetable drugs and poisons. Many of these contain methoxyl groups.

C. CHEMICAL PROPERTIES

Most esters are readily hydrolyzed by either acid or alkali. Alcohols, acetals, and ethers are stable towards alkali but generally may be cleaved by halogen acids with formation of alkyl halides. Hydrogen iodide is the most effective; hence, it is the most suitable generally for quantitative analysis.

S- and *N*-alkyl groups are not noted for their reactivity and generally require prolonged or more vigorous conditions for dealkylation. Alkylated amines generally need reaction with concentrated halogen acid at 200 to 300° for quantitative fission.

The reactivity of the *O*-alkyl and *N*-alkyl groups, however, may vary considerably with the substance and other substituted groups in the molecule. Methoxyl groups in the 7-position in isoflavones are difficult to demethylate, while the *N*-methyl bond may be so reactive that it is broken under similar conditions to an *O*-alkyl bond.

The alkyl halides formed by fission of the *O*-, *S*- and *N*-alkyl groups with halogen acids have readily replaceable halogen atoms. Since this reactivity has been utilized for quantitative estimation of the groups via the halide, some pertinent properties and reactions of the halides will be given here.

Silver nitrate reacts very slowly at room temperature with alkyl halides, indicating that they are, at most, only partially ionized, but there is a rapid precipitation of silver halide upon warming. This reaction has been used for analytical purposes—the formation of

$$C_nH_{2n+1}I + AgNO_3 \rightarrow C_nH_{2n+1}O\cdot NO_2 + AgI \quad (9)$$

silver iodide being the basis of the gravimetric determination. The iodide is more readily split than the bromide or the chloride.

The halides may be hydrolyzed to alcohols with moist silver oxide in the cold. Dilute alkali also effects a smooth hydrolysis.

$$C_nH_{2n+1}I \xrightarrow[AgOH]{} C_nH_{2n+1}OH + AgI \quad (10)$$

The mobility of the halogen atom decreases as the length of the chain increases; methyl iodide reacts faster than ethyl iodide. The order of reactivity for tertiary iodides is quite different. They are decomposed by water at room temperature, whereas the straight-chain iodide requires heating to 100°.

$$(CH_3)_3CI + H_2O \rightarrow (CH_3)_3COH + HI \quad (11)$$

IV. SEPARATION AND CONVERSION

An excellent review of the cleavage of ethers has been made by Burwell (23). In addition to cleavage by acids, which is of prime interest here, he describes cleavage by halides, anhydrides, alkylation, nucleophilic reagents alone, alkali metals, and organometallic compounds, thereby indicating mechanisms, relative ease, and alternative possibilities of fission of *O*-alkyl groups. For example, the reaction of ethers with amine salts may be used for *O*-alkyl determinations using *N*-methyl pyridinium chloride. Boron trichloride at −80°C. also offers possibilities (25).

A. MECHANISM

Although possessing quite different stabilities to acids generally, aliphatic ethers, acetals, and alcohols are all cleaved readily by hydriodic acid (sp. gr. 1.7) at room temperature (53). Alkyl aryl ethers are relatively stable and often require refluxing with hydriodic acid to remove the alkyl group as alkyl iodide and to form the phenol. This is in accord with a mechanism involving an intermediate oxonium salt formed by attraction of the basic oxygen atom for hydrogen ions. The oxonium salt decomposes by either an S_N1 or S_N2 mechanism (23).

$$ROR + H^+ \xrightarrow[\text{fast}]{} R\underset{H}{\overset{+}{O}}R \quad (12)$$

(S_N1)

$$R\underset{H}{\overset{+}{O}}R \xrightarrow[\text{slow}]{} ROH + R^+ \quad (13)$$

$$R^+ + X^- \xrightarrow[\text{fast}]{} RX \quad (14)$$

(S_N2)

$$R\underset{H}{\overset{+}{O}}R + X^- \xrightarrow[\text{slow}]{} RX + ROH \quad (15)$$

The leaving group in the S_N1 reaction determines the rate. The lower the energy of the C-leaving group bond and the greater the tendency to form an anion, the more readily does the reaction proceed by the S_N1

mechanism (92). *tert*-Butyl ethers are readily cleaved by this mechanism. In the case of aryl alkyl ethers, resonance structures such as

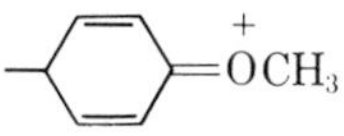

$$-\langle\,\rangle=\overset{+}{O}CH_3$$

would reduce the basicity of the oxygen atom and, hence, the tendency to form the oxonium salt. Diphenyl ether, which is only feebly basic, is not cleaved at 250° by hydriodic acid. The slow demethylation of 7-methoxy-isoflavones could also be ascribed, in part at least, to the reduction in basicity of the alkoxyl oxygen.

The great difference in the rates at which aryl ethers are hydrolyzed by the halogen acids [HCl:HBr:HI = 1.6: ∞ (45)] indicates that the rate-determining step involves the nucleophile. Displacement of RO groups by nucleophiles generally requires an acid medium to reduce the basicity of the RO groups by protonation [equation (12)] ,otherwise, they are not readily displaced. That I^- is the most effective nucleophile is due to its size; the solvation energy is lowest and the peripheral electrons are polarized most readily, which enables bonding interaction at greater internuclear distances.

B. INFLUENCE OF SUBSTITUENTS IN THE MOLECULE

The effect of nuclear substituents on the rate of fission of ethers is consistent with the above picture (45). The rates of reaction of aromatic ethers containing *o*- and *p*- directing groups are faster than those containing *m*-directing substituents. When *o*- and *p*-directing groups are substituted in the various positions in the molecule, the rates of cleavage, in descending order, are *p*-, *o*-, *m*-; for *m*-directing groups, *m*-, *o*-, *p*-. The relative effects of *o*- and *p*- directing groups, in decreasing order, are OCH_3, CH_3, OH, H, NH_2 (HX), Cl, Br; *m*-directing groups, CH_3CO, NO_2.

The rates of reaction of methoxyanisoles are quite high compared with the other rates of reaction, which has been attributed to a probability factor (45). Since there are two methoxyl groups which may be attacked, there is more of the intermediate oxonium compound and a higher reaction velocity. Thus, the introduction of methoxyl groups should facilitate the reaction of the substance with hydriodic acid, rather than hinder it as suggested earlier (35).

Phenetoles react faster than anisoles (45) presumably because the ethyl group is a more stable leaving group than the methyl group.

C. N-ALKYL FISSION

In contrast with the methods for alkoxyl fission, the choice of reagents for the dealkylation of alkimide groups is limited. The method of separation for quantitative purposes consists in gradually heating the base with hydriodic acid and ammonium iodide to 200 to 360°.

$$(C_nH_{2n+1})NR_2 + HI \xrightarrow[NH_4I]{200-360^\circ} R_2NH + C_nH_{2n+1}I \qquad (16)$$

Several repetitions of the procedure may yield quantitative recoveries of alkyl iodide. It is, however, far from being universally applicable and often gives low results. This, coupled with the length of the procedure, has limited its use. Perhaps the latter also partially explains the lack of literature concerning the method and the mechanism of the reaction.

Some dealkylation can occur in the preliminary removal of *O*-alkyl with boiling hydriodic acid. 1,2-Dimethylpyraz-3,6-dione, for example, releases one molecule of alkyl iodide under these conditions. It was suggested (48) that this is due to a rearrangement, the methyl group migrating from the nitrogen to the oxygen atoms forming a methoxyl group. Alternatively, the following mesomeric forms would allow the substance to develop full aromatic character with positive charges on the adjacent nitrogen atoms.

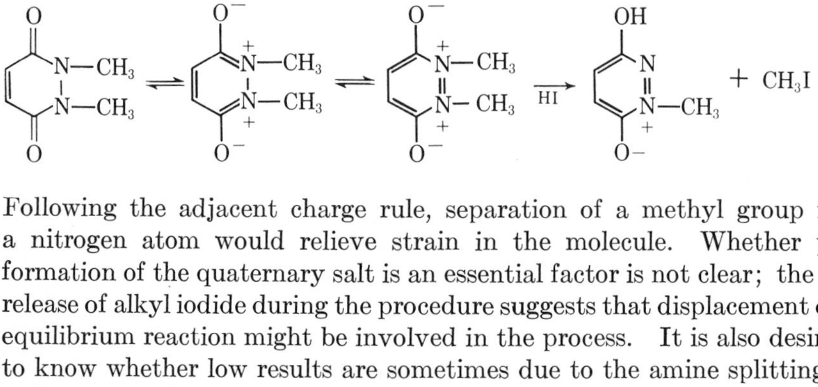

O, N—CH_3, N—CH_3, O ⇌ O^-, $\overset{+}{N}$—CH_3, $\overset{+}{N}$—CH_3, O^- ⇌ O^-, $\overset{+}{N}$—CH_3, $\overset{+}{N}$—CH_3, O^- $\xrightarrow{HI}$ OH, N, $\overset{+}{N}$—CH_3, O^- + CH_3I (17)

Following the adjacent charge rule, separation of a methyl group from a nitrogen atom would relieve strain in the molecule. Whether prior formation of the quaternary salt is an essential factor is not clear; the slow release of alkyl iodide during the procedure suggests that displacement of an equilibrium reaction might be involved in the process. It is also desirable to know whether low results are sometimes due to the amine splitting off, rather than the methyl group.

The von Braun reaction might offer an alternative separation procedure. The addition of cyanogen bromide to the tertiary base, followed by separation of alkyl bromide, would appear to be just as suitable and reliable.

$$(C_nH_{2n+1})NR_2 + CNBr \rightarrow R_2N \cdot CN + C_nH_{2n+1}Br \qquad (18)$$

Such a determination, independent of the *O*-alkyl method, could be useful for substances containing both *O*- and *N*-alkyl groups.

TABLE II
Percentage Yields of Demethylated Products for Anisole and Thioanisole with Various Reagents (60)

Reagent	Reaction conditions	Anisole, %	Thioanisole, %
Hydriodic acid, hydrobromic acid	2 hr., 130°C.	100	0
Pyridine hydrochloride	6 hr., 200°C.	100	0
Magnesium iodide	1 hr., 200°C.	58	0
Aluminum chloride	2 hr., 100°C.	100	20
Potassium hydroxide in ethanol	7 hr., 200°C.	7	0
Sodium in pyridine	5 hr., 120°C.	94	62
Sodium in ammonia	7 hr.	27	100
Sodium in ammonia	15 hr.	100	100

D. S-ALKYL FISSION

The carbon–sulfur bond is usually much more resistant to both acidic and basic reagents than the carbon–oxygen bond. Studies in the demethylation of thioanisole (60) show that cleavage is not effected readily by the usual reagents for demethylation of *O*-alkyl compounds (see Table II). However, thioanisole is readily cleaved by sodium in liquid ammonia. This reaction is also quantitative for anisole but proceeds quite slowly. It has not been developed as an analytical method for *S*-alkyl determinations. *S*-Benzyl groups are cleaved by the reagent (33).

An additional drawback to the use of the usual *O*-alkyl procedure for *S*-alkyl compounds is that the bond could break on the opposite side to the alkyl group. In a substance such as *S*-methylcysteine,

$$CH_3SCH_2CH(COOH)(NH_2)$$

fission could occur at the *S*–methylene bond, liberating the volatile thiol. This type of reaction would readily explain low values for substances regarded as difficult to demethylate. Cleavage of adenylthiomethylpentose (70) with hydriodic acid gives about two-thirds of the Me–S as methyl iodide and the other one-third as methyl mercaptan.

Sulfonium compounds resemble quaternary ammonium compounds in many respects; hence, the same procedure can be employed for separation of the alkyl group from nitrogen: namely, formation of the sulfonium compound and pyrolysis (36).

$$R_2\overset{+}{S}R'I^- \rightarrow RSR + R'I \qquad (19)$$

Since mercuric iodide and ferric chloride are necessary catalysts for the formation of sulfonium salts from the reaction of alkyl halides with disulfides, they might be of use in an analytical determination, just as gold chloride is useful in the methylimide determination.

E. STERIC EFFECTS

In routine analytical determinations, reaction times are of great importance and it is not always sufficient to know that a particular class of substance is cleaved in a particular time. The rate may be greatly diminished by atoms or groups which are not directly involved yet are close to the reactive centers of the reagent molecules. Such steric effects could well be responsible for low results in the determination. 2,6-*N*,*N*-Tetramethyl aniline, for example, does not form a quaternary ammonium salt with methyl iodide due to the two methyl groups *ortho* to the dimethylamino group (100).

F. MOLECULAR REARRANGEMENTS

Separation of the alkyl group from the O, N, or S atom does not necessarily mean that quantitative recovery of the group is assured, since the likelihood exists that it may migrate to another position in the molecule (100). This applies particularly to the *N*-alkyl determination. Trimethylanilinium iodide, for example, gives the following mixtures of compounds when heated to 250 to 280°C. and 300 to 330°C., respectively.

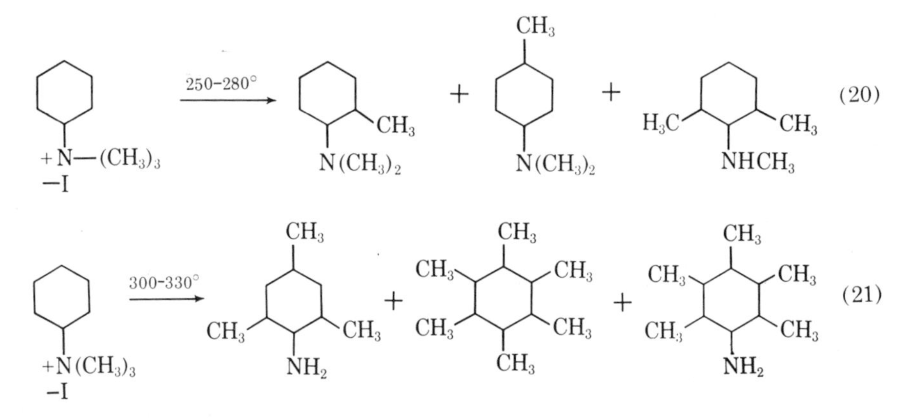

Rearrangements of alkyl aryl ethers are not as pertinent for the lower alkyl groups because the ease of rearrangement increases with the complexity of the alkyl group. However, low results might be expected for allyl ethers due to their known rearrangement to *o*- and *p*-alkyl phenols.

V. DETECTION AND IDENTIFICATION

A. PHYSICAL METHODS

Reference should be made here to the chapter on the determination of acyl groups for a general discussion on the use of spectrometric methods and for references, since they apply equally well to the *O*-, *N*-, and *S*-alkyl determinations.

1. Ultraviolet Absorption Spectroscopy

There is a low intensity band in the region below 270 mμ characteristic of ester groupings. Ethers and alcohols are transparent down to 185 mμ, hence their use as spectroscopic solvents.

2. Infrared Spectroscopy

In any CH_3X compound the individual CH frequencies are all directly related to one another and to the HX stretching frequency. When X is an element other than carbon, the position of the CH_3 symmetrical deformation frequency can shift appreciably, being dependent on the electronegativity and the position in the periodic table of the element to which the group is attached. *O*-Methyl, as exemplified by methanol, absorbs at about 1460 cm.$^{-1}$ (6.85 μ); *N*-methyl and *S*-methyl groups give absorption bands at 1420 and 1320 cm.$^{-1}$ (7.04 and 7.58 μ), respectively. However, the identification of these latter is complicated by the weakness of their vibrations and the variability of their positions. Methyl iodide absorbs at 1255 cm.$^{-1}$ (8.03 μ) and this band has been used for the quantitative determination of the methoxyl group (5), and the alkimide group (63a) after fission of the substance with hydriodic acid. Likewise, the absorption of ethyl iodide at 1215 cm.$^{-1}$ (8.30 μ) has been utilized for the ethoxyl determination.

Spectra of esters, aromatic ethers, and alcohols all show a strong absorption band between 1250 and 1200 cm.$^{-1}$ (8.00 and 8.33 μ) characteristic of the C–O linkage when the carbon is unsaturated. If the carbon is saturated, as in aliphatic ethers, this band appears at a lower frequency (1150 to 1060 cm.$^{-1}$, 8.70–9.43 μ). Methoxy and acetoxy steroids show a characteristic C–O stretching frequency at about 1100 cm.$^{-1}$ (9.09 μ) (77).

With lithium fluoride prisms or gratings the dispersion is sufficient to enable differentiation of the various types of CH groups in the 3000 cm.$^{-1}$ (3.33 μ) region. Moreover, it has been found that the intensities of CH_2 and CH_3 stretching vibrations in long-chain paraffins are directly related to the proportions of these groups present, so it is possible to measure quantitatively the relative numbers of CH_3, CH_2, and CH groups in hydrocarbons. This might also be useful in indicating the presence of different *O*-alkyl groups in the same molecule.

Methoxyl groups show a CH stretching band at 2832 to 2815 cm.$^{-1}$ (3.54–3.57 μ) which is not given by either ethoxyl or C-methyl groups (50). The intensity varies for different substances; two methoxyl groups give about double the peak height of one. *O*-Ethyl, *O*-propyl, and *O*-*sec*-butyl bands may be differentiated in the 2960 cm.$^{-1}$ (3.38 μ) region (78). Methoxyl groups in siloxane polymers have been determined at 1190 cm.$^{-1}$ (8.40 μ) (21).

3. Nuclear Magnetic Resonance Spectroscopy

The evidence accumulated suggests that nuclear magnetic resonance is suitable for identification of methoxyl groups, for estimation of the number of groups present, and for differentiation between aliphatic and aromatic methoxyl groups. The three proton signals are sharp and discrete for the group. If overlapping of signals occurs it is still a fairly straightforward matter to discriminate between signals from different protons. In chloroform solution the τ values for methoxyl attached to an aliphatic residue lie between 6.6 and 6.8, between 6.1 and 6.3 when it is linked to an aromatic residue, and between 6.2 and 6.4 in esters. These figures, of course, do not apply if there is anomalous shielding in the molecule, which would occur if a methoxyl group were constrained to a position above an aromatic ring. Pyridine is a good alternative solvent.

The signals from methylene protons α to an oxygen molecule occur at a slightly lower frequency than those from methyl protons, but the ethoxyl groups ($\cdot O \cdot CH_2 \cdot CH_3$) may be identified in aliphatics by the combination of the shifts due to the methylene group (quartet at 6.4 τ) and the methyl group (triplet at 8.8 τ). These shifts differ for aromatic protons linked to the same carbon as the alkoxyl group.

N-Alkyl and *S*-alkyl groups both produce similar paramagnetic shifts and show a band in the 7.9 τ region. *N*-Methyl groups usually give strong sharp signals which may be recognized in complex spectra but methylene groups adjacent to nitrogen, methyl ketones, and *O*- and *N*-acetates also give sharp lines in this region and confirmatory evidence may well be required. In the case of an aliphatic amine the signals can be shifted to lower frequencies by acetylation to confirm the assignment. For example, the 7.85 τ signal from methylamine shifts to 7.15 τ in *N*-methylacetamide. The quaternary ammonium salt might also be utilized in this way for identification purposes; it causes a shift of about 1.1 p.p.m.

Again, aromatic *N*-alkyl groups may be distinguished from aliphatic groups. Dimethylaniline, for example, shows a band at 7.0 to 7.3 τ.

S-Methyl groups give signals near 7.7 τ when α to aromatic protons but near 8.0 τ when α to aliphatic protons. Formation of the sulfonium salt lowers the frequency of the signal to 6.8 τ.

B. CHEMICAL METHODS

1. Methoxyl and Ethoxyl

The substance is heated with hydriodic acid to release alkyl iodide and the vapors are passed over mercuric nitrate paper (85). Formation of the orange color of mercuric iodide indicates the presence of either methoxyl or ethoxyl groups. Some butoxyl groups give the test, but it is unreliable for groups higher than ethoxyl. *S*-Alkyl groups do not give the test. Interference is given by compounds such as glycerol and epichlorhydrin, which also give volatile iodides with hydriodic acid. The colored complex formed by addition of alkaline methanolic 2,7-dihydroxynaphthalene, potassium ferricyanide, and potassium cyanide to methyl iodide absorbed in α-picoline may be used for its detection. This procedure has been developed as a submicro method for the determination of methoxyl (82).

Differentiation between methoxyl and ethoxyl groups can be made if the alkyl iodides are passed into trimethylamine in isopropyl alcohol (46). Methyl iodide forms tetramethylammonium iodide, which is practically insoluble, whereas the trimethylethylammonium iodide formed with ethyl iodide is soluble. Gas chromatography offers an excellent alternative method of detection, particularly when several alkyl iodides are present. Semiquantitative results may be obtained with very small amounts of substance.

Crystalline esters of the alkyl groups may be obtained and characterized by their melting points. For this procedure the alkyl iodides formed by hydriodic acid treatment are passed into a suspension of the silver salt of 3,5-dinitrobenzoic acid in absolute ether (36).

2. Alkyl Phenol Ethers

Phenol ethers can be dealkylated by heating to 150° with a mixture of potassium iodide and oxalic acid dihydrate (38). Gaseous hydriodic acid reacts with the ether

$$\text{ArOH} + \text{HI} \rightarrow \text{RI} + \text{ArOH} \tag{22}$$

During the heating, water is given off by the acid and facilitates the volatilization of the phenol which can be detected in the gaseous phase by the indophenol reaction with 2,6-dichloroquinone-4-chloroimine.

3. *O*- and *N*-Methyl

Oxidative cleavage with molten benzoyl peroxide produces formaldehyde from *O*- and *N*-methyl compounds (38)

$$:COCH_3 + (C_6H_5CO)_2O_2 \rightarrow :COH + (C_6H_5CO)_2O + HCHO \qquad (23)$$

$$:NCH_3 + (C_6H_5CO)_2O_2 \rightarrow :NH + (C_6H_5CO)_2O + HCHO \qquad (24)$$

Formaldehyde can be detected by the chromotropic acid test. Hexamethylenetetramine and vinyl compounds also give positive tests.

4. *O*- and *N*-Ethyl

When ethyl rather than methyl groups are attached to the oxygen or nitrogen atoms, molten benzoyl peroxide releases acetaldehyde which can be detected by the blue color upon reaction with a sodium nitroprusside–morpholine solution (38).

$$:N\ C_2H_5 + (C_6H_5CO)_2O_2 \rightarrow :NH + (C_6H_5CO)_2O + CH_3CHO \qquad (25)$$

Ethoxyl groups sometimes give difficulties and it may be better to treat these with alkali bichromate in sulfuric acid.

$$ROC_2H_5 + H_2O \rightarrow ROH + C_2H_5OH \qquad (26)$$

$$C_2H_5OH + [O] \rightarrow CH_3CHO + H_2O \qquad (27)$$

This test is very sensitive (1 μg./ml.) and may be used for detecting alcohol of crystallization or alcohol contaminant in organic compounds (32).

VI. DETERMINATION

Considering the vast range of substances that will yield an alkyl iodide upon reaction with hydriodic acid, it is perhaps remarkable that cleavage reagents other than boiling hydriodic acid have not been investigated more thoroughly for this determination. The position is in marked contrast to the acetyl determination, where the situation is, *a priori*, similar (acetic acid being substituted for alkyl iodide), yet a number of hydrolysis reagents and reaction temperatures have been proposed in order that the acetyl group may be quantitatively determined without interference from other parts of the molecule. This difference stems from the fact that the alkoxyl determination may be regarded, generally, as a simple and reliable procedure with occasional interferences, whereas the results of an acetyl determination must be accepted with caution and checked by an alternative procedure if possible.

Methods of *O*-alkyl determination have been proposed for particular classes of substances, but these have generally specified the limits of application of a method rather than the need for an alternative hydriodic acid reagent (36). While the means of separating the group has remained substantially the same, the main efforts over the years have been directed toward (*1*) overcoming problems associated with physical properties of the substance (such as volatility and solubility) which could contribute to incomplete reaction with hydriodic acid and (*2*) ensuring quantitative determination of the resultant alkyl iodide. More recently, efforts have been directed towards infrared (8)(63a) and gas chromatographic (65a)-(81a)(95) determinations of the alkyl iodide, the main advantage being that simultaneous specific determination of different alkyl iodides is possible.

The conventional determination is suitable for new personnel in a microanalytical laboratory, since its accuracy and lack of difficult techniques infuse confidence into the operator. Nevertheless, unsatisfactory results have been reported from time to time, indicating on the one hand that separation conditions were not severe enough, and on the other that groups other than the alkoxy group contributed to the result. Elimination of such results is a present need in the determination.

The general procedure for alkoxyl determination has already been outlined in the introduction. It may be broken down into a number of stages: solution of the substance in solvents, heating with hydriodic acid, separation of alkyl iodide, absorption and determination of alkyl iodide. In addition to having satisfactory conditions for each of these steps, one should also strive to have an efficient process, hence the need for a suitable form of apparatus and a carrier for the alkyl iodide in order to speed up the process without adversely affecting the reaction at each stage. Ideally, the above conditions would be independent of both the type of substance and its alkoxyl content.

These steps and conditions will be treated separately in the following discussion only for convenience of presentation. It is clear that much of the contradictory literature on this determination has occurred because of this very approach, and it cannot be stressed too strongly that the results of analyses are influenced not by one condition but by the particular combination of conditions, especially at lower levels of working.

A. EFFECT OF SOLUBILITY OF SUBSTANCE IN HYDRIODIC ACID

In many organic reactions better yields are obtained when either a solvent or a catalyst is employed, and it might be expected that some of the more inert ethers which require drastic cleavage conditions, such as heating with hydriodic acid or anhydrous aluminum chloride, would require solvents or catalysts to produce quantitative fission. Nevertheless, opinions have differed, first, as to whether solvents are advantageous, and, second, as to which are the most suitable solvents.

Phenol was first used for the macrodetermination by Weishut (99), who found that it functioned both as a solvent and as a catalyst in the reaction. He successfully analyzed substances which previously had either given low results or had required very long reaction times when heated with hydriodic acid alone, or mixed with acetic anhydride. It should be noted that Weishut used only 2 ml. of hydriodic acid with 0.2 g. of substance, whereas in the microdetermination there is usually a far greater excess of acid—2 ml. is added to only 0.005 g. of substance.

Pregl (47) dissolved the sample in phenol, or in a mixture of phenol and acetic anhydride; other workers used red phosphorus and hydriodic acid (97) or just hydriodic acid alone (66). Phenol was found to be unnecessary, although it contributed to a smoother and shorter reaction (13) while the use of solvents in general has been criticized because it sometimes leads to low results (81). On the other hand, solvents were found to be essential for highly methylated sugars (35). In this application a mixture of phenol and propionic anhydride was preferred to phenol, either alone or mixed with acetic anhydride. As a rule, most modern workers use solvents, although the question has not been resolved completely.

In the microdetermination many substances may be analyzed satisfactorily with hydroidic acid alone (62). A large number of unknown substances as well as known compounds such as narcotine, *p*-nitroanisole, 1,2,3-trimethoxybenzene, 3-(4-methoxybenzoyl)-4,5-dimethoxythionaphthene gave similar results with and without solvents and in many cases the results were slightly higher for hydriodic acid alone; furthermore, many of the samples were insoluble in hydriodic acid. The insoluble substances that give difficulty appear to be those that possess low melting points. As the acid is warmed they form an oil and creep up the sides of the flask away from the boiling hydriodic acid. Table III gives some typical results with low melting compounds. 2,6-Dimethoxybenzaldehyde is soluble and may be analyzed satisfactorily. The results for *o*-methoxytriphenylmethyl ethyl ether indicate that phenol alone is the preferable solvent for this compound because admixtures with acetic or propionic anhydrides give low results. The latter is interesting because it is initially soluble. However, it readily loses the ethoxyl group in the cold, forming the carbinol which in turn is reduced to the substituted methane. Removal of the hydroxyl group decreases the solubility to the extent that it separates as an oil in the presence of acetic or propionic anhydrides.

TABLE III

Alkoxyl Results for Low Melting Substances with Different Solvents

Substance	m.p., °C.	Solvent	Fd, %	Calc., %	Diff., %
Nerolin	72	phenol (0.7 g.)	19.6	19.6	0.0
		phenol (0.7 g.) + acetic anhydride (6 drops)	19.3	19.6	− 0.3
		none	7.7	19.6	−11.9
		none	13.9	19.6	− 5.7
p-Ethoxydiphenyl	76	phenol	22.7	22.7	0.0
		phenol + acetic anhydride	22.8	22.7	+ 0.1
		none	14.4	22.7	− 8.3
		none	17.1	22.7	− 5.6
o-Methoxytriphenylmethyl ethyl ether	81	phenol	19.4	19.5	− 0.1
		phenol	19.5	19.5	0.1
		phenol + acetic anhydride	18.6	19.5	− 0.9
		phenol + acetic anhydride	18.3	19.5	− 1.2
		phenol (few crystals) + propion anhydride	17.9	19.5	− 1.6
		phenol + propionic anhydride (6 drops)	16.7	19.5	− 2.8
		none	12.0	19.5	− 7.5
2,6-Dimethoxybenzaldehyde	71	none	37.8	37.4	+ 0.4

$$(Ar)_2\underset{\substack{\cdot\\ Ar\\ \cdot\\ OCH_3}}{C}\cdot OC_2H_5 \xrightarrow[HI]{} (Ar)_2\underset{\substack{\cdot\\ Ar\\ \cdot\\ OCH_3}}{C}\cdot H + C_2H_5I \qquad (28)$$

$$(Ar)_2\underset{\substack{\cdot\\ Ar\\ \cdot\\ OCH_3}}{C}\cdot H \xrightarrow[HI]{} (Ar)_2\underset{\substack{\cdot\\ Ar\\ \cdot\\ OH}}{C}\cdot H + CH_3I \qquad (29)$$

This, of course, is a physical phenomenon, and there is little doubt thas the reaction would be quantitative if these samples were by some means kept in contact with boiling hydriodic acid. It should be noted also that the quantity of phenol taken for these experiments is considerably greater than the "few crystals" originally recommended for microdeterminations (47).

There does not appear to be any advantage in adding propionic anhydride with phenol, since methylated sugars may be analyzed satisfactorily using phenol alone (61). There has been some criticism of the use of large amounts of solvent on the grounds that it will dilute the hydriodic acid and lead to low results, but this has not been substantiated. Propionic acid may play some part in the reaction, perhaps by acylation of hydroxyl groups, thus facilitating the fission. Hydriodic acid alone does not give 100% conversion of glycerol to isopropyl iodide unless propionic anhydride is added (19). Higher anomalous values are obtained for *tert*-butyl compounds when propionic anhydride is used instead of phenol (24).

It would seem safer to dissolve unknown substances in phenol first rather than to add a mixture of hydriodic acid and phenol. This also gives an indication of the solubility of the substance. Phenol gives a solid solution upon cooling which delays the reaction with hydriodic acid until the solution is warm, hence minimizing the risk of losing alkyl iodide upon addition of the acid.

It is surprising that so little has been written about the catalytic nature of the effect of phenol or of any other substance, since many workers boil the substance with hydriodic acid for prolonged periods to obtain complete reaction. Although Weishut (99) originally suggested that phenol acted as both a catalyst and a solvent on the macro scale, later workers (81) found that it actually retarded the reaction in some cases in the semimicrodetermination; some ethoxy compounds—for example, *p*-ethoxybenzoic acid and phenacetin—were affected. On the micro scale phenacetin gives quantitative values after only 10 minutes boiling with the reagents (62), so any retarding effect would be difficult to detect. This question of catalysis is important because the fission of some alkoxyl groups on the micro scale does not appear to follow the normal rapid pattern, and this means that the bulk of analyses on unknown samples is unnecessarily prolonged (61).

B. THE HYDRIODIC ACID REAGENT

1. *O*- and *S*-Alkyl Determinations

The reagent used in the determinations of *O*- and *S*-alkyl is the constant boiling acid (sp. gr. 1.7). In order to have a stronger reagent, or to strengthen the usual acid when it has been diluted with solvents, some workers have used fuming hydriodic acid (sp. gr. 1.94) (29,98). While this may be advantageous for fission in a sealed tube and dilution of the acid prior to distillation of the alkyl iodide, its use in the normal way is not recommended. When the acid is heated the excess hydrogen iodide is evolved until the constant boiling acid is produced. Some of this is liable to pass through the wash solution and cause high results.

The purity of the reagent has been the subject of much concern. There is a trend towards the use of the Steyermark procedure for preparation of a suitable reagent (88). This involves gentle boiling of the acid for two hours in a flask fitted with an air condenser while a slow stream of carbon dioxide or nitrogen is passed through the solution to remove volatile iodine compounds. The gas stream should be stopped at the same time as the boiling because fuming acid is produced if the gas stream is passed through cold acid. This practice gives perfect blanks and has been used for the preparation of acid for submicro determinations (15). All contact with organic matter should be avoided (26). The presence of iodine does not appear to be detrimental to the fission, although this has not been investigated thoroughly. The increase in iodine content upon refluxing the acid will increase the tendency for iodine vapor to be carried through the apparatus, with the consequent increased likelihood of a high result. This question is, of course, linked to the efficiency and choice of the wash solution.

It has been found that free iodine is advantageous for the analysis of sulfur compounds (90). These have caused difficulties, particularly in procedures which do not have a scrubber containing cadmium sulfate. However, the claim that iodine converts organic bound sulfur to the elementary form, thus preventing its interference, should be established more clearly. Some disulfides, for example, would be quite stable under these conditions, while others would be expected to produce at least some hydrogen sulfide; volatile thiols would be expected to escape from the solution and be swept into the washer with hydrogen sulfide before reacting with the iodine. This quantitative aspect of sulfur formation is most important on the micro or submicro scale. If iodine is, in fact, beneficial, some attempts should be made to define the reagent more precisely. At present it is loosely described as a dark-colored hydriodic acid. The concentration of iodine required for various types of sulfur compound would be of interest. It might also be of value for the *S*-methyl determination since formation of a volatile methyl mercaptan could be responsible for low values in this determination.

A criticism that can be made of the special grade hydriodic acid is that it may sometimes be over-strength (63). As a precautionary measure it is distilled in an all-glass apparatus, the first few milliliters of constant boiling acid being discarded. It is stored in dark bottles away from the light. If any difficulties are encountered with the reagent a pure acid may be prepared catalytically from the elements on platinized asbestos at 620°C. (61). Both procedures give a pale yellow acid which could be suitable for research into the effects of iodine concentration on the fission. A colorless hydriodic acid may be formed by addition of hypophosphorous acid (27). It gives satisfactory results and removes the necessity to separate iodine from the alkyl iodide vapors.

A reagent composed of hydriodic acid, phenol, propionic acid, and red phosphorus has also been recommended (67). The mixture was given a pretreatment similar to that described by Steyermark for hydriodic acid (88). The advantage of the extra chemicals has not been substantiated conclusively. The reagent apparently does not give satisfactory results for dimethoxydiphenylenedioxide (37,67), whereas phenol and hydriodic acid do (61), but it has been found to keep better than hydriodic acid alone (76). Results for tetraethoxysilane were some 5% low after four days, with the latter. Klimova and Zabrodina (68)(68a) prefer to use a mixture of phosphoric acid and potassium iodide at 195° for the cleavage. Reagents containing phosphorus should not be used with the infrared procedure because phosphine will vitiate results (5).

The evidence that hydriodic acid so rapidly deteriorates upon standing must be considered carefully, since it is not the first report of this nature. It was observed earlier that ethoxyl results were low for cellulose ethers once the acid strength dropped (81). Even if the acid strength were reduced, one would expect the excess water to be driven off and constant boiling acid to be formed during the analysis, but factors such as rate of boiling and flow of carrier gas through the apparatus might influence this point. Since these workers did not use solvents with the hydriodic acid their results suggest that the functions of phenol and propionic anhydride may not be solely that of solvents.

2. *N*-Alkyl Determination

The reagent for the *N*-alkyl determination is supplemented by the addition of ammonium iodide and usually one or two drops of a 5% solution of gold chloride which has been reported to give a catalytic effect (33). However, like so much of the literature on this determination, doubts have been thrown on this claim by the results of other work (35). Without gold chloride, but with freshly prepared hydriodic acid, recoveries of 95 to 100% were obtained with one distillation. The recovery varied with the quality of the hydriodic acid. Gold chloride did increase the recovery in the first distillation but did not improve the results for intractable substances. Recent work by Nakamura et al. (75a) clearly showed that aluminum is an effective catalyst.

C. SCRUBBERS

1. General

The scrubber, inserted between the boiling hydriodic acid and the absorption solution for alkyl iodide, has been the subject of much controversy, mainly because most scrubbers may be criticized on one ground or another. The fact that many competent analysts still do not agree on the suitability of various scrubbers is suggestion enough that conditions other than the composition of the scrubber might influence this choice. Before passing to a discussion of some of the recommended scrubbers, perhaps the requirements of an ideal one should be stated: It should retain hydriodic acid, iodine, and (preferably) sulfur compounds, quantitatively under optimum conditions for the analysis; it should operate effectively both when the hydriodic acid contains large amounts of iodine and when it bumps and boils unevenly, thereby upsetting the smooth flow of gas through it; it should not retain alkyl iodide. Hence, the demands on the scrubber are dependent on the substance, the carrier gas and its rate of flow, nature of the boiling, strength of the acid, concentration of iodine in the acid, and the condensing system between the boiling hydriodic acid and the scrubber.

Zeisel (102) originally used a red phosphorus suspension at 50 to 60° as a scrubber for the macrodetermination, which seems to be logical practice, since methyl iodide boils at 42° and is slightly soluble in water (see Table I). However, Pregl (47) chose a cold aqueous solution as a scrubber for the microdetermination, despite the fact that, theoretically, 1 ml. could dissolve 14 mg. of methyl iodide—or three times the amount usually present. Warming of the wash solution to 55° has been used (13) but does not appear to be necessary. It has also been criticized on the grounds that some alkyl iodide may be hydrolyzed at the elevated temperature (29). More recently, solid scrubbers have been used, although developmental work indicates that there could be difficulties resulting from their preparation.

The suitability of scrubbers for trapping hydriodic acid and iodine may be shown quite dramatically by vigorously refluxing hydriodic acid in a fast stream of carbon dioxide so that the acid condenses at the very top of the condenser (62). Table IV shows the results of various solutions with a synthetic acid in the Pregl apparatus. It is quite clear that iodine, rather than hydriodic acid mist (56), is the principal cause of the blank values under these conditions, since the alkaline solutions give high blank values. The low values obtained with carbonate solutions alone are signif-

TABLE IV

Efficiency of Various Wash Solutions with Abnormal Conditions in Pregl Apparatus

Wash solutions	Blank value
H_2O	0.483, 0.913, 1.830
Saturated $NaHCO_3$	0.640, 0.978, 1.280
10% K_2CO_3	0.180, 0.518, 0.075[a]
10% Na_2CO_3	0.090,[a] 0.083[a]
5% $Na_2S_2O_3$	0.000, 0.000
5% $Na_2S_2O_3$/10% K_2CO_3	0.000, 0.000
5% $Na_2S_2O_3$/5% $CdSO_4$	0.000, 0.000
5% $Na_2S_2O_3$/satd. NaCl	0.040, 0.035
1.5% $Na_2S_2O_3$/0.5% Na_2CO_3	0.000
Red P/H_2O	0.000
Red P/10% K_2CO_3	0.000
Red P/5% $CdSO_4$	0.000

[a] Fresh hydriodic acid of low iodine content.

icant because the acid used was fresh and had a much lower iodine content. Others have found that water is as suitable as any other wash liquid (18,57). This suggests that the conditions were less exacting than the abnormal ones above, but such contradictory conclusions serve to emphasize the earlier warning that the variables in this determination are interdependent.

2. Thiosulfate Scrubbers

Sodium thiosulfate solutions (42) replaced phosphorus suspensions, probably because the latter are efficient only while there is a thick suspension of phosphorus and become ineffective once the phosphorus begins to settle out. White (101) first drew attention to the weakness of sodium thiosulfate solutions as scrubbers. His figures showed that washing with thiosulfate alone gave results that were only 55 to 70% of those obtained with other wash solutions. This work followed a study (86) of the reaction rate of methyl iodide and sodium thiosulfate:

$$S_2O_3^{-2} + CH_3I \rightarrow {}^-S_2O_3CH_3 + I^- \qquad (30)$$

The reaction of ethyl iodide with sodium thiosulfate was found to be unimportant in an alkoxyl determination because the solubility of ethyl iodide in the wash solution is much lower and the rate of reaction is much slower. With methyl iodide the effect of the reagent is diminished upon the addition of potassium chloride, cadmium sulfate, etc., that is, neutral salts which restrict the dissociation of the thiosulfate ion; since cadmium sulfate was usually present to react with sulfur compounds, such thiosulfate scrubbers were satisfactory. In this regard, it is interesting to note in Table IV that a small blank value was obtained with a sodium thiosulfate–saturated sodium chloride scrubber. Here, presumably, the dissociation of the thiosulfate ion has been depressed to such an extent that the reaction with iodine has been affected. It would be hard to imagine a reaction with methyl iodide under these conditions. However, some years later, low results were obtained with sodium thiosulfate wash solutions in the methylimide determination (40) and also in a methoxyl determination (51). Their use has since been deprecated and workers employing them have been described as "tradition-bound chemists" (87). In view of the widespread use and satisfaction that a 5% sodium thiosulfate–5% cadmium sulfate solution has given (34,43,53,61,90), this question warrants further clarification.

The modern critics of the thiosulfate wash solution disregarded the points originally made (101) that low results with the washer are governed by two factors: first, the solubility of alkyl iodide in the scrubber solution—with a consequent dependence on the volume of the wash solution—and second, the rate of reaction of the dissolved iodide with thiosulfate. Accordingly, they not only used larger volumes of solution but also increased the time of contact of the gases with the solution by using spiral-type washers. Hence, the findings that thiosulfate washers retained alkyl iodide under these conditions could hardly be unexpected and merely vindicated White's work (101). Furthermore, it was not established that this increased volume (4 ml.) of thiosulfate solution in a spiral scrubber is essential to trap iodine or hydriodic acid quantitatively under the usual conditions of analysis. It was found independently (22,61) that both negligible losses of alkyl iodide and efficient trapping of iodine occur with smaller volumes (1 ml.) even at fast flow rates; other scrubbers have been used satisfactorily under comparable conditions. Sodium acetate, one of the reagents suggested as an alternative to a thiosulfate scrubber, was not satisfactory with fast flow rates (51). This suggests that sodium acetate is not acting as a scrubber and is only satisfactory when interfering vapors are held back in the reaction solution or in the condenser of the apparatus. If this is so, the choice lies between a reagent that can retain alkyl iodide and tends to give low results, or one which can allow iodine to pass, with the consequent tendency towards high results. But analytical chemistry abounds with methods that depend for their accuracy upon preventing undesirable side reactions. If, in the methoxyl determination, a sodium thiosulfate–cadmium sulfate wash solution is employed, the volume of the solution and the rate of gas flow are important in preventing the undesirable reaction of methyl iodide and sodium thiosulfate; likewise, if a sodium acetate solution is preferred, conditions must be controlled to ensure that interfering vapors do not reach the receiver and give high values.

The losses incurred by the use of the usual sodium thiosulfate–cadmium sulfate washer (61) become significant at the submicro level. Rather than discard such an efficient scrubber—control of blank values is extremely important at this level—the alternative is to modify it, either by increasing the concentration of cadmium sulfate or by adding sodium chloride (see Table IV). This has been done with satisfactory results (63). It should be remembered that this reduces the capacity for iodine uptake of the scrubber and, in fact, this washer should be replaced for each analysis.

3. Alternative Liquid Scrubbers

Sodium acetate (partly discussed in section C-2) (51) and sodium antimonyl tartrate solutions (40) have been proposed as alternatives to the thiosulfate mixture. In most respects the sodium antimonyl tartrate solution is a satisfactory solution and is free from controversy. It should be freshly prepared (15). It was compared with a sodium thiosulfate–cadmium sulfate mixture, each at various concentrations, using an almost colorless hydriodic acid reagent containing red phosphorus (22). The analytical results obtained for both scrubbers were dependent on the concentrations of the reagents. Under the conditions of the analysis 5% sodium antimonyl tartrate gave high values (+0.53%) whereas a 15% solution gave low values (−0.64%). The thiosulfate scrubber gave low values (−0.39%) at a concentration of 8% and high values (+0.31%) at a concentration of 3%.

Comparative tests also showed that there was little to choose between the washers in the presence of sulfur compounds (18). At the semimicro level, cadmium sulfate was found to be the reagent of choice when large amounts of sulfate were added to the hydriodic acid. The precipitate was also removed more easily from the apparatus. However, this work also showed that iodine vapors were not a problem at this level, which suggests that the scrubbing problem increases as the scale of working decreases. On the submicro scale, 8 ml. of a freshly prepared solution of sodium antimonyl tartrate in a spiral scrubber was required in order to get satisfactory blank values in the alkoxyl determination (15). For the *N*-alkyl determination this was supplemented by a 1.5-cm-long wad of silica wool soaked in sodium antimonyl tartrate and inserted in the neck of the condenser (16).

4. Solid Scrubbers

Solid absorbents such as soda asbestos (3)(39) and potassium antimonyl tartrate on kieselguhr (94) have been used successfully in the determination, but some samples of soda asbestos, and potassium antimonyl tartrate on silica gel, can retain alkyl iodide (94). The reason for this has not been ascertained.

The foregoing discussion assumes that the alkyl iodides are determined either gravimetrically as the iodide or volumetrically as the iodate. When either gas chromatography (95) or infrared techniques (3) are used, it is important to prevent water vapor from reaching the receiver (and, in the case of infrared, phosphine), hence the need for solid scrubbers. Cheronis and Ma (25) use a considerable amount of a cation-exchange resin (sulfonic acid type) moistened with 0.1*M* sodium carbonate solution for the *N*-methyl determination. When gas chromatography is used instead of the volumetric determination, this scrubber is followed by a second containing indicator-treated silica gel (81a). The latter is preferred to magnesium perchlorate, which can produce the explosive perchloric acid if hydrogen iodide passes into the drying tube.

5. Scrubbers for Sulfur Compounds

Over recent years the problems associated with the alkoxyl analysis of compounds containing sulfur have received a good deal of attention (87), although one advantage claimed for the commonly used iodometric procedure was that it was satisfactory for analysis of such compounds. To some extent this has paralleled the move away from the sodium thiosulfate–cadmium sulfate wash solution towards spiral types of washers containing larger volumes of alternative solutions. Slow passage of gas through a spiral washer, while it may insure reaction with iodine compounds, also increases the chances of undesirable side reactions.

Cadmium sulfate is an excellent scrubber for hydrogen sulfide. Other solutions are not quite as good as cadmium sulfate in this respect (18). Sodium antimonyl tartrate is also satisfactory, being only slightly less effective than cadmium sulfate (18). Sodium acetate solutions have also given satisfactory results (90) but rather because of iodine in the hydriodic acid or sufficient bromine in the receiver than because of any specific reaction in the wash solution. Soda asbestos has been used as a solid scrubber with satisfactory results (3).

D. ABSORPTION AND DETERMINATION OF ALKYL IODIDE

1. Volumetric Procedure

In contrast to the question of scrubber solutions, absorption and reaction of the alkyl iodide has been largely free from controversy. The iodometric procedure of Vieböck and Schwappach (96) has been adopted widely for routine determinations: a solution of bromine, sodium acetate, and glacial acetic acid effects thorough absorption and reaction of the iodide; upon dilution the iodine bromide so formed is oxidized to iodic acid, which is determined with sodium thiosulfate after destruction of the excess bromine with formic acid [equations (5) to (8)].

The speed and convenience of this process, plus the favorable conversion factor, make it very attractive; moreover, the presence of sulfur compounds is not important. An excess of bromine must be maintained in the solution. This may be depleted by reaction with sulfur compounds, or even by the continued passage of gas through the receiver. For microdeterminations it was found that 5 ml. of an acetic acid–potassium acetate solution containing about 1% (by volume) of bromine is desirable (51), in accord with earlier work (83), although 6 ml. of an 0.5% (by volume) solution gives satisfactory results (22).

It is also important to ensure that all bromine is removed from the solution and the flask with formic acid before the iodine is liberated. It has been found that too great an excess of formic acid can lead to low results and often imparts a yellow color to the solution (22). However, it gives identical results with the recommended procedure (section VII-A-1) (61) which requires a large excess to be used, primarily for convenience in routine determinations.

An alternative procedure (31), which uses pyridine in the receiver, has the advantage that the scrubber may be eliminated, although it is not as sensitive as the iodometric procedure. The alkyl–pyridinium iodide is titrated as a weak acid with $0.02N$ tetrabutylammonium hydroxide and azoviolet indicator. Hydriodic acid, iodine, and sulfides are first titrated as strong acids in this non-aqueous titration.

$$RI + C_5H_5N \rightarrow C_5H_5R^+I^- \qquad (31)$$

2. Gasometric Procedure

A gasometric procedure (93) can be used instead of the usual volumetric one if hydrazine is added instead of sulfuric acid. The former reacts with iodine to liberate nitrogen which is collected and measured.

$$2I_2 + N_2H_4 \rightarrow N_2 + 4HI \qquad (32)$$

This procedure does not appear to have any advantages over the iodometric titration.

3. Gravimetric Procedure

The ready solubility of alkyl iodides in organic solvents was utilized originally in the gravimetric determination when they were absorbed in an alcoholic solution of silver nitrate. To complete the determination, the contents of the receiver are transferred to a test tube using alcohol and dilute nitric acid, alternately, and boiled to ensure complete precipitation of silver iodide, which is quickly filtered and weighed. A final washing of the precipitate with concentrated nitric acid is important in eliminating errors due to silver or silver sulfide (90).

The method requires more technique than the volumetric one and is much slower. There has been a difference with respect to the need for a correction factor of 2% to the weight of silver iodide (29,42). Steyermark (90), who has used the method extensively, applies the correction and gives a detailed procedure for preparation of the reagent.

4. Separation of Alkyl Iodides

a. Gravimetric Procedure

When more than one type of *O*-alkyl group is present in the molecule, methyl iodide may be separated by passing the iodides into an alcoholic solution of trimethylamine; the tetramethylammonium iodide so formed is insoluble, whereas the higher alkyl groups form soluble salts (36). A solution of trimethylamine in isopropyl alcohol (46) was found to be superior to the ethyl alcoholic solution in giving an accurate selective determination of methoxyl from higher alkoxyl groups. Nitrobenzene has also been recommended as a solvent for the trimethylamine (72), its advantages being that the solubilities of the alkyl iodides are much different and the separation may be effected more quickly.

b. Combustion Procedure

In this gravimetric procedure, hydriodic acid fission is followed by combustion of the alkyl iodide in a tube containing a platinum catalyst (43). Iodine is absorbed on silver gauze and carbon dioxide on soda asbestos. The alkoxyl groups are characterized by the molecular ratio of the two. It has also been used for detection of alkoxyl groups by connecting the combustion tube to a solution containing potassium iodide and starch. This is a simplification of the earlier work of Friedrich (41). Fukuda (43a) employed a copper oxide catalyst instead of a platinum catalyst and adapted the procedure to a submicro determination.

c. Infrared

The alkyl iodides produced by hydriodic acid fission of *O*-alkyl and N-alkyl compounds may be collected in a special cell and determined with an infrared spectrometer (5)(63a). Methyl and ethyl iodides give quite distinct peaks at 1265 and 1215 cm.$^{-1}$, (8.93 and 8.30 λ), respectively. Phosphine and water vapor interfere, but iodine, hydrogen iodide, and hydrogen sulfide do not; hence, the usual blank value problems of the conventional determination do not arise. Instead, one must ensure that there are no phosphorus compounds in the hydriodic acid. (Hypophosphorus acid is often used as a stabilizer.)

The method requires a small amount of additional technique and separate calibration graphs—which are slightly curved—must first be obtained for ethyl and methyl iodides. Accurate analyses in the range 1.0 to 3.6 mg. of alkyl iodide may then be obtained. Besides giving positive identification of the iodides, accurate determinations of one iodide in the presence of the other may be made, unless the ratio of methoxyl to ethoxyl is greater than 4:1.

The technique has definite uses in research, as shown by the informative data obtained on the fission products of sugars and related compounds (6) which may give anomalous values. It has also been used for differentiating between ester and ether links in the substance (7) after reaction with constant boiling hydrochloric acid, which liberates only trace amounts of methyl iodide from ethers.

d. Gas Chromatography

Of the newer procedures and techniques that have been utilized in this determination, gas chromatography appears to have the attributes that could warrant using it instead of the usual iodometric determination. Methyl, ethyl, isopropyl, *n*-propyl, isobutyl, and *n*-butyl iodides can be differentiated in a sample using an octyl phthalate–Celite column at 100° (95). Only 10γ of methyl iodide is sufficient for the determination. The method is less sensitive for the higher alkyl iodides and 50γ of butyl iodide is required. It has the necessary accuracy at this level to allow simultaneous determination of the ratio of alkoxyl groups, but even at the semimicro level results were only within 5 to 10%. Chloroform was used as solvent and methylene dichloride as internal standard.

Some absorption problems have been encountered because of the small volumes of absorbent used for the alkyl iodide. Methylene dichloride, which is used as an internal standard, was lost during analyses (49) but *n*-heptane (0.5 ml.) in a spiral absorber at −80°C. was found to be satisfactory. Ethylidene dichloride is the recommended internal standard for ethyl iodide. *n*-Heptane in a Dry Ice trap has also been used in the analysis of acrylate and maleate esters in polymers (74). The column was composed of di-(2-ethylhexyl)sebacate on firebrick. Absorption on silica gel and subsequent volatilization has been recommended by Mitsui and Kitamma (75).

Kikuchi and Miki (65a) compared a variety of materials for absorption of methyl, ethyl and propyl iodides. Silica gel (30–60 mesh) coated with 5–10 per cent Apiezon L was superior to silica gel alone. Quantitative absorption and desorption occurred with both but with the former the affinity for moisture was reduced and sharper peaks were obtained on gas chromatography. Results for the gas chromatographic method were less accurate than one would expect from an alkoxyl determination but were generally satisfactory. Schachter and Ma (81a) used different conditions for collection and separation of alkyl iodides (methyl to hexyl). They simply used a cold trap to collect the alkyl iodide; tricresyl phosphate on glass beads (68a) was used for the chromatography instead of 25 per cent DOP and 5 per cent PEG 600 on Shimalite C (65a).

e. Other Methods

The procedure involving the use of sodium thiosulfate (40,101) gives a rapid distinction between methyl and ethyl iodides without the need for any elaborate equipment. If the sodium thiosulfate solution alone is placed in the scrubber of the *O*-alkyl apparatus, conditions may be adjusted so that methyl iodide is removed in the scrubber without materially affecting the yield of ethyl iodide, which is less soluble in the scrubber and slower to react.

This problem of ascertaining which alkyl group is attached to an oxygen atom naturally arises with substances of unknown structure and may be settled conclusively by other important analyses. A C-methyl analysis on the substance, for example, before and after demethylation would differ if groups higher than methyl were involved. Hence, the need for a procedure giving separation of the iodides is not often a demanding one. Furthermore, since all of the foregoing procedures are dependent on the initial reaction of the substance with hydriodic acid and assume that only *O*-alkyl groups produce alkyl iodide, they would all give misleading results if alkyl iodide was produced from some other grouping (59). Physical

TABLE V
Effect of Carrier Gas on Retention Time of Alkyl Iodides

Authors	Alkyl iodide	Time of distillation, min.	Rate of gas flow, ml./min.	Carrier gas
Inglis[a]	CH_3I	$7^1/_2$	15	CO_2
	C_2H_5I	$7^1/_2$	15	CO_2
	C_4H_9I	$7^1/_2$	15	CO_2
Houghton and Wilson[a]	CH_3I	35	6	N_2
Shaw[a]	C_4H_9I	180	6	N_2
Kirsten and Ehrlich-Rogozinsky	CH_3I	12	25	N_2
	C_3H_5I	22	25	N_2
	C_4H_9I	60	25	N_2

[a] Time of distillation includes reaction time as well; vanillin, phenacetin, and butyl alcohol, respectively, were added to the flask.

methods such as nuclear magnetic resonance (section V-A-3) could be very valuable in such cases.

E. EFFECT OF CARRIER GAS

On theoretical grounds carbon dioxide should be a more efficient carrier gas than nitrogen, and the results in Table V would confirm this. At slower flow rates it gives quicker quantitative transfer. In the submicrodetermination, carbon dioxide was found to produce an aerosol which gave high blank values, even at slow rates of flow, whereas nitrogen was satisfactory. As noted earlier, such differences may be connected with other variables in this determination. Certainly the Rigakos (79) type reaction flask, in which the carrier gas is actually passed through the acid (as in Fig. 1), would be more likely to produce aerosols than other types.

The rate of flow of carrier gas has generally been defined quite poorly as the number of bubbles per second issuing from an orifice which itself is not critically defined. The practice continues even though it is much more informative to relate this to the actual flow of gas through the apparatus. Fast flow rates have been criticized in the determination (4), the reason being that slow flow rates (of the order of 6 ml./minute) minimize tendencies towards inadequate gas scrubbing and incomplete alkyl iodide retention in the receiver. This criticism prompted speculation that anomalous values reported in the literature for polyhydric alcohols (80) might be caused by iodine which is formed during the reaction and not retained by the scrubber. This, however, is highly unlikely since the method

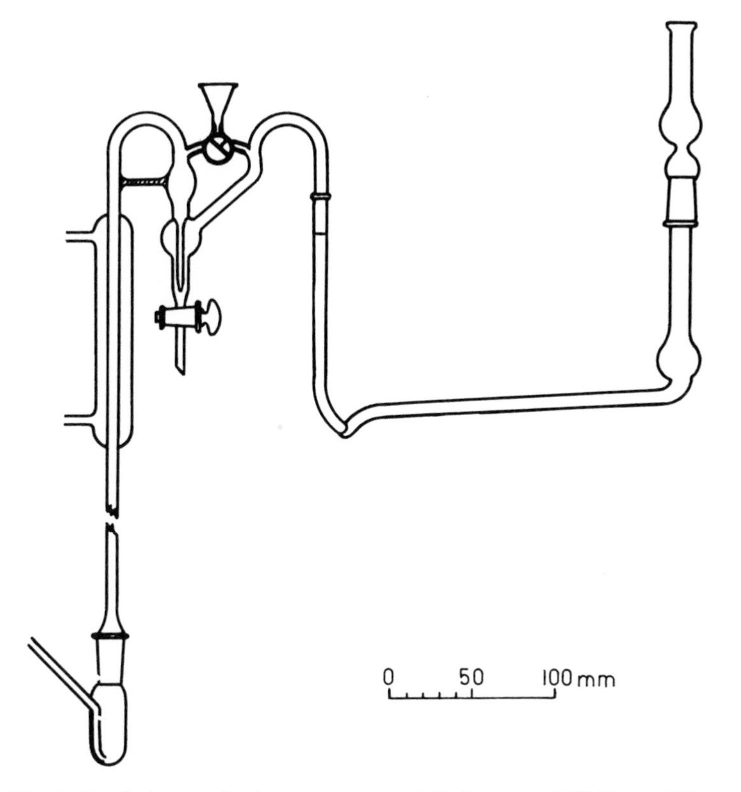

Fig. 1. Semimicro and micro apparatus of Hoffman and Wolfrom (53).

(53) used by Rudloff (80) recommended a scrubber containing sodium thosulfate, which is excellent for removing iodine (see Table IV) and the apparatus, Fig. 1, has a relatively long condensing system.

Procedures have been proposed (22,53,61,67,90) in which either carbon dioxide or nitrogen is passed through the apparatus at rates of from 12 to 25 ml./minute. Table VI shows the conditions used by these workers who all used different forms of the apparatus (compare Figs. 1 to 5). Using carbon dioxide at 15 ml./minute, which is certainly as fast as one would need for efficient transfer of alkyl iodide to the receiver (Table V), the analysis is not prolonged unnecessarily, particularly for the analysis of alkyl groups higher than methyl, and there is less chance of side reactions in the scrubber. This also means that duplicate determinations can be carried out with different reaction times, which gives an indication of whether the compound is behaving normally.

TABLE VI

Conditions Used with Fast Flow Rates of Carrier Gas

	A. C. S. Method	Budesinsky and Korbl	Hoffman and Wolfrom	Inglis	Kirsten and Ehrlich-Rogozinsky
Carrier gas	CO_2	N_2	CO_2	CO_2	N_2
Flow rate, ml./min.	15	25	25	12–15	25
Reagent	HI C_6H_5OH	HI C_6H_5OH Red P	HI	HI C_6H_5OH	HI C_6H_5OH Red P C_2H_5COOH
Scrubber	25% NaOAc	10% Na Sb tartrate	Red P/5% $CdSO_4$ or 5% $Na_2S_2O_3$/5% $CdSO_4$	5% $Na_2S_2O_3$/ 5% $CdSO_4$	30% $KHCO_3$
Apparatus	micro and semimicro Fig. 4	micro Fig. 2	micro and semimicro Fig. 1	micro Fig. 5	micro Fig. 3

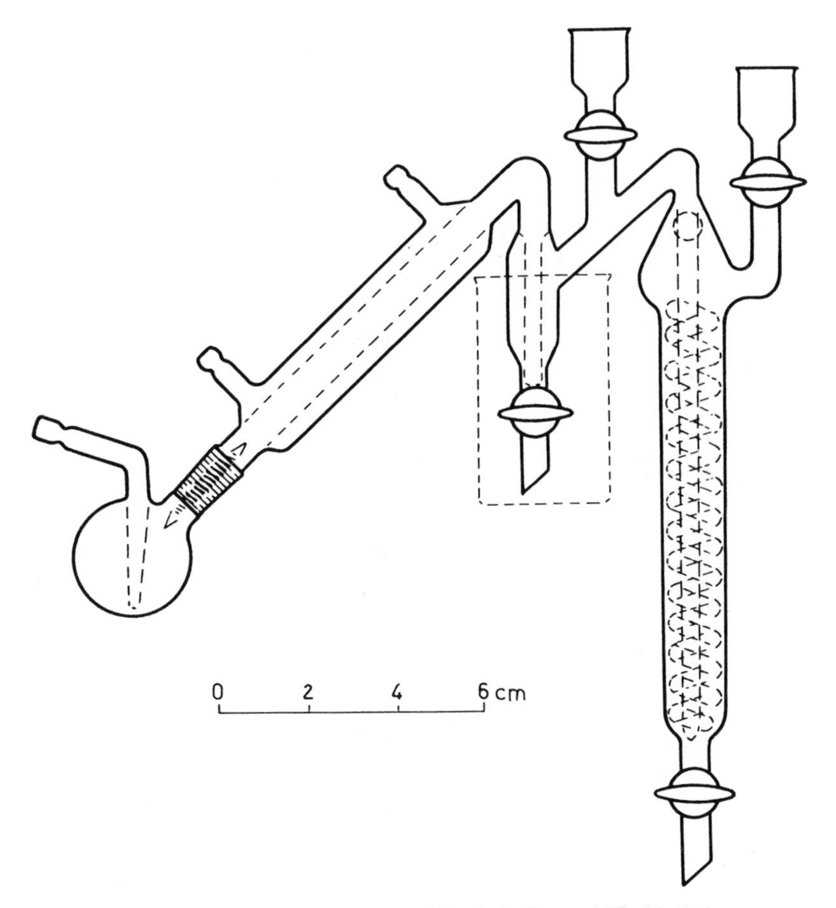

Fig. 2. Micro alkoxyl apparatus of Budesinsky and Korbl (22).

F. ANALYSIS OF DIFFICULT SUBSTANCES

The discussion in Section III on separation of the group has already indicated that a method for the determination of alkoxyl groups may not be universally applicable, and, indeed, there have been numerous examples both of low results due to incomplete reaction and of anomalous high results.

Often, the difficulty in a routine analytical laboratory is to obtain sufficient amounts of the so-called difficult substances to develop the method further, since they are often research compounds of value. Moreover, until the structure and the purity of such compounds is established unequivocally, there is need for reservation about the results. It is important that difficult compounds be supplied as standard substances because the

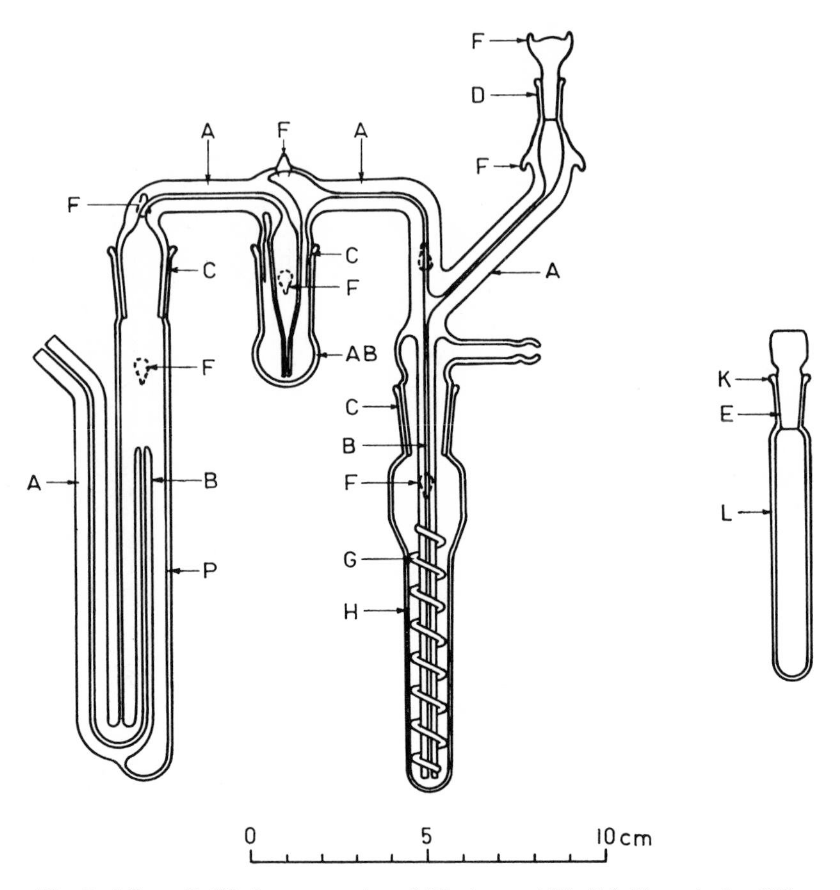

Fig. 3. Micro distillation apparatus of Kirsten and Ehrlich-Rogozinsky (67).

usual standards, such as vanillin and phenacetin, are readily cleaved, and while they are satisfactory for testing most parts of the procedure they do not certify that the cleavage conditions are ideal.

1. Substances which Give Low Results

a. Methoxyl and Ethoxyl Determinations

Several substances which have been reported to give low results gave no difficulty when the fission was carried out in boiling hydriodic acid containing substantial amounts of phenol (61); in fact, the reaction is very fast under these conditions, even for highly methylated substances, such as tetramethylmethylglucoside, which were previously given prolonged diges-

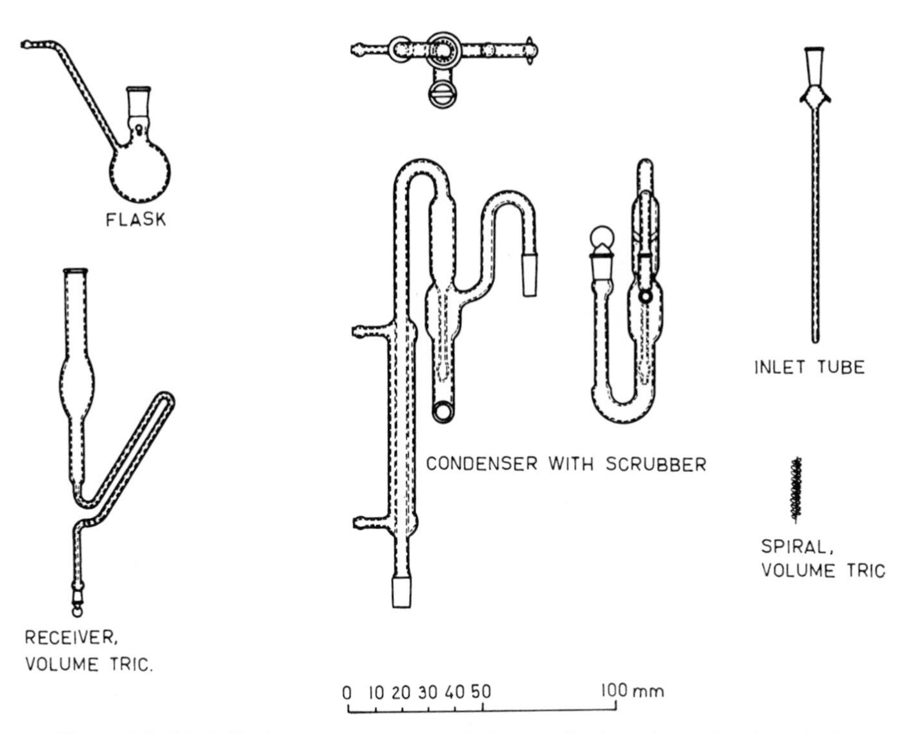

Fig. 4. Modified Clark apparatus for semimicro- and micro-determinations (89).

tion times with small amounts of phenol and propionic acid as an additional solvent (35). 9,10-Dimethoxy-1,2,5,6-dibenzanthracene (10) and dimethoxytriphenylene dioxide which had earlier given zero values were also satisfactory. The latter substance required $2^1/_2$ hours for quantitative fission in a reaction mixture containing red phosphorus, phenol, and propionic acid in addition to hydriodic acid (67). However, it is probable that the difference is due to a temperature, rather than a reagent effect. The fission was carried out in a sealed tube at 100° (not 127°) and it has been shown that the reaction rate usually decreases quite markedly with decreasing temperature (61).

Some isoflavones with methoxyl groups in the 7-position and the substance melicopicine require prolonged reaction times (61). These particular compounds have in common a carbonyl group in the *p*-position to the methoxyl and an additional benzene ring, both of which would be expected to deactivate the group. If this were the case, modification of the carbonyl group might be effective in shortening the reaction time for these substances. Compounds with a nitro group substituted in either the 6-position

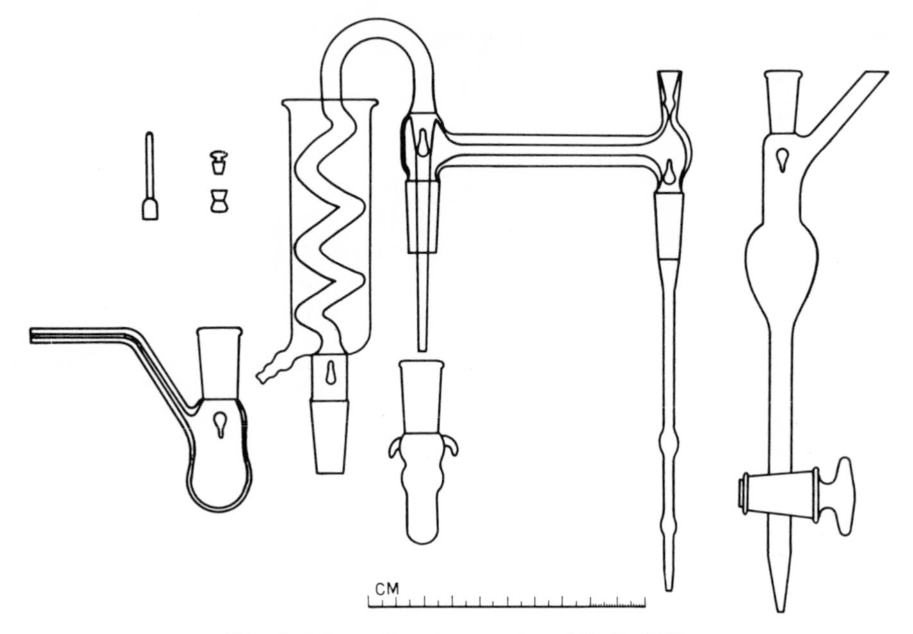

Fig. 5. Micro alkoxyl apparatus of Inglis (61).

of 5,8-dimethoxyquinolines, or the 5-position for 8-methoxyquinaldines gave very low recoveries of methyl iodide (77a). Quantitative yields were obtained using the more vigorous conditions employed for N-alkyl determinations.

b. Higher *O*-Alkyl Determinations

Isopropyl-β-glucoside tetraacetate was found to be intractable (67); results were 1.7% low after 30 minutes at 100°; extension of the reaction time to $2^1/_2$ hours produced lower results. Steric hindrance was considered to be the cause of the difficulty of analysis, but such results would also be consistent with the occurrence of a secondary reaction of isopropyl iodide with the reaction mixture in the sealed tube.

Determination of higher groups such as *O*-butyl, -amyl, and -allyl has received very little attention. Since the iodides have higher boiling points, precautions must be taken to ensure their complete removal from the apparatus either by heating the condenser jacket (61) or the scrubber (22,94). Fission of some butoxyl and allyl ethers (63) appears to be straightforward, at least, but *tert*-butyl groups give very low recoveries of *tert*-butyl iodide (24). This is undoubtedly due to the reactivity of the latter. It has been shown that interference of the *tert*-butyl group in a methoxyl determination might be eliminated by appropriate choice of solvents and scrubber solution (24). Kretz (69) used a sealed tube at 120 to 130°C. for 2 to 4 hours for fission of semimicro amounts of alkoxy groups.

c. *N*- and *S*-Alkyl Determinations

The *N*-alkyl determination may give low results even after several repetitions of the procedure (90). Since the substances are often unknown research compounds, it is impossible to know whether the low values are due to failure to form the quaternary ammonium salt, volatility of the compound, migration of the group, or steric hindrance in the molecule. Thermal decomposition of methyl and ethyl iodides can be a source of error (40).

Simply heating the substance with hydriodic acid in a sealed tube at 250–300°C for 2 hours is sufficient to demethylate some alkaloids (63a) but results for morphine hydrochloride are not quantitative.

A much simpler indirect submicrodetermination (17) has been developed for tertiary nitrogen contents of organic compounds as an alternative to the tedious Herzig and Meyer method. The quaternary ammonium compound is formed by reaction with excess methyl iodide; the halide is replaced by hydroxyl using an anion-exchange resin in the form of the free base, and the quaternary ammonium hydroxide is titrated with sulfuric acid using a mixed indicator (methyl red–methylene blue). Excellent results were obtained for alkaloids. This method, too, has limitations, in that the methylation is not always quantitative. *o*-Phenylenediamine, for example, reacts with only one mole of methyl iodide, whereas the *p*-compound reacts with two (20).

The determination of *S*-alkyl groups has not been widely applied to known substances. A method (12) which is essentially a prolonged Zeisel determination, and was developed for the amino acid methionine, gives low results for other amino acids, such as *S*-methylcysteine, even with variations of the procedure (1). The determination might require much more drastic conditions: the alkimide conditions, for example (36). However, methods should be developed which cover the possibility that fission does not occur primarily at the sulfur–methyl linkage because formation of the volatile methyl mercaptan would lead to low results with the usual method. A method has been developed which determines alkoxyl, *S*-methyl, and thiol groups (54).

d. Volatile Substances

Volatile substances obviously present an additional problem since they are liable to escape from the hydriodic acid before the alkyl group is split off. Fortunately, the very volatile esters, ethers, and alcohols react very rapidly, and the main problem is to bring them into contact with the acid underneath its surface (53). Sealed glass capillaries must be used for weighing the very volatile ones. They should be made in relatively long capillaries and centrifuged to the tail; when they are dropped tip first into the reaction flask the volatile liquid is above the level of the hydriodic acid. The liquid is only ejected when the acid is warm and reacts immediately.

It should be remembered with highly volatile compounds, such as ethyl ether, that the weighing is extremely important because the alkoxyl content is very high, and only small sample weights are required for the analysis.

Small glass bottles, either stoppered (61) or covered with a film of tartaric acid (53) may be used for less volatile liquids.

To guard against the possibility of these substances escaping from the reaction solution, some workers prefer to use two reaction flasks (44,90). Anisole was found (29,61) to be particularly troublesome and to warrant special apparatus. On the other hand, satisfactory results were obtained merely by adding it in a stoppered glass bottle to a mixture of phenol and hydriodic acid (4).

Volatility of the compound has been suggested (16) as the reason for low *N*-methyl results on *N*-methyl acetanilide.

2. Substances which Give High Results

High results may be expected in *O*-alkyl determinations whenever the substance also has an *N*-alkyl group present, since the latter may be so labile as to rupture with boiling hydriodic acid (36). 1,2-Dimethylpyraz-3,6-dione gives a result corresponding to one methoxyl group (48). *S*-Alkyl groups react in a similar manner (11). It may sometimes be possible to vary the reaction time and so distinguish the groups, although a total determination of *O*- and *N*-, or *O*- and *S*-alkyl groups would be more satisfactory.

Trimethylammoniumpyrimidine chloride (65) and *p*-nitrosodimethylaniline (3) slowly liberate methyl iodide on boiling with hydriodic acid. 2,3-Dimethyl-2,3-bis(*p*-hydroxyphenyl)butane (58) also gives positive results, but under abnormal conditions.

tert-Butyl phenols give positive results for the alkoxyl determination (8). The extent of their contribution in the methoxyl determination can be minimized by using phenol as solvent and a scrubber containing sodium thiosulfate–cadmium sulfate (24). Presumably the thiosulfate reacts with the labile iodine of *tert*-butyl iodide. A hydrochloric acid reagent allows differentiation between *tert*-butyl and *tert*-butoxyl groups (8).

The microdetermination of glycerol (19) may be carried out in an alkoxyl apparatus, isopropyl iodide being determined after reaction of the substance with hydriodic acid. Therefore, it is to be expected that substances with similar groupings will form volatile iodides with hydriodic acid. The major product of hydriodic acid reaction with ethanediol and with 1,2-propanediol was found to be ethyl iodide (80), and compounds with adjacent hydroxyl groups would be expected to react as follows:

$$HOCH_2CH_2OH \xrightarrow[HI]{} ICH_2CH_2I \longrightarrow CH_2{=}CH_2 \longrightarrow CH_3CH_2I \quad (33)$$

Likewise, in the case of glycerol, isopropyl rather than propyl iodide is produced due to favored Markownikoff addition of hydriodic acid to the intermediate olefin. This work suggests that a true methoxyl value might be obtained by using a method involving separation of methyl and ethyl iodides. This work also confirmed that sorbitol and mannitol gave high alkoxyl values (9)—12 to 14% "methoxyl" content—which accounted for high values for the hexamethylated substances. It should be possible to eliminate this interference either by modifying the substance prior to reaction or by changing the cleavage reagent. The reaction may be modified somewhat by the addition of solvents. Excellent results have been reported for hexamethylsorbitol (35) although the main difference in the determination was that phenol and propionic anhydride were used as solvents with the hydriodic acid. Values of 0.2% "methoxyl" were obtained for sorbitol after a reaction time of 30 minutes with phenol and hydriodic acid (63). It has been shown by infrared analysis that sugars and related compounds yield 2,5-dimethylfuran, which is a cause of anomalous values (6). These products, however, are generally minute and only significant for very low alkoxyl contents. This work also showed that α-methyl-D-glucoside gave satisfactory values in disagreement with an earlier finding (80).

Polyoxyalkalenes, $[-(CH_2CH_2O)_x]$, may give a stoichiometric amount of iodine with hydriodic acid (84). Dioxane and ethylene glycol, for example, may be determined in this way. The presence of this grouping in a substance would, therefore, be responsible for high alkoxyl values.

Cyclic ethers would also be expected to yield volatile iodides with hydriodic acid. A modification of the hydrogen iodide reagent—phosphoric acid and potassium iodide (91)—is effective in splitting substances such as tetrahydrofuran:

$$\begin{matrix} CH_2\text{—}CH_2 \\ | \quad\quad | \\ CH_2 \quad CH_2 \\ \diagdown \; \diagup \\ O \end{matrix} \rightarrow \begin{matrix} CH_2\text{—}CH_2 \\ | \quad\quad | \\ CH_2 \quad CH_2 \\ | \quad\quad | \\ OH \quad\; I \end{matrix} \rightarrow \begin{matrix} CH_2\text{—}CH_2 \\ | \quad\quad | \\ CH_2 \quad CH_2 \\ | \quad\quad | \\ I \quad\quad I \end{matrix} \qquad (34)$$

Diiodobutane is quite stable in boiling hydriodic acid.

3. Specific Methoxyl Determination

A sensitive colorimetric determination has been developed for the methoxyl group (73). The substance is hydrolyzed with sulfuric acid; the methanol formed is separated by distillation and oxidized to formic acid with permanganate; excess permanganate is removed with sodium bisulfite and the formaldehyde is determined colorimetrically after reaction with chromotropic acid.

The method is applicable to submicro quantities but is not as attractive as the general method because it is longer and more involved. But it is largely specific for the methoxyl group—glycolic acid, as well as other methylenedioxy or labile methylene groups, give high values—and, therefore, valuable in confirming anomalous values due to glycerols and labile *N*-alkyl groups. It has been applied to the methoxyl determination in alkaloids (71) since these often contain both *O*-alkyl and *N*-alkyl groups.

Alcohol formed on acid hydrolysis of alkoxyl in borohydrides (2) and organoaluminum (30) compounds can be determined either titrimetrically (2) or colorimetrically (30) by oxidation with excess standard ceric ammonium nitrate solution. The use of reagents other than hydriodic acid for fission of the group has been quite neglected and the usefulness of this particular alternative should stimulate further work in this direction.

G. APPARATUS

The apparatus, more than any other feature of the procedure, serves to indicate the important influence of quite small details on the results of the determination. It is remarkable how many papers, irrespective of the aspect they cover, include an apparatus modified in some way—admittedly sometimes only because of personal preference, but more often motivated by erratic results in practice. This has led to a great variety of forms of apparatus, including a tendency toward overelaboration by the addition of double scrubbers and double absorption tubes for the alkyl iodide (18). The various features of the apparatus will be discussed briefly here, first with reference to the simple apparatus (Fig. 6) with which Pregl originally determined micro quantities of alkyl groups in 20 minutes.

A regulated flow of carbon dioxide is passed down the side tube *A* and carries the alkyl iodide through the apparatus to the receiver *B*. It passes through the air condenser *SR* and the washer *W* which removes interfering vapors such as iodine.

1. Reaction Flask

Pregl, of course, did not have excellent ground glass joints available. Modern practice is to use joints between the condenser *SR* and flask *SK* and at the washer *W* so that series determinations are facilitated. This also simplifies cleaning the apparatus. However, there has also been a trend in the United Kingdom towards keeping the reaction flask and condenser in one piece with the joint at the top of the condenser (Fig. 7). This is presumably to eliminate losses at the hot joint, although the reaction flask is not as convenient for adding the substance.

Smooth boiling is important for optimum results. Various boiling aids have been recommended, perhaps the most satisfactory being platinum tetrahedra and boiling tubes. Rigakos (79) led the carrier gas directly to the bottom of the flask. This is a good means of getting even boiling but may give small losses due to sample remaining unreacted in the inlet tube (61). The flasks are often placed in electrically heated air baths for uniform and reproducible conditions.

2. Condensing System

Apart from the free use of ground joints, it is interesting to note the close resemblance of the modern apparatus of Bethge and Carlson (Fig. 8) to the Pregl apparatus (Fig. 6). A significant difference, however, is the increased volume and length of the condenser in the apparatus. There is ample safety margin for condensing acid and iodine vapors and, accordingly, it was found that there is no need to scrub for iodine vapors at the semimicro level of work (18). In the Pregl form this portion is so short that any uneven boiling of the hydriodic acid or variation in gas flow is immediately accompanied by a sudden surge of bubbles through both the washer and the receiver with the consequent danger of erratic results. In addition, the condenser is not efficient enough for prolonged boiling periods. It is this weakness in the apparatus for micro determinations that has led to slow flow rates, work on aids for smooth boiling of the acid, water condensers above the reaction flask, and scrubbers and receivers modified to give complete absorption of the appropriate vapors.

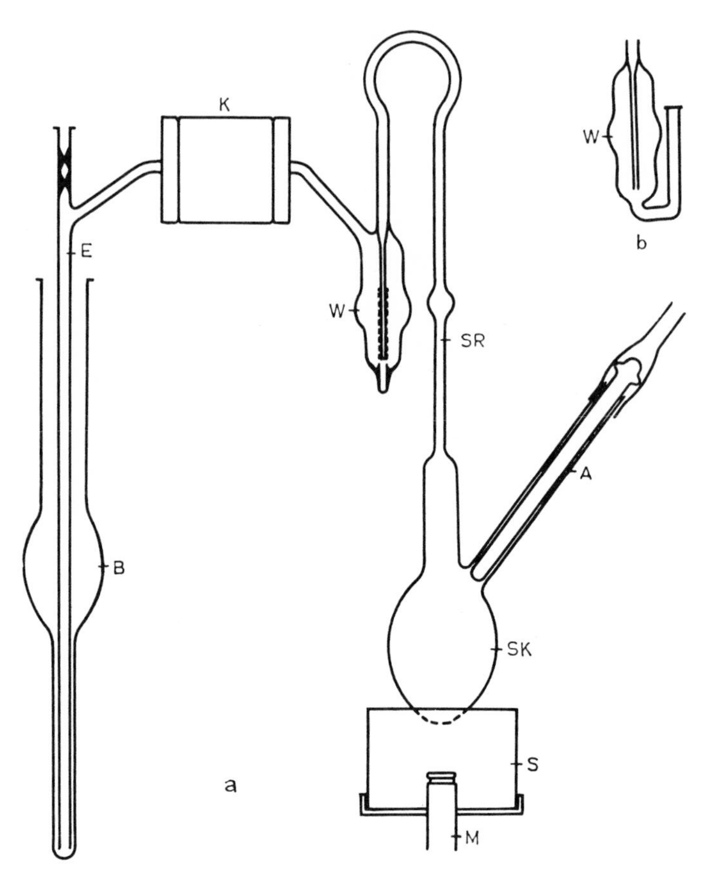

Fig. 6. Pregl apparatus (47).

Water condensers plus slow gas flow rates have been recommended (14) (see Fig. 7). This may be contrasted with the use of an air condenser with the fast flow rate of 25 ml./minute (67) (Fig. 3). However, the apparatus in Fig. 3 is used only for the distillation of the alkyl iodide and at the same time the problem of removing iodine is mullified because red phosphorus is present in the reaction mixture. A suitable compromise consists of a simple jacket with an opening at the bottom (Fig. 5). It may be filled with water at any temperature and emptied readily. The tube inside the jacket is sinuous and not the usual vertical one. It should result in more turbulent flow, hence minimizing effects due to erratic boiling of the hydriodic acid and to aerosols. This has not been proved directly, although zero blank values may be obtained with this system at carbon dioxide flow rates of 15 ml./minute.

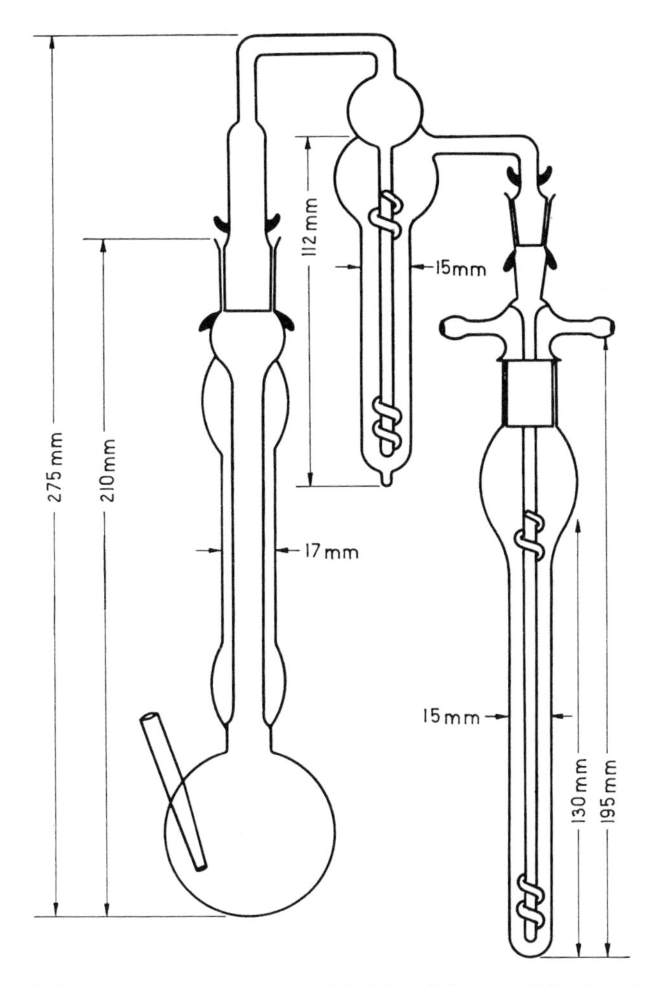

Fig. 7. Semimicro and micro apparatus of Belcher, Fildes, and Nutten (14).

Minor objections to water condensers are that the alkyl iodide may be cooled unnecessarily and that there is increased likelihood of breakage when dismantling the apparatus for cleaning. If higher *O*-alkyl groups are to be determined, hot rather than cold water jackets are desirable, otherwise quantitative recoveries of the higher boiling iodides will not be obtained in reasonable times. Since the determination can still be free from other iodide interferences under these conditions (61) it seems illogical to use cold-water jackets at all. However, they do enable more methoxyl and ethoxyl determinations to be made without cleaning the apparatus.

As an alternative, the wash solution may be heated with water at 95°C. (22) (Fig. 2), the receiver being shielded from the heat with asbestos.

3. Scrubbers

The design of the scrubber is linked to the speed of reaction of the solution therein with hydriodic acid and iodine. The simple scrubber used by Pregl (Fig. 6) is effective with a good scrubber solution (see Table III). Glass wool or sintered glass (90) at the inlet to the solution breaks up the gas stream and increases the efficiency of the washing. Spiral scrubbers (Fig. 7), which increase the time of contact of the gas stream with the solution, have also been recommended (14).

4. Receivers

There has been a trend towards standardization on spiral-type receivers (35)(64) (Figs. 4 and 7) which allow a long period of contact with the gas, rather than the simpler type of receiver exemplified in either Fig. 6 or Fig. 8, although collaborative studies have shown that the spiral is unnecessary (89). The criticism that may be made of both these types is that large bubbles of gas pass quickly through the solution when any sudden increase in gas flow occurs. The receiver in Fig. 5 does not suffer from this disadvantage, while it retains the simplicity and ease of rinsing of the Pregl unit. In the event of a sudden increase in gas flow, a gas pocket builds up in the bottom of the receiver while small bubbles continue to be pressed up through the solution. Complete absorption was confirmed by addition of a second receiver to the system (62).

5. Modifications for *N*-Alkyl Determinations

The method for *N*-alkyl determinations assumes that no methyl iodide is released under the conditions of the *O*-alkyl determination and is usually preceded by the latter to prevent interference. It is, therefore, convenient to use the same apparatus for both determinations. In this case, however, since the method depends upon pyrolytic decomposition of the quaternary ammonium salt at 360°, there must be provision for collecting the hydriodic acid before it reaches the scrubber and for its return to the reaction flask for a repetition of the procedure. The acid is usually drawn back into the reaction flask upon stopping the gas flow and cooling the flask; consequently, the design must be such that the scrubber solution is not also affected similarly.

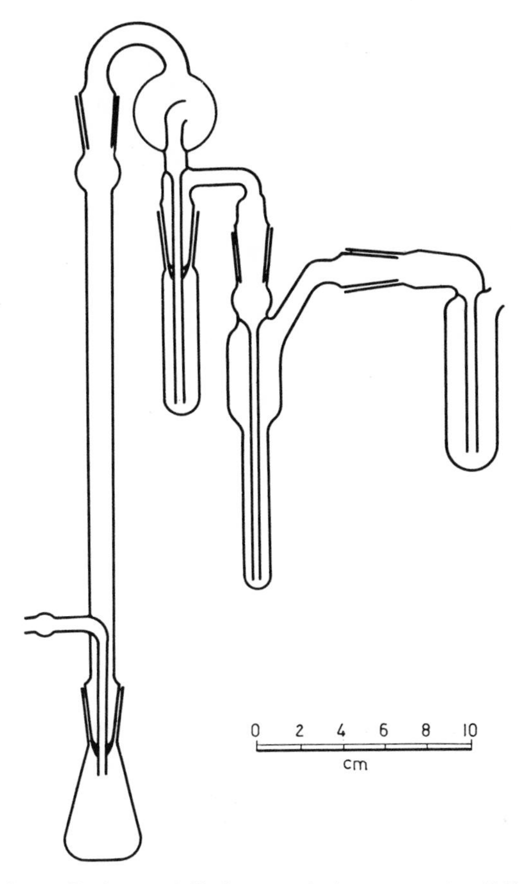

Fig. 8. Bethge and Carlson semimicro apparatus (18).

The apparatus of Friedrich (Fig. 9) was primarily designed for *N*-alkyl determinations and has been used with success on tractable substances (90). The tube *BB′* allows the alkyl iodide to pass to the washer unimpeded while the hydriodic acid condenses and collects in *AA′*. The wash solution is prevented from being sucked back into *AA′* by including the bulb *C* in the apparatus and making the stopcock connecting *BB′* three-way. As the reaction flask is cooled, air is sucked in via *B′*–*A′*–*A* and forces the acid back into the reaction flask. Fig. 10 illustrates a methoxyl apparatus, also suitable for determination of higher alkoxyl groups and alkimide groups (81a). The unit is primarily designed for collection of alkyl iodides prior to quantitative gas chromatography, hence the solid scrubbers and the special collector unit immersed in a cold bath.

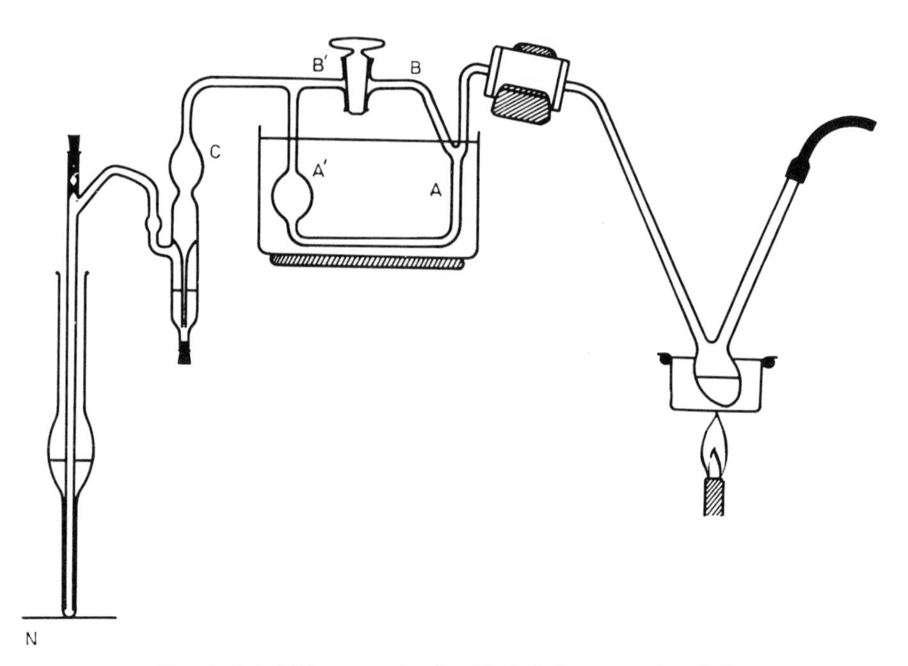

Fig. 9. Friedrich apparatus for *N*-alkyl determination (42).

6. Modifications for Volatile Liquids

Although the need for specialized apparatus has been questioned for the analysis of substances that may escape from the hydriodic acid solution prior to reaction (see section F-1-d), relatively convenient forms have been proposed and, indeed, found to be essential (44) to ensure complete reaction. Very volatile ethers and esters have been analyzed in an apparatus with two reaction flasks (44). A modern form (90) of this apparatus is shown in Fig. 11.

The same principle may be utilized with a conventional type of alkoxyl apparatus (61) (Fig. 5) but with a modified reaction flask (Fig. 12). The small ground-glass adaptor is inserted in the reaction flask after addition of the reagent and forms an upper chamber. Additional reagent is placed in this to trap substances that escape from the reaction flask.

Both units also satisfy the requirements for *N*-alkyl determinations, since the second reaction flask may be left empty to act as the receiver for hydriodic acid distilled from the first. They therefore have the advantage of versatility. With the apparatus in Fig. 11 the problem of preventing transfer of the scrubber solution is overcome simply by lowering the second reaction flask just enough to break the seal at the joint, whereas the

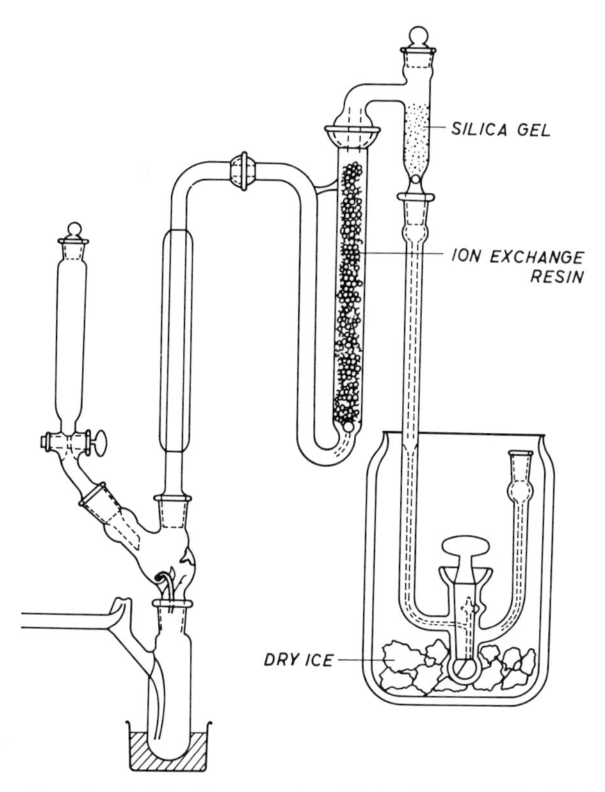

Fig. 10. Micro alkoxyl and alkimide apparatus of Schachter and Ma (81a) for collection of alkyl iodides prior to gas chromatography.

vessel containing the scrubber solution (see Fig. 5) would be lowered if the reaction flask in Fig. 12 were used.

VII. RECOMMENDED METHODS

A. *O*-ALKYL DETERMINATION

1. Micro Method (61)

Apparatus

The apparatus shown in Fig. 5 is used routinely. The two bulbs on the delivery tube should just check the gas flow, forming small pockets of gas at these points. The reaction flask is connected to a source of carbon dioxide. Dry Ice packed into a Thermos flask fitted with a mercury pressure head is suitable. Pass the gas through a bubbler of concentrated sulfuric acid. Check the flow rate at the bubbler.

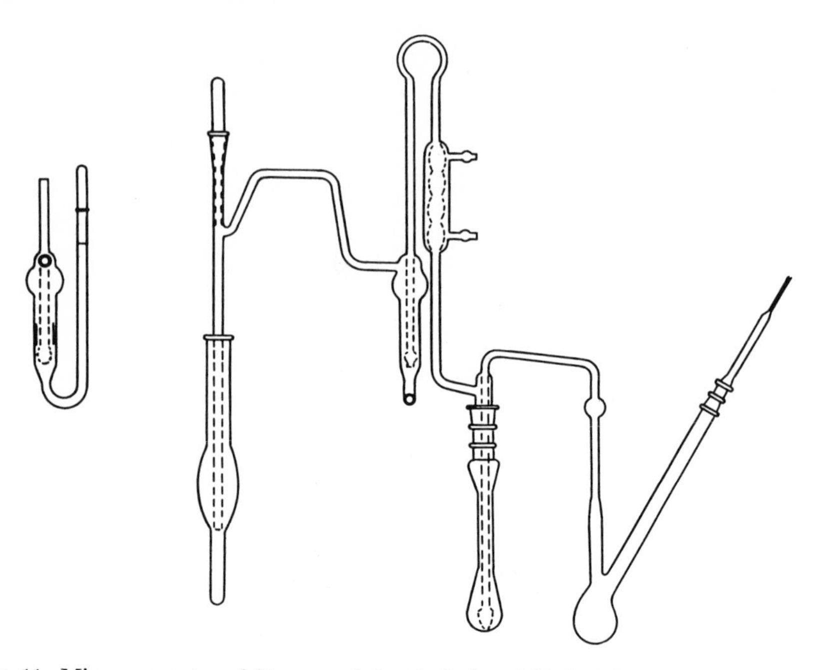

Fig. 11. Micro apparatus of Steyermark for *O*-alkyl and *N*-alkyl determinations (90).

The reaction vessel shown in Fig. 12 is used for volatile liquids. It can also be used for *N*-alkyl determinations.

Reagents

All reagents are analytical or microanalytical grade.

Hydriodic acid (sp. gr. 1.7) (Note **1**).

Phenol.

Cadmium sulfate 5%.

Sodium thiosulfate 5%. The solution contains a small amount of sodium carbonate (0.02%).

Sodium acetate (hydrated) 20%.

Bromine–sodium acetate–acetic acid. Dissolve 40 g. sodium acetate in 400 ml. glacial acetic acid and add 4 ml. bromine.

Formic Acid 95%.

Potassium iodide.

Sulfuric acid 20%.

Starch 1%. Make 1 g. of soluble starch into a paste with cold distilled water and add with stirring to 100 ml. of boiling distilled water. Add a few milligrams of mercuric iodide as a preservative.

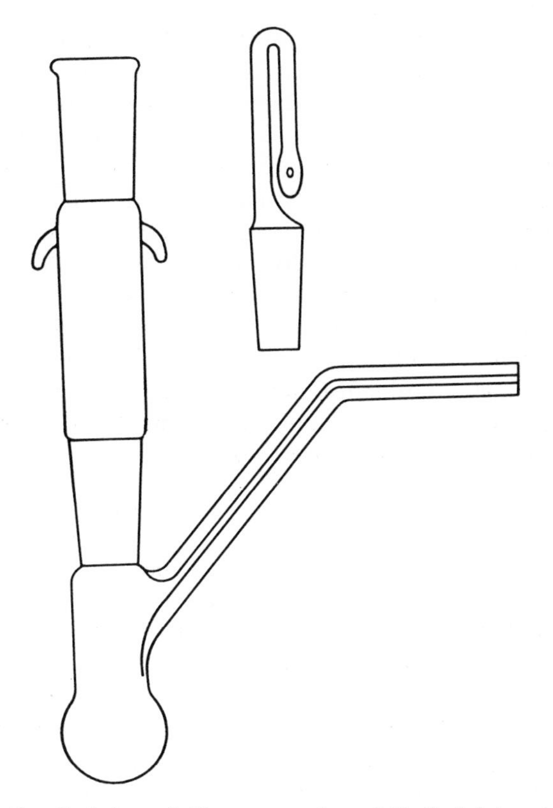

Fig. 12. Reaction flask for volatile compounds and *N*-alkyl determinations (61).

Sodium thiosulfate 0.02*N*. Dissolve 0.2 g. anhydrous sodium carbonate in about 600 ml. of distilled water; add 5 g. of sodium thiosulfate and make up the solution to 1 liter.

Potassium iodate 0.02*N*. Weigh exactly 0.7134 g. of dried potassium iodate. Transfer quantitatively to a 1-liter flask, "A" grade; fill to the mark with distilled water at 20°. This is used to standardize the sodium thiosulfate solution.

Procedure

To prepare the apparatus for an analysis add 0.5-ml. portions of cadmium sulfate and sodium thiosulfate to the washer and connect it to the apparatus with springs, sealing the joint with a drop of distilled water. Seal the top of the delivery tube with a drop of distilled water and close it with a silicone rubber tubing containing a glass rod. Half-fill the receiver with bromine–sodium acetate–acetic acid solution and connect it to the delivery tube. For series determinations close the outlet of the jacket around the sinuous tube and add cold water to it. The washer is usually satisfactory for five analyses but should be replaced if there are obvious signs of reaction in it. After five analyses, clean the apparatus by rinsing with water and then twice with acetone. Remove most of the acetone with a water pump, dry the apparatus in an oven at 120° for 5 minutes, and, finally, attach it to the water pump to remove the last traces of acetone.

For alkyl groups higher than ethyl, fill the water jacket with boiling water and use the apparatus for only one analysis.

The suitability of the conditions generally, and for series determinations, should first be ascertained by running in succession a blank determination, three analyses of a standard substance, and then another blank determination. After subtraction of the blank value from the titres of the samples, the results should all be within 0.5% (relative).

1. Solids

Weigh 3 to 5 mg. of substance in either a short-handled weighing tube or a small glass bottle. The latter is preferable for low-melting substances. Add 0.7 g. of phenol (taken in a small scoop) to the flask, dissolve the substance by gently warming the flask over a micro flame, cool until the phenol crystallizes, then add 2 ml. of hydriodic acid and a boiling tube. Connect the flask to the apparatus, sealing the joint with a drop of hydriodic acid, and connect the side arm of the flask to the carbon dioxide supply. The flow rate of the latter should be adjusted to 12 to 15 ml./minute. Place the flask on an electric heater and boil vigorously for 45 minutes (Note **2**). Remove the apparatus from the heater and lower the receiver, rinsing the inside and outside of the delivery tube with distilled water. Open the top of the receiver to allow the contents to run into a titration flask containing 10 ml. of sodium acetate. After rinsing the receiver with distilled water, destroy the bromine by adding 2 ml. of formic acid, swirling and blowing lightly across the top of the flask to ensure complete removal of bromine vapors. Liberate iodine with 0.1 g. potassium iodide and 5 ml. of sulfuric acid and titrate to the starch end point with 0.02*N* sodium thiosulfate solution, adding the starch when the solution becomes a pale straw color.

2. Liquids

The modified reaction flask (Fig. 12) is safer for liquids. Weigh them in a small glass bottle with a ground glass stopper and drop the bottle into the reaction flask containing 1 ml. of hydriodic acid, 0.7 g. of phenol, and the boiling tube. Insert the inner cone (Note **3**) and add the same amounts of phenol and hydriodic acid to the upper chamber. This forms the second reaction solution. Connect the flask to the apparatus, but before connecting the carbon dioxide line reduce the flow rate to a minimum. Heat the reaction flask for 10 minutes, then increase the flow rate to the normal fast rate of 12 to 15 ml./minute and heat for an additional 30 minutes. Complete the analysis as in (*1*).

3. Volatile Liquids

Weigh volatile liquids in small glass capillaries with fine tips. The length of the capillary should be about 2 cm. and the length of the solid glass tail about 1 cm. Gently warm the capillary and immerse the tip in the liquid until a suitable amount has been drawn into the cooling capillary. Centrifuge the liquid to the tail of the capillary, seal the tip, cool, and weigh. Cool the capillary, if necessary; scratch and break the tip and drop the capillary tip first into the flask, which should already contain the reagents and the boiling tube. The length of the capillary should be such that the liquid is above the surface of the hydriodic acid. As in the procedure for liquids, the gas flow should be negligible for the initial heating period (Note **4**), after which it may be increased to the usual 12 to 15 ml./minute to sweep the methyl iodide into the receiver.

Calculation for methoxyl

$$1 \text{ ml. } 0.02N \text{ Na}_2\text{S}_2\text{O}_3 = 0.6204 \text{ mg. OCH}_3$$

$$\% \text{ methoxyl} = \frac{(\text{Analysis titre} - \text{blank titre}) \times 0.1034 \times 100}{\text{mg. sample}}$$

The foregoing procedure should give a value of 20.4 ± 0.1% for vanillin.

2. Submicro Method

A similar procedure can be used for the analysis of 30 to 200 γ of sample by making the final titration with a micrometer syringe (15,63). The following changes should be made:

1. Apparatus. A slightly smaller apparatus than the one shown in Fig. 5 is preferred with ₮ 10/20 ground glass joints for the reaction flask and the washer.

2. Reagents. The wash solution must be replaced for each analysis. For 0.5 ml., use 0.25 ml. 5% cadmium sulfate, 0.25 ml. 5% sodium thiosulfate and 10 mg. of sodium chloride.

Take only 0.16 g. of phenol, 0.5 ml. of hydriodic acid, 1 ml. of sodium acetate–acetic acid–bromine solution, 0.5 ml. of sodium acetate solution,

6 to 7 drops of formic acid, 8 drops of 20% sulfuric acid, 25 mg. of potassium iodide and 25 mg. Thyodene.

3. Reaction time; 30 minutes.

Small blank values are difficult to eliminate but relatively easy to reproduce. A blank value of 0.015 ml. $0.02N$ sodium thiosulfate can be tolerated. Results should be within 1% (relative).

3. Semimicro and Micro Method

The semimicro apparatus and method of Clark (27,28) were improved by a collaborative study (89) so that they are suitable for both semimicro and micro quantities. For semimicro work, weigh 20 to 30 mg. of sample and titrate with $0.1N$ sodium thiosulfate.

Apparatus

Modified Clark apparatus as shown in Fig. 4.

Reagents

Acetic acid–potassium acetate–bromine solution. Dissolve 10 g. of potassium acetate in enough acetic acid to make 100 ml. and add 3 ml. of bromine.

Sodium acetate (hydrated), 25%.

Starch indicator, 2%. Mix 2 g. of finely powdered potato starch with cold water to a thin paste; add 200 ml. of boiling water, stirring constantly, and immediately discontinue heating. Add 1 ml. of mercury, shake, and let the solution stand over the mercury.

Sodium thiosulfate solution, $0.02N$. Bring 1 liter of distilled wa er to boil to remove carbon dioxide and cool while loosely covered. Dissolve 4.96 g. of sodium thiosulfate in the water and dilute to 1 liter. Transfer to a brown rubber-stoppered bottle. Add 1 ml. of cloroform as preservative and shake for a few minutes. Standarize every few days if a pool of chloroform is not present, otherwise monthly.

Hydriodic acid. Place 250 ml. of constant boiling hydriodic acid (sp. gr. 1.7) in a 500-ml. round-bottom flask connected by a joint to an air condenser and reflux for 2 hours with a stream of carbon dioxide or nitrogen bubbling through the acid. Do not let the acid vapors come in contact with organic material. As soon as the refluxing stops, discontinue the gas flow. Cool and store in a dark glass-stoppered bottle.

Phenol.

Potassium iodide.

Sulfuric acid 10%.

Procedure

Half-fill the scrubber with sodium acetate solution and two-thirds fill the receiver with acetic acid–potassium acetate–bromine solution. Weigh enough sample in a platinum boat to require about 8 ml. of sodium thiosulfate solution for the titration and place in the bottom of the reaction flask. Add 2.5 ml. of melted phenol from a wide-tip pipet, then add 5 ml. of hydriodic acid and connect the flask to the apparatus. Attack the source of carbon dioxide to the side arm of the flask and pass carbon dioxide through at the rate of 15 ml./minute. With a suitable heater (manteled microburner or electric heater), boil the liquid so that it condenses not more than halfway up the condenser. After 30 minutes drain the circulating water from the condenser and continue the reaction for another 30 minutes. Disconnect the flask, remove the receiver, and rinse the delivery tube and contents of the receiver into a 125 ml. Erlenmyer containing 5 ml. of sodium acetate. Adjust the volume to about 50 ml. and add formic acid dropwise until excess bromine is destroyed. Remove any bromine vapors by blowing air over the liquid, then add 0.5 g. of potassium iodide and 5 ml. of sulfuric acid. Swirl the solution to dissolve the potassium iodide, mix the contents, and titrate the liberated iodine with the sodium thiosulfate solution using starch solution as indicator.

Determine the blank value by making a determination without sample and subtract this titre from the analysis titre.

Notes 1. See Section VI-B-1 for discussion of this vital reagent. It should give a negligible blank value, although small blank values can be tolerated.

2. Most substances give quantitative results after 20 minutes. If sufficient sample is available, do two analyses with reaction periods of 20 minutes and 45 minutes. This often gives an indication of the reliability of the analysis.

3. This inner cone should be lightly greased with silicone grease which helps to seal it as well as to hold it in place.

4. Once the liquid has been expelled from the capillary the flow rate may be increased.

B. *N*-ALKYL DETERMINATION (Note **1**)

1. Microchemical Method

The following microchemical procedure (Note **2**) can be recommended with reservations (Note **3**) for *N*-alkyl determinations with the apparatus used for determination of *O*-alkyl in liquids (Figs. 5 and 12). For the *N*-alkyl determination the upper chamber of the reaction flask is left empty and is required for collection of hydriodic acid distilled from the bottom chamber.

Reagents

Since alkyl iodide is determined, the reagents and apparatus can be prepared exactly as for the *O*-alkyl determination. Two additional reagents are required:

Aluminum foil.

Ammonium iodide.

Procedure

Weigh 3 to 5 mg. of substance into the reaction flask, dissolve in 0.7 g. of phenol with gentle warming, cool, and add 50 mg. of ammonium iodide, 7–10 mg. aluminum foil, and 1 ml. of hydriodic acid. Pack the flask with glass beads (Note **4**), connect the reaction flask to the apparatus and the source of carbon dioxide, and proceed as for the *O*-alkyl determination (Note **5**). At the end of the analysis, fill the receiver with fresh absorption solution and replace it on the apparatus.

Before beginning the separation of the alkyl groups by pyrolysis, reduce the flow of carbon dioxide through the apparatus until it is minimal and maintain this adjustment when necessary during the subsequent heating. Gradually raise the temperature of the reaction flask to 360°, maintaining it at 290 to 300° for 30 minutes and at 350 to 360° for 1 hour. During the last 30 minutes of the pyrolysis, increase the carbon dioxide flow rate to 12 to 15 ml./minute. Complete the analysis as for an *O*-alkyl determination.

At the end of this first pyrolysis, raise the apparatus from the heating bath, disconnect the carbon dioxide supply, and remove the washer, When the flask has cooled to room temperature apply mild suction to the side arm of the reaction flask to draw the hydriodic acid back into the flask (Note **6**).

Repeat the pyrolysis procedure at least twice more to be certain that all the *N*-alkyl bonds have been cleaved.

Calculation for N-methyl (Note 7)

$$1 \text{ ml. } 0.02N \; Na_2S_2O_3 = 0.3005 \text{ mg. of } CH_3$$

$$\% \text{ methyl} = \frac{(\text{analysis titre} - \text{blank titre}) \times 0.05012 \times 100}{\text{mg. sample.}}$$

Notes 1. Both this and the *O*-alkyl procedure with prolonged reaction time have been used for *S*-alkyl determinations.

2. Reference (90) gives a similar procedure with different apparatus. The apparatus of Schachter and Ma (81a) is preferable for collection of the alkyl iodides prior to gas chromatography.

3. The method is time-consuming and liable to give low results. Addition of aluminum (75a) is an improvement in this respect.

4. Although this is not usually recommended for micro-determinations, it was found to be essential for the submicro-determination (16).

5. Substances with labile *N*-alkyl or *S*-alkyl groups will contribute to the *O*-alkyl analysis at this stage.

6. Rinsing of the upper chamber is not usually necessary. If the substance is suspected of being volatile, disconnect the flask from the apparatus and wash by adding 0.5 ml. of hydriodic acid to the upper chamber and return it to the lower chamber as before.

7. The results are usually expressed as per cent alkyl for *N*-alkyl determinations in contrast to the per cent *O*-alkyl for alkoxyl determinations.

REFERENCES

1. Alicino, J. F., in N. D. Cheronis, Ed., *Microchemical Techniques,* Interscience, New York, 1962, p. 567.
2. Alexander, A. P., P. Y. Bourne, and D. S. Littleball, *Anal. Chem.*, **27**, 105 (1955).
3. Anderson, D. M. W., and J. L. Duncan, *Chem. Ind.* (*London*), **1959**, 1151.
4. Anderson, D. M. W., and J. L. Duncan, *Talanta,* **7**, 70 (1960).
5. Anderson, D. M. W., and J. L. Duncan, *Talanta,* **8**, 1 (1961).
6. Anderson, D. M. W., and J. L. Duncan, *Talanta,* **8**, 241 (1961).
7. Anderson, D. M. W., and J. L. Duncan, *Talanta,* **9**, 661 (1962).
8. Anderson, D. M. W., J. L. Duncan, M. A. Herbich, and S. S. H. Zaidi, *Analyst,* **88**, 353 (1963).
9. Araki, T., and Y. Hasi, *J. Chem. Soc. Japan,* **61**, 99 (1940).
10. Badger, G. M., J. W. Cook, and P. A. Ongley, *J. Chem. Soc.*, **1950**, 867.
11. Baernstein, H. D., *J. Biol. Chem.*, **97**, 663 (1932).
12. Baernstein, H. D., *J. Biol. Chem.*, **106**, 451 (1934).
13. Bailey, A. J., *Ind. Eng. Chem.*, *Anal. Ed.*, **14**, 181 (1942).
14. Belcher, R., J. E. Fildes, and A. J. Nutten, *Anal. Chim. Acta,* **13**, 16 (1955).
15. Belcher, R., M. K. Bhatty, and T. S. West, *J. Chem. Soc.*, **1957**, 4480.
16. Belcher, R., M. K. Bhatty, and T. S. West, *J. Chem. Soc.*, **1958**, 2393.
17. Belcher, R., M. K. Bhatty, and T. S. West, *J. Chem. Soc.*, **1960**, 2473.
18. Bethge, P. O., and O. T. Carlson, *Anal. Chim. Acta,* **15**, 279 (1956).
19. Bradbury, R. B., *Mikrochemie,* **38**, 114 (1951).
20. Brown, H. C., and K. L. Nelson, *J. Am. Chem. Soc.*, **75**, 24 (1953).
21. Brown, P., and A. L. Smith, *Anal. Chem.*, **30**, 549 (1958).
22. Budesinsky, B., and J. Körbl, *Mikrochim. Acta,* **1960**, 369.
23. Burwell, R. L., Jr., *Chem. Rev.*, **54**, 615 (1954).
24. Campbell, A. D., and V. J. Chettleburgh, *Analyst,* **84**, 190 (1959).
25. Cheronis, N. D., and T. S. Ma, *Organic Functional Group Analysis by Micro and Semimicro Methods,* Interscience, New York, 1964.
26. Christensen, B. E., L. Friedman, and Y. Sato, *Ind. Eng. Chem.*, *Anal. Ed.*, **13**, 276 (1941).
27. Clark, E. P., *Ind. Eng. Chem.*, *Anal. Ed.*, **10**, 677 (1938).

28. Clark, E. P., *Semimicro Quantitative Organic Analysis,* Academic Press, New York (1943).
29. Colson, A. F., *Analyst,* **58,** 594 (1933).
30. Crompton, T. R., *Analyst,* **86,** 652 (1961).
31. Cundiff, R. H., and P. C. Markunas, *Anal. Chem.,* **33,** 1028 (1961).
32. Dirscherl, A., *Mikrochim. Acta,* **1962,** 153.
33. DuVigneaud, V., *Science,* **123,** 968 (1956).
34. Edlbacher, S., *Z. Physiol. Chem.,* **101,** 278 (1918).
35. Elek, A., *Ind. Eng. Chem., Anal. Ed.,* **11,** 174 (1939).
36. Elek, A., "Determination of Alkoxyl Groups" in J. Mitchell, Jr., I. M. Kolthoff, E. S. Proskauer, and A. Weissberger, Eds., *Organic Analysis,* Vol. 1, Interscience, New York, 1953, pp. 67–126.
37. Erdtman, H., *Proc. Roy. Soc. (London),* **A143,** 234 (1933).
38. Feigl, F., *Spot Tests in Organic Analysis,* 6th Eng. Ed., Elsevier, New York, 1960.
39. Filipovic, L., and Z. Stefanac, *Croat. Chem. Acta,* **30,** 149 (1958).
40. Franzen, F., W. Disse, and K. Eysell, *Mikrochim. Acta,* **1953,** 44.
41. Friedrich, A., *Mikrochemie,* **7,** 185 (1929).
42. Friedrich, A., *Die Praxis der quantitativen organischen Mikroanalyse, Deuticke,* Leipzig and Vienna, 1933.
43. Fukida, M., *Mikrochim. Acta,* **1960,** 448.
43a. Fukuda, M., *Bunseki Kagaku,* **15,** 1360 (1966).
44. Furter, M., *Helv. Chim. Acta,* **21,** 873 (1938).
45. Ghaswalla, R. P., and F. G. Donnan, *J. Chem. Soc. (London),* **1936,** 1341.
46. Gran, G., *Svensk Papperstid.,* **57,** 702 (1954).
47. Grant, J., *Quantitative Organic Microanalysis,* 4th Ed., Churchill, London, 1945, p. 160.
48. Gysel, H., *Mikrochim. Acta,* **1954,** 743.
49. Haslam, J., J. B. Hamilton, and A. R. Jeffs, *Analyst,* **83,** 66 (1958).
50. Henbest, H. B., G. D. Meakins, B. Nicholls, and A. A. Wagland, *J. Chem. Soc.,* **1957,** 1462.
51. Heron, A. E., R. H. Reed, H. E. Stagg, and H. Watson, *Analyst,* **79,** 671 (1954).
51a. Herzig, J., and H. Meyer, *Chem. Ber.,* **27,** 319 (1894).
52. Hodgman, C. D., *Handbook of Chemistry and Physics,* 41st Ed., Chemical Rubber Publ. Co., Cleveland, 1959.
53. Hoffman, D. O., and M. L. Wolfrom, *Anal. Chem.,* **19,** 225 (1947).
54. Holasek, A., H. Lieb, and W. Merz, *Mikrochim. Acta,* **1956,** 1216.
55. Houben-Weyl-Müller, *Methoden der organischen Chemie,* 4th Ed., Vol. 2, 1953, p. 401.
56. Houghton, A. A., and H. A. B. Wilson, *Analyst,* **69,** 363 (1944).
57. Hozumi, K., and K. Hazami, *J. Pharm. Soc. Japan,* **81,** 1298 (1961).
58. Huang, R. L., and F. Morsingh, *Anal. Chem.,* **24,** 1359 (1952).
59. Huang, R. L., and K. T. Lee, *Anal. Chem.,* **27,** 1030 (1955).
60. Hughes, G. K., and E. O. P. Thompson, *J. Proc. Roy. Soc. N. S. Wales,* **83,** 269 (1949).
61. Inglis, A. S., *Mikrochim. Acta,* **1957,** 677.
62. Inglis, A. S., M. Sc. thesis, Melbourne 1954.
63. Inglis, A. S., Unpublished work.
63a. M. A. Khan, S. S. H. Zaidi, S. Parveen, and N. Ahmad, *Pakistan J. Sci. Ind. Res.,* **10,** 24 (1967).
64. Kahovec, L., *Mikrochemie,* **14,** 341 (1934).

65. Karpitschka, N., *Mikrochim. Acta,* **1961,** 738.
65a. Kikuchi, N., and T. Miki, *Bunseki Kagaku,* **17,** 1102 (1968).
66. Kirpal, A., and T. Bühn, *Montash. Chem.,* **36,** 853 (1915).
67. Kirsten, W. J., and S. Ehrlich-Rogozinsky, *Mikrochim. Acta,* **1955,** 787.
68. Klimova, V. A., and K. S. Zabrodina, *Izv. Akad. Nauk Arm. SSSR,* **1961,** 2234.
68a. Klimova, V. A., K. S. Zabrodina and N. L. Shitikova, Izv. Akad. Nauk. SSSR, *1965,* 178.
68b. Kratzl, K., and K. Gruber, Mh. Chem. **89,** 618 (1958).
69. Kretz, R., *Z. Anal. Chem.,* **176,** 421 (1960).
70. Kuhn, R., L. Birkofer, and F. W. Quackenbush, *Chem. Ber.,* **72B,** 407 (1939).
71. Langejan, M., *Pharm. Weekblad,* **92,** 667 (1957).
72. Makens, R. F., R. L. Lothringer, and R. A. Donia, *Anal. Chem.,* **31,** 1265 (1959).
73. Mathers, A. P., and M. J. Pro, *Anal. Chem.,* **27,** 1662 (1955).
74. Miller, D. L., E. P. Samsel, and J. G. Cobler, *Anal. Chem.,* **33,** 677 (1961).
75. Mitsui, T., and Y. Kitamma, *Microchem. J.,* **7,** 141 (1963).
75a. Nakamura, K., Y. Utsüi, and T. Maeda, *Bunseki Kagaku,* **16,** 535 (1967).
76. Nessonova, G. D., and E. K. Pogosyants, *Zavodsk. Lab.,* **24,** 953 (1958).
77. Page, J., *J. Chem. Soc.,* **1955,** 2017.
77a. Pietrogrande, A., F. Bordin, and G. Dalla Fini, Mikrochim. Acta, *1966,* 1156.
78. Pozefsky, A., and N. D. Coggeshall, *Anal. Chem.,* **23,** 1611 (1951).
79. Rigakos, D. R., *J. Am. Chem. Soc.,* **53,** 3903 (1931).
80. Rudloff, E. von, *Anal. Chim. Acta,* **16,** 294 (1957).
81. Samsel, E. P., and J. A. McHard, *Ind. Eng. Chem., Anal. Ed.,* **14,** 750 (1942).
81a. Schachter, M. M., and T. S. Ma, Microchim. Acta, *1966,* 55.
82. Schole, J., *Z. Anal. Chem.,* **193,** 321 (1963).
83. Shaw, B. W., *J. Soc. Chem. Ind. (London),* **66,** 147 (1947).
84. Siggia, S., A. C. Starke, J. J. Garis, Jr., and C. R. Stahl, *Anal. Chem.,* **30,** 115 (1958).
85. Shriner, R. L., R. C. Fuson, and D. Y. Curtin, *The Systematic Identification of Organic Compounds: A Laboratory Manual,* 4th Ed., Wiley, New York, 1956, p. 116.
86. Slator, A., *J. Chem. Soc.,* **85,** 1286 (1904).
87. Stephens, W. I., *Proceedings of the International Symposium on Microchemistry,* 1958, Pergamon, New York, 1960, p. 163.
88. Steyermark, Al, *Anal. Chem.,* **20,** 368 (1948).
89. Steyermark, Al, H. K. Alber, V. A. Aluise, E. W. D. Huffman, E. L. Jolley, J. A. Kuck, J. J. Moran, and C. L. Ogg, *Anal. Chem.,* **28,** 112 (1956).
90. Steyermark, A., *Quantitative Organic Microanalysis,* 2nd Ed., Academic Press, New York, 1961.
91. Stone, H., and H. Schechter, *J. Org. Chem.,* **15,** 491 (1956).
92. Sykes, P., *A Guidebook to Mechanism in Organic Chemistry,* Longmans Green, London, 1961.
93. Takuyra, K., Y. Takino, and S. Hareda, *J. Pharm. Soc. Japan,* **76,** 1328 (1956).
94. Věcéra, M., and A. Spěvak, *Chem. Listy,* **52,** 1520 (1958).
95. Vertalier, S., and F. Martin, *Chim. Anal.,* **40,** 80 (1958).
96. Vieböck, F., and A. Schwappach, *Chem. Ber.,* **63,** 2818 (1930).
97. Vieböck, F., and C. Brecher, *Chem. Ber.,* **63,** 3207 (1930).
98. Ware, G. M., *Mikrochemie,* **8,** 352 (1931).
99. Weishut, F., *Monatsh.,* **33,** 1165 (1912).
100. Wheland, G. W., *Advanced Organic Chemistry,* 2nd Ed., Wiley, New York, 1954, p. 561.
101. White, E. P., *Ind. Eng. Chem., Anal. Ed.,* **16,** 207 (1944).
102. Zeisel, S., *Monatsh. Chem.,* **6,** 989 (1885).

Part II
Section B-2

DETERMINATION OF ETHERS AND EPOXIDES

BY ROBERT T. HALL AND ROBERT D. MAIR,
Research Center, Hercules Incorporated,
Wilmington, Delaware

Contents

Contents (*continued*)

I. INTRODUCTION

Inertness, not reactivity, is the characteristic most generally associated with ethers. Broadly speaking, this is true for typical ethers R—O—R and R—O—R′, where the R groups can be either aliphatic or aromatic in character. The two R groups may be linked to give cyclic ethers, and these also are relatively inert as long as the ring formed is five- or six-membered and is, therefore, free from strain. Small-ring cyclic ethers such as alpha-epoxides, on the contrary, are quite reactive. Intermediate in reactivity between the inert types and the reactive cyclic ethers are those containing the vinyl linkage. It is therefore convenient to consider the ethers as a graded sequence of compounds based on their chemical reactivity. By doing so a more fundamental, cohesive, and selective treatment of their analytical chemistry is possible, and this has been our guiding principle in organizing this chapter.

Tabulations of some analytically useful properties and suggested methods for a number of ethers and epoxides are given in Meites' handbook (56). The detection and estimation of these compounds have been discussed also in a recent advanced treatise on the chemistry of the ether linkage (34a).

II. ANALYSIS OF ETHERS AND EPOXIDES

A. INERT OR NONCYCLIC ETHERS (ROR AND ROR′)

1. Physical and Chemical Properties

This class includes such compounds as ethyl ether, isopropyl ether, phenyl ether, etc. Physical properties of some ethers of this type are shown in Table I.

TABLE I

Typical Inert or Noncyclic Ethers—Physical Constants

	Boiling point, °C/760 mm	Sp. gr., 20/20°C	Refractive index, n_D^{20}
Methyl ether, CH_3OCH_3	−24.9	0.661	
Ethyl methyl ether, $C_2H_5OCH_3$	7	0.697	
Ethyl ether, $C_2H_5OC_2H_5$	34.6	0.7146	1.3527
Isopropyl ether, $(CH_3)_2CHOCH(CH_3)_2$	68.3	0.7244	1.3679
n-Hexyl ether, $(n\text{-}C_6H_{13})O(n\text{-}C_6H_{13})$	208.8	0.7950	1.4204
Phenyl ether, $C_6H_5OC_6H_5$	259	1.0728	1.5763^{30}
Anisole, $C_6H_5OCH_3$	154	0.9940	1.5179

In these compounds the oxygen is attached to two carbon atoms, and the carbon–oxygen bond can be broken only under rather special conditions. Ordinarily, they are little affected by acids, alkalies, oxidizing agents, and acid chlorides. At high temperature and pressure there is reaction with some reagents; in fact, ethers react as rapidly as alcohols with reagents which do break the C—O bond. The apparent nonreactivity of ethyl ether arises in large measure from its volatility, which lowers its concentration at higher temperatures and atmospheric pressure.

Because of the absence of a hydroxyl group, and the resulting lack of association between molecules, the physical properties of ethers are nearer to those of the corresponding hydrocarbons than to the isomeric alcohols.

2. Chemical Reactivity

The relatively inert simple ethers possess fundamental properties that lead to some chemical reactivity, as discussed in the next section; for instance, because their oxygen atom has two unshared electron pairs that can interact with electron-deficient groups, such ethers may act as Lewis bases to form oxonium salts, undergo coordination reactions, and form complexes with such acids as boron trifluoride. Although seldom useful for bond cleavage, these reactions allow the ready separation of ethers from hydrocarbons and alkyl halides.

For simple ethers the nature of the R groups determines the strengths of the C—O bond. The aromatic C—O bond is very resistant to cleavage, so that a compound such as phenyl ether is unusually inert, and the cleavage of mixed aralkyl ethers with halogen acids always yields a phenol and an alkyl halide:

$$C_6H_5OR \xrightarrow{HX} C_6H_5OH + RX \tag{1}$$

For substituted ethers reactivities are frequently enhanced through inductive effects. Thus in aromatic ethers nitro and nitroso groups substituted in the *ortho* or *para* position exert a strong activating effect on the alkoxyl group toward replacement reactions.

Substituent groups such as ester, nitro, and cyano in the beta position activate the adjacent C—O bond, as illustrated by the reaction:

$$ROCH_2CH_2CN \underset{\text{acid or base}}{\rightleftharpoons} ROH + CH_2{=}CHCN \tag{2}$$

Also, unsaturation in the alpha position activates the C—O linkage, and is an analytically important reaction for vinyl ethers (see discussion in Sections II-B-2 and -3).

Another typical reaction of ethers is their tendency to form explosive peroxides (70) on storage. Interestingly enough, tests (28,55,87) have been devised for detection of ethers based on their ability to form peroxides readily. This same ability also creates a potential hazard that must always be considered in handling ethers. Many fires and explosions have been reported during the evaporation or distillation of easily autoxidizable solvents such as ethyl and isopropyl ethers, dioxane, and tetrahydrofuran. Before carrying out such operations, a qualitative test for peroxide should always be applied. We especially recommend the reports of Noller (62) and his associates concerning the safe handling of ethers and other compounds which tend to form organic peroxides.

3. Chemical Methods of Analysis

The most important analytically useful reaction of aliphatic ethers is cleavage of their carbon–oxygen bond, usually by reaction with halogen acids, especially hydriodic acid. This reaction is the basis of the widely used Zeisel procedure and has been highly developed as a general method for the quantitative determination of ethers and alkoxyl groups. A previous chapter of this volume is concerned with this method and it is therefore not treated here. Qualitative tests for ethers are also based on cleavage with hydriodic acid (17). An interesting new method for the cleavage of ethers has been described by Long and Freeguard (54). By this method, which employs an iodine–diborane reagent, a rapid, smooth and complete cleavage can be effected, usually at room temperature or even below. In preparative work, this new approach has been applied to symmetric, unsymmetric, aliphatic, aromatic, or cyclic ethers. It might be quite interesting to investigate the analytical possibilities of this new cleavage reagent.

For some aryl ethers a colorimetric reaction with formaldehyde (2) has been developed; a more general reaction for forming suitable derivatives for identifying aromatic ethers is chlorosulfonation followed by ammonolysis to give the substituted benzenesulfonamide. A good discussion of this latter and other qualitative tests, and of the preparation of derivatives suitable for identification of both aliphatic and aromatic ethers, is given by Cheronis and Entrikin (18) and by Feigl (34).

One of the most noteworthy and extensive investigations of methods for the identification of aliphatic ethers is the work of Jurecek *et al.* (47), who made a study of various cleavage reagents, including hydriodic acid, 3,5-dinitrobenzoyl chloride with anhydrous zinc chloride catalyst, and 3,5-dinitrobenzoic anhydride with tin(IV) as catalyst. They determined optimum reaction conditions for cleaving various ethers, and developed paper chromatographic techniques for separating and identifying the 3,5-dinitrobenzoate derivatives. They concluded that combination of the cleavage and the subsequent paper chromatography is a generally satisfactory micro method for identifying aliphatic ethers. The cleavage reaction is best carried out in a sealed tube to prevent loss of ether and to get more complete reaction. The R_f values of a number of the 3,5-dinitrobenzoates have been conveniently tabulated (11,12,58,79). Only the identification of asymmetric ethers with isomeric alkyls presents a problem because some such compounds yield derivatives with similar R_f values and give only a single elongated spot.

A more recent method, also involving cleavage, for the classification and identification of ethers has been reported by Waszeciak and Nadeau (85). It consists of reacting the ether with acetyl chloride in the presence of a strong Lewis acid salt (ferric chloride) and examining the products by gas–liquid chromatography. The procedure has been successfully applied to both simple aliphatic and polymeric ethers. This method would likely provide added versatility and effectiveness compared to previously used techniques for examining the cleavage products. Although we have had no personal experience with this method, it is clearly necessary to run known compounds and establish optimum conditions for the system being studied. Even so, the method may turn out to be one of the better qualitative ones, especially for polyethers. Furthermore, based on this same reaction, a colorimetric method for the quantitative determination of polyethers in low concentrations seems possible.

4. Physical and Instrumental Methods

As is the case with so many other functional groups, infrared is probably the most widely useful physical technique for detection and determination of the ether linkage (5,61). The characteristic feature of its infrared spectrum is the strong, broad absorption arising from the single bond C—O— stretching vibration; formals and acetals are different, giving several sharper bands in this region. Bellamy (5) gives for alkyl ethers, $—CH_2—O—CH_2—$, 1150–1060 cm^{-1} (8.70–9.44 μ) and for aryl ethers and others with the group, =C—O—, 1270–1230 cm^{-1} (7.88–8.13 μ). Unfortunately the latter range also includes esters, O=C—O—. Davison (26) reports a band at about 960 cm^{-1} (10.42 μ) for vinyl ethers.

Numerous applications of infrared to the determination of specific ethers have been published, e.g., determination of cineol (23), 2,2′-dihydroxyethyl ether (66), *cis*- and *trans*-anethole (50), and phenyl ether (49). Of somewhat broader interest is the work of Saier and Hughes (68), who developed a group-type analysis for determination of acids, alcohols, aldehydes, ketones, esters, and ethers in a mixture. Specific conditions for determining ethers in such a mixture of oxygenated compounds were established and the method proved to be more rapid than chemical procedures. Erley (32) and Stewart and Erley (77) have used long-path infrared cells to detect and determine ethers in air or exhaled breath, after collection on silica gel and release by heat.

Although infrared is probably one of the most useful physical techniques for application to compounds containing the ether linkage, it lacks both sensitivity and specificity to any high degree, and generally would not be suitable for determining traces, or even appreciable amounts, of ethers in the presence of other functionalities absorbing in the infrared. More recently, the near-infrared technique has become of interest for the determination of the more reactive types of ethers such as epoxides; it will be considered in Section C-4-a.

Nukada (63) has made a study of the application of nuclear magnetic resonance (NMR) to aliphatic compounds containing oxygen atoms, especially ethers. He determined the chemical shifts for a number of ether linkages, and shows their NMR spectra. As applied to ethers, NMR has been used chiefly in proof-of-structure investigations rather than as a means of analysis. There appears to be no reason for this limitation, especially with the continued improvement and increasing availability of instrumentation specifically designed for analytical purposes (7).

The mass spectrometer has been applied successfully to the identification, structure determination, and analysis of aliphatic ethers (57). The mass spectra of saturated ethers are similar to those of other compounds with electron-donating functional groups such as alcohols and amines. *gem*-Diethers (compounds with two alkoxyl groups on the same carbon), however, exhibit anomalous behavior (53) in the mass spectrometer. These compounds can give two different spectra depending on the type of filament; e.g., on bare tungsten they decompose into alcohols and vinyl ethers, and their mass spectra are a summation of these. A carbonized or rhenium filament, which prevents decomposition, should be used for reliable analytical work.

Because of their solubility properties and ability to form oxonium salts and various complexes, ethers can be separated from hydrocarbons and various other oxygenated compounds by a variety of chemical treatments and extraction techniques. In addition, their inertness often makes them readily separable from more reactive organic functionalities by regular column adsorption chromatography. Solubilization chromatography (71) has also been applied to the separation of mixtures of aliphatic and aromatic ethers.

Gas chromatography has, of course, been widely applied to the separation and analysis of ethers (16,81). It provides a rapid, simple method for determining trace and major quantities of lower alcohols, ethers, and acetone in aqueous solutions. A wide range of compositions and mixtures of these components have been so analyzed, with acceptable precision and accuracy (8); neither mass spectrometric techniques nor distillation techniques were as suitable.

B. VINYL ETHERS

1. Physical Properties

Ethers containing the vinyl linkage ($ROCH{=}CH_2$) are much more reactive than the inert or noncyclic ethers. Some typical vinyl ethers, along with some of their physical constants, are shown in Table II.

TABLE II

Typical Vinyl Ethers—Physical Constants

	Boiling point, °C/760 mm	Sp. gr. 20/20°C	Refractive index, n_D^{20}
Ethyl vinyl ether, $CH_2{=}CHOC_2H_5$	35.5	0.7602	1.3774
Isobutyl vinyl ether, $CH_2{=}CHOCH_2CH(CH_3)_2$	83.4	0.7659	1.3961
Vinyl ether, (divinyl oxide) $CH_2{=}CHOCH{=}CH_2$	39	0.7744	1.3989
Butoxyethyl vinyl ether, $CH_2{=}CHOCH_2CH_2OC_4H_9$	decomp.	0.8659	1.4213
2-Methoxyethyl vinyl ether, $CH_2{=}CHOC_2H_4OCH_3$	108.8	0.8967	1.4103

2. Analytical Reactions

The reactions of chief analytical significance are illustrated by:

1. Hydrolysis:

$$ROCH{=}CH_2 + H_2O \xrightarrow{H^+} ROH + CH_3CHO$$

2. Iodine absorption:

$$R{-}OCH{=}CH_2 + I_2 + CH_3OH \rightarrow ROCH(OCH_3){-}CH_2I + HI$$

3. Mercuric acetate addition:

$$R{-}OCH{=}CH_2 + Hg(O{-}\overset{O}{\overset{\|}{C}}{-}CH_3)_2 + CH_3OH \rightarrow R{-}OCH(OCH_3){-}CH_2{-}HgO{-}COCH_3 + CH_3COOH$$

3. Chemical Methods of Analysis

Methods based on reaction (1), which are also applicable to acetals and ketals, differ chiefly in the variations used in carrying out the hydrolysis and in the methods for estimating the acetaldehyde. A reliable procedure of this type is described by Siggia (74), and is suitable for many vinyl ethers. It consists of hydrolyzing them, using hydroxylamine hydrochloride as the acid catalyst and also as the agent to remove the resulting aldehyde by forming the oxime, thus driving the reaction to completion; the liberated hydrochloric acid is titrated with standard alkali. A refinement of this technique involves the addition of triethanolamine, after hydrolysis in the acidified solution of hydroxylamine hydrochloride, to increase the pH of the solution and help push the oximation reaction to completion (20). The use of hydrobromic acid has been recommended (14) to effect the hydrolysis of the compound to the corresponding carbonyl which is then oximated, and the excess hydroxylamine determined by titration with potassium cyanoferrate(III). Because rather high concentrations of hydrobromic acid may be used, this method is worth considering for vinyl ethers that are difficult to hydrolyze.

The hydrolysis procedure has been refined to make it applicable to trace amounts of vinyl ethers and compounds containing combined acetaldehyde (13). In this case the acetaldehyde, generated when the compound is hydrolyzed with sulfuric acid, is steam distilled (which serves to separate it from a number of interferences) into an aqueous solution of sodium bisulfite and reacted with *p*-phenylphenol and cupric sulfate in concentrated sulfuric acid to give a violet compound absorbing at 572 mμ. This route has been used specifically in the determination of sesamex, the pyrethrin and carbamate synergist, but is generally useful for the determination of combined acetaldehyde in very low concentrations. The method is simple and is reported (13) to be reasonably accurate—better than 5% in the microgram range—and reproducible.

Most of the other chemical methods for vinyl ethers are based on the formation of addition products with various reagents; e.g., the consumption of iodine by the vinyl ether linkage and back-titration of the excess iodine, as developed by Siggia (75). The presence of alcohol is required as indicated in reaction (2), since the end product of the reaction is the iodo acetal. This method is especially indicated for samples containing acetaldehyde or acetals, which would interfere in the hydrolysis methods. Based on its application to a variety of pure compounds, the average deviation from theory was estimated to be 0.2%. In cases where the sample contains free alkali, it should be neutralized before addition of the methanol–iodine reagent, or else the vinyl ether will form iodoform. Acids, unless present in amounts large enough to lower the pH of the analysis solution below 2, cause no problem. The method does not work on vinyl ethers having too limited a solubility in the reagent.

The mercuration procedure illustrated by reaction (3) is especially useful for analyzing vinyl ethers. It is based on determining the unsaturation of the terminal double bond. In fact, it is essentially restricted to determination of terminal or isolated double bonds with *cis* configuration and, accordingly, other such unsaturated compounds will interfere. Some methods measure the acetic acid liberated by methoxy-mercuration, whereas others measure the unconsumed mercuric acetate. One of the more satisfactory versions, described by Johnson and Fletcher (44), involves reaction of the sample with methanolic mercuric acetate reagent;

the acetic acid liberated from the addition reaction is then determined by titration. The interference that would result from the presence of excess mercuric acetate in the titration is avoided by adding sodium bromide, which converts it to mercuric bromide. Average deviation from theory on applying this method to a range of vinyl ethers was found to be less than 0.2%. Here, too, samples should first be neutralized, as the measurement is based on an acidimetric titration.

4. Physical and Instrumental Methods

Infrared is of some value for the qualitative detection of the vinyl ether linkage (26). The absorptions at about 962 cm^{-1} (10.40 μ) are somewhat typical of the vinyl group but they are relatively insensitive. In this laboratory, however, the near infrared has been found far more useful for the quantitative determination of vinyl ethers. An excellent discussion of the technique and its application to the analysis of compounds giving terminal methylene absorption is given by Goddu (36). The two bands found most useful for quantitative measurement of terminal double bonds are the combination band at 4740 cm^{-1} (2.11 μ) and the first overtone of the C—H stretching band at about 6170 cm^{-1} (1.62 μ). Table III gives typical data on a variety of vinyl ethers.

TABLE III

Effect of Structure on 6170 cm^{-1} (1.62 μ) Terminal Methylene Absorption in Carbon Tetrachloride (36)

			Molar absorptivity, ϵ, liter/mole-cm		
Compound	Max, cm^{-1}	λ max, μ	Cary Model 14	Beckman DK-2	Spectracord 4000A
Allyl vinyl ether	6135 (allyl)	1.630 (allyl)	0.24		
Allyl vinyl ether	6203 (vinyl)	1.612 (vinyl)	0.40		
n-Butyl vinyl ether	6192	1.615		0.35	
Isobutyl vinyl ether	6192	1.615		0.36	
tert-Butyl vinyl ether	6192	1.615	0.45	0.35	0.35
α-Terpinyl vinyl ether	6192	1.615		0.37	
Isobornyl vinyl ether	6200	1.615		0.38	

Goddu has made some generalizations as to the effect of structure on the wavelength of the absorption maximum. Vinyloxy compounds (CH_2=CH—O—) absorb at the shortest wavelength, about 6190 cm^{-1}

(1.615 μ). Compounds with the group $CH_2{=}C{-}C{=}O$ absorb at about 6170 cm^{-1} (1.62 μ); unsaturated hydrocarbons generally absorb above 6135 cm^{-1} (1.630 μ). The excellent resolution obtainable in the 6170 cm^{-1} (1.62 μ) region is illustrated by the spectrum of allyl vinyl ether, shown in Fig. 1.

As a consequence of this resolution it is often possible to make both qualitative and quantitative determinations on an unknown unsaturated compound; for example, the presence of vinyl ether could be predicted from the position of the short-wavelength band. Few methods can differentiate such closely related functional groups so rapidly and completely. We strongly recommend it.

The near-infrared technique was found to be uniquely suited for process control in following the vinylation of an alcohol (36):

$$ROH + HC{\equiv}CH \xrightarrow[\text{donor solvent}]{\text{catalyst}} \underset{\text{Vinyl ether}}{ROCH{=}CH_2} \tag{3}$$

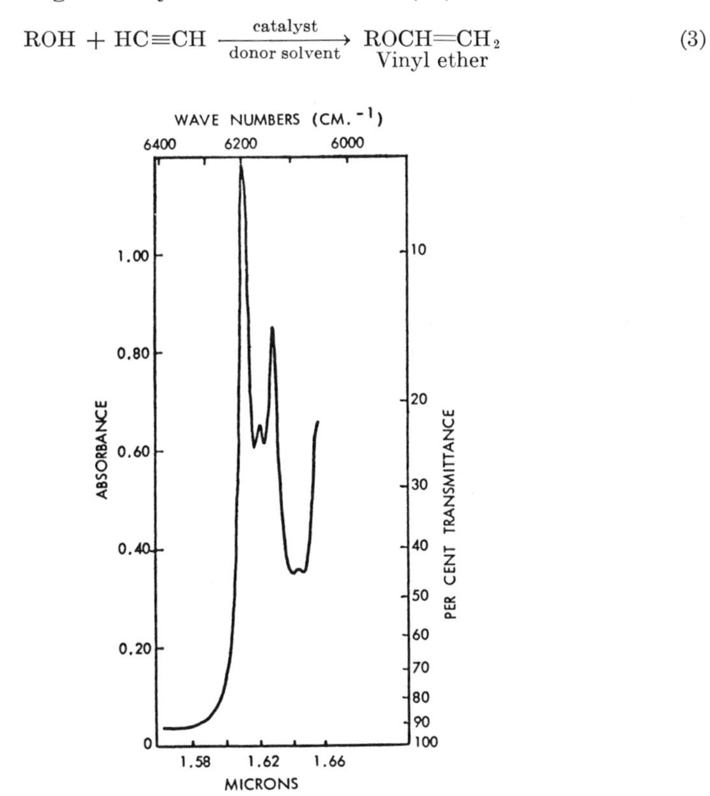

Fig. 1. Near-infrared absorption spectrum of allyl vinyl ether in the 6300 cm^{-1} (1.6 μ) region on the Cary Model 14 recording spectrophotometer; concentration 2.40 M in carbon tetrachloride; 1-cm cell.

The vinyl ether content of the reaction mixture is readily determined at 6190 cm^{-1} (1.615 μ); furthermore, the isolated vinyl ether may be analyzed for purity either directly or, more accurately, by using high absorbance compensation techniques.

A further point illustrated by the data in Table III is how relatively constant the molar absorptivities are for a range of vinyl ethers; it is thus possible to make reasonably good determinations even when a standard is unavailable. One of the chief interferences in the near infrared is water, but this is not too severe a limitation, as most vinyl ethers can be dissolved in carbon tetrachloride or methylene chloride, both especially suitable solvents for the near infrared region.

C. SMALL RING ETHERS—α-EPOXIDES OR OXIRANES

1. Physical and Chemical Properties

In these cyclic ethers, the oxygen and two adjacent carbon atoms form a three-membered ring. Because of their chemical reactivity, these compounds have achieved wide commercial importance as reagents, intermediates, and end products, and numerous chemical and physical methods have been developed for their detection and determination. Their chemistry is broadly discussed by Winstein and Henderson (86). Some typical compounds containing the reactive α-epoxide linkage are shown in Table IV.

TABLE IV

Typical Epoxides—Physical Properties

	Boiling pt. °C/760 mm	Sp. gr., 20/20°C	Refractive index, n_D^{20}
Ethylene oxide, CH_2—O—CH_2 (ring)	10.4	0.8711	1.3597[7]
Propylene oxide, CH_3CH—O—CH_2 (ring)	34.0 (35)	0.8304	1.3657
Styrene oxide, C_6H_5CH—O—CH_2 (ring)	191	1.052	1.5350
Epichlorohydrin, $CH—CH_2Cl$, —O—CH_2 (ring)	115.2 (116.5)	1.1761	1.4359
Allyl 1-(2,3-epoxy)propyl ether, $CH_2{=}CH—CH_2—O—CH_2—CH—CH_2$ (CH—CH_2 bridged by O)	153.9	0.969	1.4348

2. Chemical Reactivity

One of the most characteristic reactions of the 1,2-epoxide linkage is ring opening, accompanied by formation of addition compounds. Many nucleophilic reagents, e.g., halogen acids, thiosulfates, bisulfites, and amines, will react in this manner, which is the basis for numerous analytical methods as illustrated in the following reaction (HX being a nucleophilic reagent):

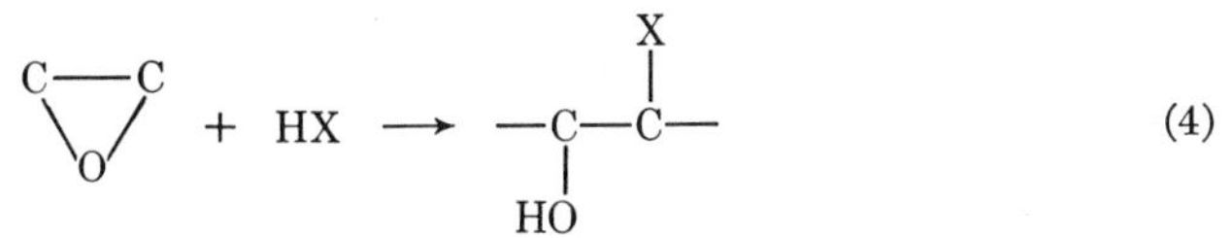

There is, of course, no universally applicable procedure. Such matters as the reactivity of the particular epoxide compound being determined, its side reactions with the reagent, its solubility in the particular reaction medium, the types of interferences that may be present, whether it is a terminal or internal epoxide, and whether one is determining traces or major amounts—all these influence the selection of the method. An excellent discussion of the pros and cons of the classical, hydrochlorination technique, as well as methods based on other nucleophilic reagents, is available (46). The more widely applicable of these methods as well as some developed more recently have been critically reviewed by Critchfield (21).

A problem encountered in hydrochlorination and in similar methods is the tendency of a very reactive epoxide linkage to undergo hydrolysis to a glycol because of water in the solvent or reagent, thereby interfering with the desired reaction. This interference is usually eliminated by operating in relatively anhydrous media or, for the case of a water-soluble epoxide, by using a large excess of salt (calcium chloride or magnesium chloride) to drive the reaction to completion. Another annoying side reaction causing difficulty, but limited mainly to epoxides containing a tertiary carbon atom, is isomerization under acid conditions to the equivalent aldehyde. How to overcome or minimize these and other interferences will be noted in the following discussions of various types of procedures.

3. Chemical Methods of Analysis

a. Hydrohalogenation

One of the best and most useful of the methods for determining oxirane oxygen is based on its reaction with anhydrous hydrobromic acid in glacial acetic acid medium to form the corresponding bromohydrin. Many epoxides react so readily that they can be titrated directly to the crystal violet endpoint with the hydrobromic acid–glacial acetic reagent (29). Chlorobenzene or chlorobenzene–chloroform (1:1) mixture are used as solvents. This direct titration method has been adopted as a tentative standard by both ASTM (3) and the American Oil Chemists' Society (9). Amines, if present, are titrated quantitatively but can be corrected for (30) by adding excess mercuric acetate to the solution after titrating to the crystal violet endpoint with hydrobromic acid. This reagent regenerates the interfering amine acetate, which is then titrated with 0.1N $HClO_4$ to the same crystal violet endpoint.

Less reactive epoxides, or those that react too slowly to be determined by direct titration with the hydrobromic acid reagent, are often determined by the indirect hydrogen bromide method. Here the epoxide compound is allowed to react with an excess of the hydrobromic acid–glacial acetic reagent for a period of time varying from 15 to 60 min and the unreacted hydrobromic acid is titrated with a standard solution of sodium acetate in acetic acid using crystal violet indicator. One of the main problems with the indirect method is the interferences that stem from the use of such a reactive reagent under these conditions, particularly where reaction times longer than 15 min are required. For example, nonepoxy compounds such as 1,2-glycols and tertiary alcohols interfere by consuming hydrobromic acid depending on the reaction time; some unsaturated compounds and peroxides may also consume reagent. Obviously, some of these interferences are minimized or even eliminated when the epoxide to be determined is sufficiently reactive that the direct titration method can be used. By making suitable modifications and additions the hydrogen bromide titration methods have been extended to a number of specific situations, e.g., determination of adsorbed epoxides in gum rubber (38a).

In the course of recent work in this laboratory, neither the direct nor indirect hydrogen bromide methods could be satisfactorily applied to the determination of oxirane oxygen in epoxy resin-finished glass fiber. The direct method did not give complete reaction. In the indirect procedure, the glycol that was present interfered and caused high results because of the unusually long reaction periods required for diffusion of reagent to difficultly accessible sites. This interference by glycol was eliminated (40) by substituting the somewhat less active reagent, hydrochloric acid, for hydrobromic acid in glacial acetic. The glass fiber is treated directly with this reagent, as the epoxide compound cannot be completely removed from it by extraction. The excess hydrochloric acid is not titrated with sodium hydroxide because of the presence of amines in the epoxide coatings. Instead, the chloride ion is determined by addition of excess silver nitrate and back-titration with potassium thiocyanate using ferric alum indicator. Aside from the unique glass fiber analysis, where the epoxide cannot be extracted and brought into solution, direct potentiometric titration with silver nitrate is quite satisfactory, in fact, preferable. Any free halide present does, of course, interfere in an argentimetric procedure, and a correction must be made.

Vorobjov (83) also used hydrochloric acid to cleave epoxy resins. He used a 15-min treatment in dioxane followed by direct titration of excess chloride by mercuric nitrate to a diphenylcarbazone endpoint.

The argentimetric technique is particularly valuable for the determination of α-epoxides in the presence of free acid or base and of readily hydrolyzable materials. Other successful applications (51,76) have been reported on epoxide samples containing large amounts of methyl formate, ethyl acetate, benzoyl peroxide, butylamine, and other compounds that would have interfered with the regular hydrohalogenation methods.

In addition to the above, many variants of the hydrochlorination reagent have been developed and applied satisfactorily in specific situations, using hydrochloric acid *per se* in solvents such as dioxane (46) and methyl ethyl ketone (45), and as the hydrochloride complex (21), e.g., in pyridine and dimethyl formamide. In fact, hydrochlorination in pyirdine is still widely used, especially for the more reactive oxiranes; 1,2-glycols and unsaturated compounds are unlikely to interfere. But there is no doubt that this reagent has its limitations. It is much less reactive than the hydrobromic acid–glacial acetic acid reagent, and methods employing it usually require heating at elevated temperatures, sometimes for extended periods of time.

The most important advances toward getting shorter and more generally applicable titration methods for epoxides appear in two procedures involving *in situ* generation of hydrobromic acid. For example, Dijkstra and Dahmen (27) found the direct titration method inconvenient for the determination of glycidyl ethers because the anhydrous hydrobromic–acetic acid reagent reacted too slowly with them. As a consequence of their studies on the mechanism of hydrobromic acid addition, a new rapid method was evolved, based on titration of the glycidyl ethers with perchloric acid in the presence of cetyl trimethylammonium bromide in acetic acid medium to a crystal violet endpoint. Because the endpoint shown by the indicator does not coincide exactly with the potentiometrically determined equivalence point, it is necessary to run a blank titration and correct accordingly. This method gave good results also on epichlorohydrin and ethylene oxide, and was suggested as being especially suitable for determining α-epoxids in gas streams.

Another variation of this approach, also involving the *in situ* generation of hydrobromic acid, was proposed by Jay (43), who titrated with perchloric acid in the presence of tetraethylammonium bromide. These reagents have the advantage of being more stable and easier to handle than anhydrous hydrobromic acid–acetic acid and, in addition, give sharper endpoints and more rapid titrations. Such modifications appear to be worthwhile improvements of the standard hydrobromic acid titration procedure, and are recommended.

b. Special Types of Oxirane Linkages

A limitation common to all of the hydrohalogenation procedures is their inapplicability to epoxides in which the oxirane oxygen is attached to a tertiary carbon atom. Such epoxides are usually readily hydrated, isomerize easily to form an aldehyde, or form easily hydrolyzable chlorohydrins. Although not fitting into this more general category, styrene oxide also acid hydrolyzes to an aldehyde and thus cannot be determined by the regular methods. Durbetaki (31) has developed a method designed specifically for such compounds including styrene oxide. His procedure consists of converting the α-epoxide to its corresponding aldehyde by catalytic isomerization with zinc bromide in benzene; the resulting aldehyde is determined gravimetrically as the 2,4-dinitrophenylhydrazone. α-Pinene oxide, camphene oxide, α-methylstyrene oxide, and 1,2-diisobutylene oxide have all been analyzed successfully by this procedure.

Another, more recent, method (21) in which epoxide compounds that isomerize to aldehydes present no problem is one employing morpholine as the reagent. This reacts with the 1,2-epoxy compound to form the corresponding tertiary amine, and excess reagent is acetylated to the amide with acetic anhydride. The tertiary amine is titrated with perchloric acid in methyl Cellosolve medium using thymol blue–xylene cyanol mixed indicator. The morpholine method is restricted to the more reactive epoxides. Many, α,β-unsaturated compounds react to consume this reagent and thus interfere. Nevertheless, the method can sometimes be used to circumvent interferences that would be encountered in the more generally applicable methods.

As a special case, epichlorohydrin can be determined by periodate cleavage to formaldehyde, which is determined by one of the colorimetric procedures—acetylacetone, fuschsin, or chromotropic acid (65).

c. Use of Other Nucleophilic Reagents

In addition to the foregoing methods, which include the most generally applicable ones as well as those most useful for avoiding certain specific interferences, methods based on numerous other nucleophilic reagents have been used for epoxide determinations. These include procedures based on sodium sulfite (80) and sodium thiosulfate (78). Such methods are applicable mainly to water-soluble, terminal epoxides. The sodium thiosulfate method gave good results on styrene oxide, and should work well for other terminal epoxides; a variation of the method, especially designed for basic α-epoxides, has been developed by Leary (52).

One of the most important of the newer methods for determining epoxides is that described by Gudzinowicz (38) employing dodecanethiol as the reagent. This reagent was first (4) developed for the determination of α,β-unsaturated carbonyl compounds and nitriles. With epoxides reaction (5) occurs. The excess mercaptan is then titrated iodometrically.

$$\underset{\text{(epoxide)}}{H_2C\overset{O}{-}CH_2} + RSH \xrightarrow[\text{excess, RSH}]{OH^-} \underset{SR}{CH_2}-\underset{OH}{CH_2} \tag{5}$$

This novel method can be applied in media that are rich in benzene or alcohols. It has been used successfully on the simpler epoxides as well as on a number with higher molecular weight; a precision and accuracy of 1–2% is claimed. Aldehydes, such as formaldehyde and acetaldehyde, when present in relatively large amounts (~50%), affect the accuracy of the method, causing slightly high results. Of course, α,β-unsaturated compounds present in mixtures with epoxides are included quantitatively and must be determined independently by other methods to derive the epoxide content by difference.

We have had occasion to compare the Gudzinowicz method with those using hydrochloric–acetic acid (argentimetric modification) and have found good agreement with the exception of the results on some resin-coated fiber glass samples. (As already noted, their analysis is a unique problem because of the slow diffusion of reagents into the resin and consequent slow or incomplete reaction.) The dodecanethiol method does not give correct results on epichlorohydrin or in other cases where the carbon alpha to the epoxide is substituted with a group that makes it possible for ring closure to occur in the presence of excess base, as indicated in the reaction series of equation (6). On the other hand, there were

$$\underset{\text{(epoxide)}}{H_2C\overset{O}{-}CH}-\underset{}{\overset{Cl}{CH_2}} \xrightarrow[OH^-]{RSH} \overset{RS}{H_2C}-\overset{OH}{CH}-\overset{Cl}{CH_2} \xrightarrow[OH^-]{\text{excess}}$$

$$\overset{RS}{H_2C}-CH\overset{O}{-}CH_2 \xrightarrow[RSH]{OH^-} \overset{RS}{H_2C}-\overset{OH}{CH}-\overset{SR}{CH_2} \tag{6}$$

indications that the dodecanethiol method combined with the direct hydrobromic acid or other methods would offer the possibility of distinguishing among certain types of epoxide linkages, based on reaction rate differences.

A disadvantage of dodecanethiol or other mercaptan reagents is their tendency to oxidize. According to Gudzinowicz (38), this requires a rather involved correction. However, we found it necessary to run only one blank titration, thus simplifying the procedure and calculations.

To try to overcome the oxidation problem encountered with dodecanethiol, monothioglycol (67) was tested in this laboratory as a reagent, on the expectation that it would have less tendency to oxidize. Such did not prove to be the case; in fact, it oxidized more rapidly than the dodecanethiol. It also reacted faster with epoxide, so the net effect was to make it a slight improvement over dodecanethiol; perhaps the addition of an antioxidant should be considered. The development of better mercaptan reagents appears to be an area worthy of future work.

d. Colorimetric Methods

Several colorimetric methods for detection of α-epoxides have been suggested; included is one based on their reaction with lepidine (4-methylquinoline) in the presence of diethylene glycol at 170°C to form an intense blue dye (39). The sensitivity of the reaction is primarily dependent upon the number of substituents on the cyclic carbon atoms. Ammonia and simple amines interfere in the test. The colorimetric procedure of Critchfield (22) is good for the specific determination of ethylene oxide in ppm quantities. It is based on hydrolysis of the epoxide to ethylene glycol and cleavage of the latter with sodium periodate to yield formaldehyde, which is then determined by chromotropic acid. Schaefer (69) has made a study of the detection of epoxides by paper chromatography. More recently, Urbanski and Kainz (82) developed a colorimetric method based on reaction of epoxy compounds (in dioxane) with 2,4-dinitrobenzenesulfonic acid. The resulting dinitrobenzenesulfonic ester gives a yellow color with piperazine. This procedure is claimed to be capable of detecting and determining very low concentrations of epoxides. We applied this method to various epoxy-containing polymers and to some other compounds containing the epoxy group. In the case of the polymers there were solubility problems which were overcome to some extent by making minor modifications in the procedure. With different compounds containing the epoxide linkage, the absorptivity per epoxide equivalent varied considerably, so we found it necessary to prepare calibration standards for each epoxy compound. For qualitative purposes, however, and especially for application to the simpler types of epoxy compounds, the method might be useful.

The colorimetric procedure of Fioriti et al. (33), which is based on reaction of the epoxide with picric acid, appears to be the most generally applicable and reliable. In a thorough study, these authors found that maximum color development requires a considerably longer time than by the Urbanski and Kainz procedure; but they demonstrated that their method gave reproducible and reliable results on a wide variety of oxirane compounds. The method has been extended to the detection of epoxides in heated oils (33a); in this case thin-layer chromatography was used to separate mixtures of the picrated epoxidized methyl esters. We recommend the Fioriti method.

4. Physical and Instrumental Methods

a. Infrared and Near Infrared

The application of infrared to cyclic ethers, and especially to the identification and determination of the epoxide group, has been extensively studied (6). A good discussion on the correlation of the infrared spectra with the structure of a wide range of oxirane compounds is given by Bomstein (10); he includes an overall evaluation of his own data and that of previous investigators (41,64,72). A number of bands in the 900–800 cm^{-1} (11–12 μ) region have been found suitable for quantitative analyses, and are used routinely. The successful application of infrared to the specific problem of the analyses of cured epoxy resins has been described by Dannenberg and Harp (24). Experience with their procedure in this laboratory has been generally satisfactory. The band at 9175 cm^{-1} (10.90 μ) due to the terminal epoxy group in condensation products from the reaction of Bisphenol A and epichlorohydrin is used to monitor the curing of these resins. The band diminishes as curing proceeds. Interfering absorption from reaction products makes this method unreliable at low epoxy levels. Other specific applications of infrared have been developed for the identification of 1,2-epoxybutane (15) in a mixture of four epoxybutanes, and for the determination of impurities in technical 1-chloro-2,3-epoxypropane (50). An interesting infrared method (75a) has been proposed based on spectral changes associated with the opening of the epoxide ring by reaction with gaseous hydrogen chloride; differences in spectra before and after treatment are determined. Measurements are made in the range of fundamental frequencies (3650–3450 cm^{-1}; 2.74–2.90 μ) of the hydroxyl groups formed in the reaction. The method is claimed to be applicable to all types of epoxides and to be capable of determining as little as 1×10^{-3} mmole/ml with an accuracy of 1–2%; phenolics and other hydroxyl components in high concentrations interfere.

Although the near infrared does not have the general utility of the middle infrared for qualitative purposes, it is perhaps the best of all methods for the quantitative determination of terminal epoxides; other oxygen rings or ethers do not interfere. Goddu (37) used the bands in the near-infrared region at 4550 cm^{-1} (1.65 μ) and 6050 cm^{-1} (2.20 μ) for the rapid determination of epoxide in a variety of mixtures. The range covered is from 100 ppm to pure epoxides; the accuracy and precision over most of the range are ±1–2% of the amount present. An example of a near-infrared absorption spectrum of an epoxide is that of epichlorohydrin shown in Fig. 2.

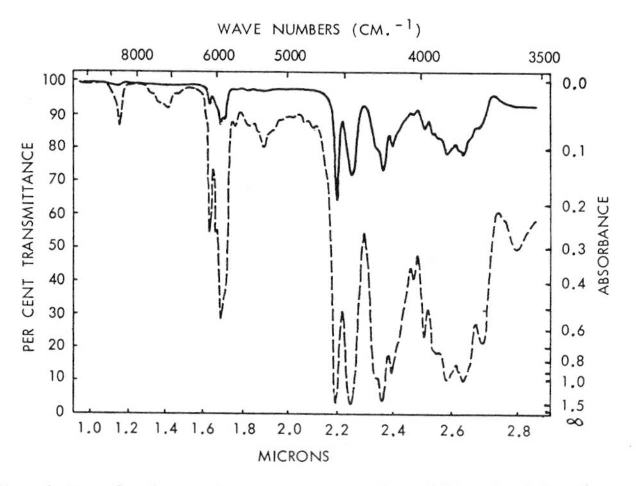

Fig. 2. Near-infrared absorption spectrum of epichlorohydrin in carbon tetrachloride. Solid line, 1% solution in 1-cm cell; broken line, 10% solution in 1-cm cell.

More recently the use of the near-infrared technique for the analyses of epoxy resins has been reported (25). Epoxy values are determined at 4535 cm^{-1} (2.205 μ) or else at 8630 cm^{-1} (1.159 μ). Near-infrared spectroscopy was found to be more accurate and freer from interference than the previously used infrared method, and was applied quite successfully to curing studies on epoxy resins. The technique of attenuated total reflectance should be uniquely valuable in examining epoxy-containing surfaces of molded pieces, thick films, or coatings where it is impossible to use ordinary infrared or near-infrared procedures.

b. Other Methods

Differential thermal analysis (DTA) is becoming increasingly important for the study of epoxide reactions, particularly ones involving such processes as polymerizations and curing (1). DTA thermograms have been used to characterize various epoxide systems. Perhaps of much more importance, though, especially in the protective coatings field, is the combined application of DTA and infrared (19,35). In this laboratory we have found this combination useful in elucidating the mechanism of curing processes involving epoxy resins. For volatile epoxides, gas chromatography must always be considered.

D. OXETANES AND OTHER LARGER RING COMPOUNDS

1. Chemical Reactivity and Chemical Methods

The oxetanes or β-epoxides are considerably less reactive than the oxiranes because of the increased stability of the oxetane ring. Trimethylene oxide is much less reactive than ethylene oxide but more reactive than tetrahydrofuran. The gamma and delta compounds are usually still more resistant, with the exception of unsaturated substances such as furan and pyran. Broadly speaking, the same types of methods have been applied to these compounds as to oxiranes, but they have been generally less effective because of the increased stability. For example, in the determination (48) of the oxetane content of 3,3-bis(chloromethyl)oxetane, refluxing 3 hr with a pyridinium chloride–anhydrous pyridine reagent is required to open the ring and convert it to 1,3-chlorohydrin (reaction (7)). The excess

$$(ClCH_2)_2C\begin{matrix} CH_2 \\ \diagup \quad \diagdown \\ \diagdown \quad \diagup \\ CH_2 \end{matrix}O + HCl \xrightarrow[\text{pyridine}]{} (ClCH_2)_2C\begin{matrix} CH_2OH \\ \diagup \\ \diagdown \\ CH_2Cl \end{matrix} \qquad (7)$$

pyridinium chloride is back-titrated with standard aqueous alkali. The procedure can be shortened to 1 hr by carrying out the reaction in sealed tubes at elevated temperature (170°C) and pressure. These were applied satisfactorily to a range of β-epoxides; the relative standard deviation, using either modification, is about 0.5%.

Sometime ago the analysis of 3,3-bis(chloromethyl)oxetane and related compounds was extensively investigated in this laboratory, incident to the commercial development of the chlorinated polyether Penton.* Gas chromatography (42) proved generally superior to chemical, infrared, or

*Registered trademark of Hercules Incorporated.

near-infrared techniques for analysis of the monomer and various process streams.

The gamma and delta epoxides are less reactive than the alpha and beta type, yet they are far from inert; for example, the ether ring of tetrahydrofuran is fairly readily opened by treatment with benzoyl chloride in the presence of zinc chloride catalyst (reaction (8)). With 3,5-dinitro-

$$\text{(tetrahydrofuran ring: } H_2C{-}CH_2,\ H_2C{-}O{-}CH_2) + C_6H_5COCl \xrightarrow{ZnCl_2} C_6H_5COOCH_2CH_2CH_2CH_2Cl \qquad (8)$$

benzoyl chloride, the corresponding solid 3,5-dinitrobenzoates are formed; they are valuable for characterization purposes.

2. Physical and Instrumental Methods

As with the smaller ring ethers, infrared is one of the more useful techniques for the detection and determination of the larger ring cyclic ethers. It also has the same drawbacks, i.e., lack of sensitivity and interference problems. Infrared spectra for 15 compounds of the furan series have been obtained (60). The absorption between 1580 and 1620 cm^{-1} (6.17–6.33 μ) is attributed to the double bond in the dihydrofuran nucleus. Bellamy (6) gives 1150–1060 cm^{-1} (8.70–9.44 μ) for six-membered rings, 1100–1075 cm^{-1} (9.09–9.30 μ) for five-membered ones, 980–970 cm^{-1} (10.20–10.30 μ) for four-membered ones, and 890–830 cm^{-1} (11.23–12.05 μ) for three-membered ones.

Mixtures of furan and tetrahydrofuran homologs have been separated by gas chromatography (73) using tritolyl phosphite on diatomite as the stationary phase and hydrogen as the carrier gas.

III. SUMMARY OF METHODS FOR ETHERS AND EPOXIDES

For the detection and identification of inert ethers, the procedures of Cheronis (18) involving derivative formation, and of Jurecek et al. (47) involving cleavage and paper chromatography, are the most suitable. The recently developed method of Waszeciak and Nadeau (85), involving gas–liquid chromatography, appears to have the greatest future potential; it has been applied successfully to both simple aliphatic and polymeric ethers. Infrared, NMR, and mass spectroscopy are useful in certain specific instances, but none are broadly applicable. Lack of sensitivity, interference problems, and other considerations limit them both qualitatively and quantitatively.

Of the methods discussed for vinyl ethers, those involving hydrolysis and estimation of the liberated acetaldehyde are quite useful except when the sample also contains acetaldehyde or acetals. The hydrolysis type of procedure with appropriate modifications is the best also for determining trace amounts of vinyl ethers (13). For specificity in determining vinyl ethers, the iodometric method of Siggia (75) is suggested. It is simple and will give results of acceptable precision and accuracy; it is limited to lower molecular weight vinyl ethers, or those soluble in the methanol–iodine reagent. The modified mercuric acetate method is particularly indicated where terminal or isolated double bonds with *cis* configuration are involved (44). Near-infrared spectroscopy (36) is widely applicable, much faster, and generally superior to chemical methods for vinyl ethers; it is highly recommended. Water in the sample or solvent does interfere because it absorbs in the near-infrared, but so many vinyl ethers can or must be examined in nonaqueous systems that this is not a practical limitation.

There are so many methods for epoxides that it is a problem to select the one best suited for the particular system; there is no universally applicable procedure. The most widely useful is that involving reaction of the epoxide linkage with hydrobromic acid, both the direct and indirect modifications (3,9,29). The new development of generating the hydrobromic acid *in situ* as described by Jay (43) is a worthwhile improvement of this already widely accepted method and is recommended. Some of the older hydrochlorination methods (21), although less useful than the hydrobromic acid techniques, are still often chosen for application to the more reactive oxirane linkages, especially where it is desired to minimize interference from 1,2-glycols or unsaturated compounds. None of the hydrohalogenation procedures are reliable in the particular instance where the oxirane oxygen is attached to a tertiary carbon atom. Special methods, involving catalytic isomerization (31) or the use of morpholine (21) as a reagent, must be used.

Of the numerous other nucleophilic reagents used to determine epoxides, the procedure employing dodecanethiol (38) deserves favorable mention. It has its limitations, but combinations of this method with other appropriate ones does, in some instances, offer a means of distinguishing among certain types of epoxide linkages.

For detection of epoxides the colorimetric method based on their reaction with picric acid (33) is suggested. Infrared and near-infrared have proven valuable in both identification and determination of epoxides; but for the quantitative determination of terminal epoxide linkages, near infrared is the best of the available methods (37). For specific situations such as elucidating the mechanism of curing and polymerization reactions involving epoxide linkages, the combination of infrared or near infrared and DTA has been quite effective (19,35).

For analysis of the oxetanes (β-epoxides) and other larger ring compounds, many of the same methods used for the previous sections are applicable, but because of the increased stability of these epoxides more rugged reaction conditions are required. For the specific determination of the oxetane content of 3,3-bis(chloromethyl)oxetane, a gas chromatographic method is recommended.

IV. RECOMMENDED LABORATORY PROCEDURES

A. ALIPHATIC ETHERS—DETECTION AND IDENTIFICATION

1. Paper Chromatography after Cleavage with 3,5-Dinitrobenzoyl Chloride and Zinc Chloride (47)

A sample is reacted with 3,5-dinitrobenzoyl chloride and anhydrous zinc chloride, the reaction mixture washed to remove the excess reagents, and the derivatives identified by paper chromatography.

Procedure

Introduce 0.2 ml of ether, 1.1 g of 3,5-dinitrobenzoyl chloride, and 0.4 g of freshly fused anhydrous zinc chloride catalyst into a small tube, seal, and place in an oven at 115°C for 30 min. After cooling, cautiously open the tube and take up the reaction mixture in 5 ml of benzene. Shake the benzene solution with 30% sodium hydroxide to remove zinc chloride and excess aryl chloride, and discard the aqueous layer. Wash the benzene solution with water, discard the washings, and dry it over anhydrous calcium sulfate. Chromatograph the benzene solution of the alkyl 3,5-dinitrobenzoate and at the same time run a 1% benzene solution of the appropriate known ester as a standard. Spot 15–60 μg of the alkyl 3,5-dinitrobenzoate onto paper impregnated with dimethylformamide. Use cyclohexane (containing no aromatics) as the solvent, making sure that the chromatograph chamber is saturated with the vapors of both solvents. Use a 1% solution of α-naphthylamine as a developer; the spots are brownish red, easily recognized under ultraviolet light. See references 11, 12, and 46 for R_f values.

2. Gas Chromatography after Reaction with Acetyl Chloride and Ferric Chloride (85)

A sample is reacted with acetyl chloride and anhydrous ferric chloride and the resulting derivatives identified by gas chromatography. A typical procedure used for identification of polypropylene glycols follows.

Procedure

Mix about 1.5 g of sample with 1.5 g of ferric chloride and 1.5 g acetyl chloride at 0°C in a small flask. Allow to come up to room temperature. Chromatograph a 5-μl portion of the reaction mixture using a 1-m column of Dow-Corning Silicone Oil DC-500 on Chromosorb W, inlet port 250°C, column 75°C for 9 min, programmed to 200°C. The major peaks from a sample of polypropylene glycol, mol wt about 3000, will be acetyl chloride, acetic acid, acetic anhydride, chloropropyl acetate, and propylene glycol diacetate.

The relative retentions and response factors for other ethers must be determined by chromatographing pure materials.

B. VINYL ETHERS—QUANTITATIVE MEASUREMENT

1. Hydrolysis–Aldehyde Measurement (20,74)

A sample is reacted with neutralized aqueous hydroxylamine hydrochloride solution (pH 3.1) at either room or reflux temperature and the liberated hydrochloric acid titrated potentiometrically with standard alkali.

Depending on the nature of the compound being determined, methyl alcohol is added to the reaction mixture to aid solution, and, if necessary, the mixture is refluxed to promote completeness of reaction.

Procedure

Weigh a sample containing about 12 meq of vinyl ether into a 250-ml Erlenmeyer flask with ground joint. If the sample is appreciably acid or alkaline, dissolve in a small volume of methanol and adjust to a pH of 3.1 as measured by a pH meter equipped with glass—calomel electrodes.

Add, with a graduate, 50 ml of 0.5*M* aqueous hydroxylamine hydrochloride solution which has been adjusted to a pH of 3.1. If necessary, add methanol to obtain a one-phase solution. Keep the amount of methanol at a minimum because it slows hydrolysis of the sample.

Allow the reaction mixture to stand at room temperature until reaction is complete. If a reaction time of over 2 hr is required, reflux the mixture for the required time, 30 min to 2 hr.

After hydrolysis titrate potentiometrically with 0.5*N* aqueous sodium hydroxide solution. If methanol was added to dissolve the sample, add an equal volume of water before titrating. Observe the volume of titrant at the inflection point which is usually about pH of 3.1.

Calculation

$$\frac{\text{ml NaOH} \times N \times \text{mol. wt.} \times 100}{\text{grams of sample} \times 1000} = \% \text{ Vinyl ether}$$

Precision and Accuracy

Duplicate determinations by the same analyst should agree within 0.5%. Siggia states that this method is accurate to ±5% (absolute).

2. Iodine Addition (75)

This method is applicable to all vinyl ethers soluble in the methanol–aqueous iodine reaction mixture used.

A sample is added to a mixture of iodine solution and methanol, allowed to stand for a short while, and the excess iodine titrated with standard sodium thiosulfate solution. The reaction is

$$ROCH{=}CH_2 + I_2 + CH_3OH \rightarrow \underset{\displaystyle OCH_3}{ROCH}{-}CH_2I + HI \qquad (9)$$

Many of the vinyl ethers are somewhat volatile; therefore, the sample is weighed in a glass bulb.

Procedure

Weigh a sample containing about 25 meq of vinyl ether in a small glass bulb. Cool the bulb with Dry Ice to draw up the sample and seal the capillary tip of the bulb with a small flame.

Add 25 ml of glass beads, 50 ml of methanol, and 50.00 ml of 0.1N standard iodine solution to a 500-ml wide-mouth glass-stoppered bottle. Add the bulb with sample, grease the ground area of the stopper, and insert the stopper in the bottle.

Shake the bottle vigorously to break the glass bulb and crush it. For low-boiling samples, place the bottle on a mechanical shaker for 10 min to absorb the vapors. Higher boiling materials may be allowed to stand for 10 min with only occasional swirling.

After shaking, open the bottle and rinse down the stopper and inside walls of the bottle with water and titrate with 0.1N sodium thiosulfate solution to the disappearance of the yellow color.

Make a blank titration on 50.00 ml of the iodine solution carried through the complete determination as described for a sample.

Calculation

$$\frac{(B - S) \times N \times {}^1/_2 \text{ (mol wt)} \times 100}{\text{g sample} \times 1000} = \% \text{ Vinyl ether}$$

where B = ml of sodium thiosulfate required for the reagent blank, S = ml of sodium thiosulfate required for the sample, and N = normality of the sodium thiosulfate used.

Precision

Siggia shows data wherein duplicate determinations by the same analyst agree within 0.5%.

3. Mercuric Acetate Addition (44)

A sample is reacted with a methanolic solution of mercuric acetate stabilized with a small amount of acetic acid. Both mercuric acetate and methanol add to the double bond, liberating acetic acid. Sodium bromide is added to convert the excess mercuric acetate to mercuric bromide and the acetic acid is then titrated with methanolic potassium hydroxide solution using phenolphthalein indicator.

Reagents

(*1*) *Mercuric acetate in methanol*, approximately 0.12*M*. Dissolve 40 g of mercuric acetate in anhydrous methanol, dilute to 1 liter, and add 3–8 drops of glacial acetic acid.

Procedure

Pipet 25.00 ml of the mercuric acetate solution into a 250-ml Erlenmeyer flask with ground joint. Cool the solution to about −10°C. Add a weighed sample containing 20 meq of vinyl ether, stopper, swirl to mix, and allow to stand in a −10°C bath for 10 min. (See reference 76 for recommended reaction temperatures for other unsaturated compounds.)

After reaction, add 1 ml of 1% alcoholic phenolphthalein indicator solution and titrate with 0.1*N* methanolic potassium hydroxide solution to a pink endpoint. Keep the temperature of the solution below 15°C during the titration. Make a blank titration on 25.00 ml of the reagent after treating in the same manner as for the sample.

Calculation

$$\frac{(S - B)N \times \text{mol wt} \times 100}{\text{g sample} \times 1000} = \% \text{ Vinyl ether}$$

Precision

Siggia (74) shows data wherein duplicate determinations by the same analyst agree within 0.3%.

4. Near-Infrared (36)

A sample is dissolved in carbon tetrachloride and its infrared absorbance at about 6300 cm^{-1} (1.6 μ) is determined as a measure of the vinyl ether content. Other compounds which absorb at this wavelength will interfere.

Calibration of Spectrophotometer

Calibrate the spectrophotometer by obtaining the absorbance of several carbon tetrachloride solutions of varying amounts (mg/ml) of pure vinyl ether of the type to be determined in the same manner as for a sample. Plot absorbance versus concentration (mg/ml) to give a calibration curve.

Procedure

Prepare a carbon tetrachloride solution of the sample. Fill a 1-cm silica cell with the solution. Obtain the spectrum over the region of 6330–6020 cm^{-1} (1.58–1.66 μ) against carbon tetrachloride in the reference beam using a Beckman DK-2, Cary Model 14, or equivalent instrument. Read the absorbance from the peak for the analytical wavelength to the selected baseline. Using the observed absorbance, refer to the calibration curve and read the corresponding amount of vinyl ether present.

Calculation

$$\frac{\text{mg/ml vinyl ether} \times 100}{\text{mg/ml of sample in soln}} = \%\ \text{Vinyl ether}$$

Precision

Duplicate determinations by the same analyst should agree within 1%. (Estimated by authors of this chapter.)

C. EPOXIDES—QUANTITATIVE MEASUREMENT

1. Titration with Hydrogen Bromide (3,9)

a. Direct

A sample is dissolved in chlorobenzene and titrated with a standard solution of hydrogen bromide in glacial acetic acid to a crystal violet blue-green color. The direct titration approach has been adopted by ASTM and AOCS and thus has obtained wide acceptance for use. For less reactive epoxides, the indirect hydrobromic acid method is often applicable. If amines are present, they will be titrated by either procedure; however, they can be corrected for by use of mercuric acetate reagent (30).

Reagents

(*1*) *Crystal violet indicator solution*, 0.1% in glacial acetic acid.

(*2*) *Hydrogen bromide solution*, 0.1*N*, in glacial acetic acid. Add 6.5 ml of reagent grade hydrogen bromide to 1 liter of glacial acetic acid. Add 1.0-g increments of reagent grade phenol until the solution becomes light straw color. Usually about 10 g of phenol is required. Then add 1.0 g in excess. Mix the solution after each addition of phenol and then allow to stand for 12 hr before using. Standardize against primary standard grade potassium acid phthalate dissolved in glacial acetic acid using crystal violet indicator. Store the reagent in an automatic buret assembly and protect from atmospheric moisture with drying tubes containing indicating Drierite. This reagent has a high coefficient of expansion with temperature. Either standardize at the temperature at which it is to be used or keep in a constant-temperature room.

Procedure

Add 5 ml of chlorobenzene to a 50-ml Erlenmeyer flask. Weigh a sample containing about 20 meq of epoxy compound into the flask. Add 5 drops of crystal violet indicator solution and a Teflon-coated magnetic stirrer bar. Close the neck of the flask with a one-hole rubber stopper with a small groove cut in one side as an air vent.

Insert the tip of the buret through the hole in the stopper in the flask. Stir the solution with a magnetic stirrer and titrate to a blue-green color which is permanent for 1 min. Some of the epoxides react at a relatively slow rate. As the endpoint is approached, add the titrant in small increments, allowing a few seconds stirring to make certain that the blue color does not return.

Calculation

$$\frac{\text{ml} \times N \times \text{mole wt.} \times 100}{\text{g sample} \times 1000} = \%\ \alpha\text{-Epoxide}$$

Precision

ASTM states that average difference between duplicate runs may approximate 0.7% of the epoxy content of the resin tested.

b. Indirect

Procedure same as in Section IV-C-1-a above, except that a measured volume of the hydrogen bromide solution is added to the sample in an iodine flask. After reaction (usually 15–30 min), both the sample and a reagent blank are titrated with $0.1N$ sodium acetate in acetic acid to a crystal violet endpoint.

2. In situ Hydrobromic Acid Method (43)

A sample is dissolved in chloroform and titrated with a standard solution of perchloric acid in acetic acid or dioxane to a crystal violet endpoint in the presence of an excess of a soluble quaternary ammonium bromide. This method appears equal or superior to the older hydrobromic acid methods and has the advantages of more stable reagents, sharper endpoints, and more rapid titrations.

Procedure

Weigh a sample containing 0.6–0.9 meq of oxirane into a 50-ml Erlenmeyer flask and dissolve in 10 ml of chloroform, acetone, benzene, or chlorobenzene. Add 10 ml of tetraethylammonium bromide in glacial acetic acid (100 g in 400 ml) and 2 or 3 drops of crystal violet indicator solution. Titrate with $0.1N$ perchloric acid in glacial acetic acid using a 10-ml microburet.

Calculation

$$\frac{\text{ml} \times N \times \text{mol wt} \times 100}{\text{g sample} \times 1000} = \%\ \alpha\text{-Epoxide}$$

Precision

Jay (43) showed variations for duplicate determinations not exceeding 2 units of epoxy equivalent weight in the region of 400 (i.e., butyl epoxystearate).

3. Hydrochlorination with Pyridine–Hydrochloric Acid (21)

A measured volume of hydrogen chloride in pyridine is reacted with the sample to form the corresponding chlorohydrin. The unreacted hydrogen chloride in the sample, and also in a reagent blank, is titrated with standard alkali.

Reagents

Pyridine hydrochloride solution, 0.2N. Cautiously add 17 ml of reagent grade hydrochloric acid to 1 liter of reagent grade pyridine and mix thoroughly.

Procedure

Accurately measure 25 ml of the pyridine hydrochloride solution into a 250-ml Erlenmeyer flask with ground joint. Add a weighed sample containing 25 meq of α-epoxide and swirl to dissolve. Connect the flask to a reflux condenser and reflux for 20 min.

After refluxing, cool the flask and contents, add 6 ml of distilled water and about 0.5 ml of 1% phenolphthalein indicator solution, and titrate with 0.1N alcoholic potassium hydroxide solution to the first permanent pink color.

Carry out a blank determination on 25 ml of the pyridine hydrochloride solution in the same manner.

Calculation

$$\frac{(B - S) \times N \times \text{mol wt} \times 100}{\text{g sample} \times 1000} = \% \ \alpha\text{-Epoxide}$$

where B = ml alcoholic potassium hydroxide solution for the blank, S = ml alcoholic potassium hydroxide solution for the sample, N = normality of the alcoholic potassium hydroxide solution used.

Precision

Siggia (74) gives an estimated reproducibility of replicate determinations of 0.8%.

4. Colorimetric (33)

An ether solution of the sample is reacted with picric acid for 12–48 hr. Aliquots of this solution are diluted with basic ethanol solution and the absorbance measured at 490 mμ using a photometer calibrated with known solutions.

Procedure

Weigh a sample containing 0.04–0.4 meq of oxirane into a 10-ml volumetric flask. Dissolve in about 5 ml of ether. Add 2.0 ml of 0.25M picric acid in ethanol, make up to the mark with ether, and mix. Allow to stand at 24 ± 2°C until maximum color develops, usually 12–48 hr. Then withdraw 1.0-ml aliquots and dilute to 50.0, 100.0, and 200.0 ml with basic ethanol solution (20 ml 5% sodium hydroxide and 80 ml 95% ethanol). Read the absorbance in a 1-cm cell at 490 mμ within an hour after dilution. Run a blank on the picric acid solution under the same conditions. Using the absorbance of the dilution having an absorbance in the midrange of the scale, corrected for the blank, read the concentration of oxirane (moles/liter × 10^5) from the calibration curve.

Calculation

$$\frac{\text{Oxirane concentration} \times \text{dilution ratio} \times 100}{\text{mol wt of oxirane}} = \%\ \text{Oxirane}$$

Precision

The repeatability for duplicate determinations by the same analyst is not given by Fioriti.

Different methods agree within 4% absolute at the 25–30% oxirane level.

5. Near Infrared (37)

The same procedure used for vinyl ethers (Section IV-B-4) applies for epoxides provided suitable reference standards are available for calibration of the spectrophotometer.

REFERENCES

1. Anderson, H. C., *Anal. Chem.*, **32**, 1592 (1960).
2. Ashworth, M. R. F., *Anal. Chim. Acta*, **18**, 330 (1958).
3. ASTM D 1652-67.
4. Beesing, D. W., W. F. Tyler, D. M. Kurtz, and S. A. Harrison, *Anal. Chem.*, **21**, 1973 (1949).
5. Bellamy, L. J., *The Infrared Spectra of Complex Molecules*, 2nd ed., Wiley, New York, 1958, p. 115.
6. Bellamy, L. J., reference 5, pp. 118–120.
7. Bhacca, N. S., L. F. Johnson, and J. N. Shoolery, *NMR Spectra Catalog*, Varian Associates, 1962.
8. Bodnar, S. J., and S. J. Mayeux, *Anal. Chem.*, **30**, 1384 (1958).
9. Bolley, D. S., "Report of Epoxidized Oils Subcommittee on Oxirane Oxygen," 1963, *J. Am. Oil Chem. Soc.*, **41**, 86 (1964); see also AOCS Cd 9–57.
10. Bomstein, J., *Anal. Chem.*, **30**, 544 (1958).
11. Borecky, J., and J. Gasparic, *Collection Czech. Chem. Commun.*, **25**, 1287 (1960).
12. Borecky, J., and J. Gasparic, *J. Chromatog.*, **4**, D25–6 (1960).

13. Bowman, M. C., M. Beroza, and F. Acree, Jr., *Anal. Chem.*, **33,** 1053 (1961).
14. Budesindky, B., and J. Korbl, *Mikrochim. Acta,* **1960,** 697.
15. Carrington, R. A. G., *Anal. Chem.*, **31,** 1117 (1959).
16. Carruthers, W., R. A. W. Johnstone, and J. R. Plimmer, *Chem. Ind. (London),* **1958,** 331.
17. Cheronis, N. D., "Micro and Semimicro Methods," in *Techniques of Organic Chemistry,* Vol. VI, A. Weissberger, Ed., Interscience, New York, 1954, pp. 454 and 459.
18. Cheronis, N. D., J. B. Entrikin, and E. M. Hodnett, *Semimicro Qualitative Organic Analysis,* 3rd ed., Interscience, New York, 1965, pp. 541–547.
19. Clampitt, B. H., D. E. German, and J. R. Galli, *J. Polymer Sci.*, **27,** 515 (1958).
20. Critchfield, F. E., *Organic Functional Group Analysis,* Macmillan, New York, 1963, p. 65.
21. Critchfield, F. E., reference 20, pp. 125–140.
22. Critchfield, F. E., and J. B. Johnson, *Anal. Chem.*, **29,** 797 (1957).
23. Cross, A. H. J., A. H. Gunn, and S. G. E. Stevens, *J. Pharm. Pharmacol.*, **9,** 841 (1957).
24. Dannenberg, H., and W. R. Harp, Jr., *Anal. Chem.*, **28,** 86 (1956).
25. Dannenberg, H., *SPE Trans.*, **3,** 78 (1963).
26. Davison, W. H. T., and G. R. Bates, Jr., *J. Chem. Soc.*, **1953,** 2607.
27. Dijkstra, R., and E. A. M. Dahmen, *Anal. Chim. Acta,* **31,** 38 (1964).
28. Dugan, P. R., *Anal. Chem.*, **33,** 1630 (1961).
29. Durbetaki, A. J., *Anal. Chem.*, **28,** 2000 (1956).
30. Durbetaki, A. J., *Anal. Chem.*, **30,** 2024 (1958).
31. Durbetaki, A. J., *Anal. Chem.*, **29,** 1666 (1957).
32. Erley, D. S., *Am. Ind. Hyg. Assoc. J.*, **23,** 288 (1962).
33. Fioriti, J. A., A. P. Bentz, and R. J. Sims, *J. Am. Oil Chemists' Soc.*, **43,** 37 (1966).
33a. Fioriti, J. A., A. P. Bentz, and R. J. Sims, *J. Am. Oil Chemists' Soc.*, **43,** 387 (1966).
34. Feigl, F., R. Belcher, and W. I. Stephen, "Progress in Qualitative Organic Analysis," in *Advances in Analytical Chemistry and Instrumentation,* Vol. 2, C. N. Reilley, Ed., Interscience, New York, 1963, p. 18.
34a. Fritz, J. S., "Detection and Estimation of Ethers," in *The Chemistry of the Ether Linkage,* Patai, Ed., Interscience, New York, 1967, p. 669.
35. German, D. E., B. H. Clampitt, and J. R. Galli, *J. Polymer Sci.*, **38,** 433 (1959).
36. Goddu, R. F., "Analytical Applications of Near-Infrared," in *Advances in Analytical Chemistry and Instrumentation,* Vol. I, C. N. Reilley, Ed., Interscience, New York, 1960, p. 364–6, 417.
37. Goddu, R. F., and D. A. Delker, *Anal. Chem.*, **30,** 2013 (1958).
38. Gudzinowicz, B. J., *Anal. Chem.*, **32,** 1520 (1960).
38a. Gunther, D. A., *Anal. Chem.*, **37,** 1172 (1965).
39. Gunther, F. A., *et al.*, *Anal. Chem.*, **23,** 1835 (1951).
40. Haden, W. W., Hercules Incorporated, Research Center, private communication (1964).
41. Henbest, H. B., *et al.*, *J. Chem. Soc.*, **1957,** 1459.
42. Hudy, J. A., Hercules Incorporated, Research Center, private communication (1960).
43. Jay, R. R., *Anal. Chem.*, **36,** 667 (1964).
44. Johnson, J. B., and J. P. Fletcher, *Anal. Chem.*, **31,** 1563 (1959).
45. Jung, G., and W. Kleeberg, *Kunstoffe,* **51,** 714 (1961).

46. Jungnickel, J. L., E. D. Peters, A. Polgar, and F. T. Weiss, "Determination of the Alpha-Epoxy Group," in *Organic Analysis*, Vol. I, J. Mitchell, Ed., Interscience, New York, 1953, p. 127–54.
47. Jurecek, M., M. Hubik, and M. Vecera, *Collection Czech. Chem. Commun.*, **25**, 1458 (1960).
48. Keen, R. T., *Anal. Chem.*, **29**, 1041 (1957).
49. Kiley, L. R., *Anal. Chem.*, **30**, 1303, 1577 (1958).
50. Kolbasov, V. I., S. B. Bardenshtein, and R. V. Dzhagatspanyan, *Zavodsk. Lab.*, **26**, 1120 (1960); *Anal. Abstr.*, **8**, 2003 (1961).
51. Krull, L., *Farbenchemiker*, **61**, 23 (1959); *Anal. Abstr.*, **6**, 4971 (1959).
52. Leary, J. B., *J. Am. Pharm. Assoc., Sci. Ed.*, **49**, 606 (1960).
53. LeBlanc, R. B., *Anal. Chem.*, **30**, 1797 (1958).
54. Long, L. H., and G. F. Freeguard, *Nature*, **207**, 403 (1965).
55. Ma, T. S., and M. Gutterson, *Microchem. J.*, **5**, 416 (1961).
56. Meites, L., *Handbook of Analytical Chemistry*, McGraw-Hill, New York, 1963.
57. McLafferty, F. W., *Anal. Chem.*, **29**, 1782 (1957).
58. Micheel, F., and W. Schminke, *J. Chromatography*, **1**, Tables XVI and XVII, p. XIV (1958); from *Angew. Chem.*, **69**, 334 (1957).
59. Naves, Y. R., *Compt. Rend.*, **246**, 1734 (1958).
60. Nahum, R., *Compt. Rend.*, **240**, 1898 (1955).
61. Nakanishi, K., *Infrared Absorption Spectroscopy*, Holden-Day, San Francisco, 1962, p. 36.
62. Noller, D. C., and D. J. Bolton, *Anal. Chem.*, **35**, 887 (1963).
63. Nukada, K., *Bull. Chem. Soc. Japan*, **33**, 1606 (1960).
64. Patterson, W. A., *Anal. Chem.*, **26**, 823 (1954).
65. Pimenova, Z. M., and S. O. Khamaza, *Novoe v Oblasti Sanit. Khim. Analiza* (*Raboty po Prom-Sanit. Khim.*), **1962**, 113; *Chem. Abstr.*, **59**, 2094e (1963).
66. Pinchas, S., *Anal. Chem.*, **31**, 1742 (1959).
67. Price, C. C., University of Pennsylvania, private communication (1964).
68. Saier, E. L., and R. H. Hughes, *Anal. Chem.*, **30**, 513 (1958).
69. Schäfer, W., W. Nuck, and H. Jahn, *J. Prakt. Chem.*, **11**, 1 (1960).
70. Shanley, E. S., in *Kirk-Othmer, Encyclopedia of Chemical Technology*, Vol. 10, A. Standen, Ed., Interscience, New York, 1953, p. 70.
71. Sherma, J. A., Jr., and W. Rieman III, *Anal. Chim. Acta*, **18**, 214 (1958).
72. Shreve, O. D., and M. R. Heether, *Anal. Chem.*, **23**, 277 (1951).
73. Shuikin, N. I., B. L. Lebedev, and V. V. An, *Zavodsk. Lab.*, **27**, 1868 (1962).
74. Siggia, S., *Quantitative Organic Analysis via Functional Groups*, 3rd ed., Wiley, New York, 1963, pp. 399–403.
75. Siggia, S., *Quantitative Organic Analysis via Functional Groups*, 3rd ed., Wiley, New York, 1963, pp. 403–406.
75a. Soucek, J., and J. Vasatkova, *Collection Czech. Chem. Commun.*, **31**, 2860 (1966).
76. Stenmark, G. A., *Anal. Chem.*, **29**, 1367 (1957).
77. Stewart, R. D., and D. S. Erley, *J. Forensic Sci.*, **8**, 31 (1963).
78. Sully, B. D., *Analyst*, **85**, 895 (1960).
79. Sundt, E., and M. Winter, *Anal. Chem.*, **29**, 851 (1957).
80. Swan, J. D., *Anal. Chem.*, **26**, 878 (1954).
81. Tenney, H. M., *Anal. Chem.*, **30**, 2 (1958).
82. Urbanski, J., and G. Kainz, *Mikrochim. Acta*, **1**, 60 (1965).
83. Vorobjov, V., *Chem. Prumysl*, **13**, 381 (1963); *Anal. Abstr.*, **11**, 4420 (1964).

84. Ward, G. A., Hercules Incorporated, Research Center, private communication (1964).
85. Waszeciak, P., and H. G. Nadeau, *Anal. Chem.*, **36**, 764 (1964).
86. Winstein, S., and R. B. Henderson, "Ethylene and Trimethylene Oxides," in *Heterocyclic Compounds*, Vol. I, W. Elderfield, Ed., Wiley, New York, 1950, pp. 1–60.
87. Wolfe, W. C., *Anal. Chem.*, **34**, 1328 (1962).

Part II

Section B-2

DETERMINATION OF ORGANIC PEROXIDES

By Robert D. Mair and Robert T. Hall,
Research Center, Hercules Incorporated, Wilmington, Delaware

Contents

Contents (*continued*)

Contents (*continued*)

I. INTRODUCTION

Organic peroxides are compounds in which oxygen is bonded directly to oxygen. An unusually broad and varied category, their complexity arises in part from a dual nature: Besides possessing the definitive —O—O— linkage, they are also closely related to their nonperoxidic analogs in many physical and chemical properties (e.g., hydroperoxides are related to alcohols, peroxy esters to esters). The peroxide functionality thus challenges the analytical chemist with an extreme range of reactivities and a wide variety of analytically useful properties.

Consequently, the analytical chemistry of organic peroxides is so complex that chemists have failed to devise any general methods applicable to all members of this category. They have discovered many reactions for determining particular types of peroxides, and procedures have proliferated from these reactions; discovery of a single general-purpose method has remained elusive, however, and is probably unattainable. A more realistic approach, well illustrated by the work of Horner and Jürgens (82), is to use a series of coordinated methods and techniques (see Section IV-A-2-c).

The literature on peroxide analysis is extensive and cumbersome—a large body of knowledge arising from diverse fields of research that have little overlap of basic interests. Fortunately, there are a number of books and reviews that provide convenient access to and guidance through this literature; several recommended general references are listed at the end of our text.

Some readers may be interested in a broad treatment of the whole field of peroxide analysis. Others may have more specific objectives, such as recommended methods for one peroxide type or a particular mixture of types, fresh ideas for the development of specific process control methods, or a survey of methods for trace peroxide analysis in various systems. In recognition of these varying needs we have made an effort in this chapter to discuss briefly some characteristic properties of the peroxide classes, to consider the techniques by which certain peroxides can be separated from a mixture of other compounds (including other peroxides), to review the many reactions that are available for detecting and determining organic peroxides, and to consider the relative merits and limitations of particularly effective procedures.

II. STRUCTURE, OCCURRENCE, AND STABILITY OF ORGANIC PEROXIDES

A. REVIEW OF STRUCTURE AND PROPERTIES

As organic peroxides are a broad and complex category, we shall begin this section with an orienting review of their structures. Table I shows eleven classes whose characteristic traits will now be considered briefly. More detailed treatments of the peroxide types and their reactions and properties are available, particularly those of Milas (125) and Davies (35).

One structural feature, common at least to the noncyclic peroxides, is a skewed conformation about the —O—O— grouping. This derives from the structure of hydrogen peroxide, whose two —O—O—H linkages lie in planes that intersect at a dihedral angle of 120° (151); the —O—O—H angle is 95°. Peroxy acids, diacyl peroxides, dialkyl peroxides, alkyl hydroperoxides, and peroxy esters all have adaptations of this hydrogen peroxide skew (110,158,193). The chief source of information about the molecular geometries of these organic peroxides has been dipole moment data on their solutions.

The conformation of cyclic peroxides has been studied less. In one example, however, the disparity between observed and calculated dipole moments in ascaridole threw doubt on an assumed *cis*-planar configuration for the transannular C—O—O—C group (158).

1. Peroxy Acids

Swern and Silbert have reviewed the structural properties of peroxy and carboxylic acids (190). X-ray diffraction and infrared absorption data show that in the solid state both acid classes exist as hydrogen-bonded dimers. The strength of this hydrogen bonding is weaker for the peroxy acids, lying between that of carboxylic acids and solid alcohols, but closer to the latter.

In solution there are greater differences. Carboxylic acids show a concentration-dependent equilibrium between monomer, with a free hydroxyl group, and a dimer that has hydrogen-bonded hydroxyls. Peroxy acids, in contrast, remain strictly monomeric in solution, and their hydroxyl is hydrogen bonded intramolecularly; this has been shown clearly both by infrared absorption and molecular weight measurement.

Diagram 1 summarizes this structural comparison of these acid classes.

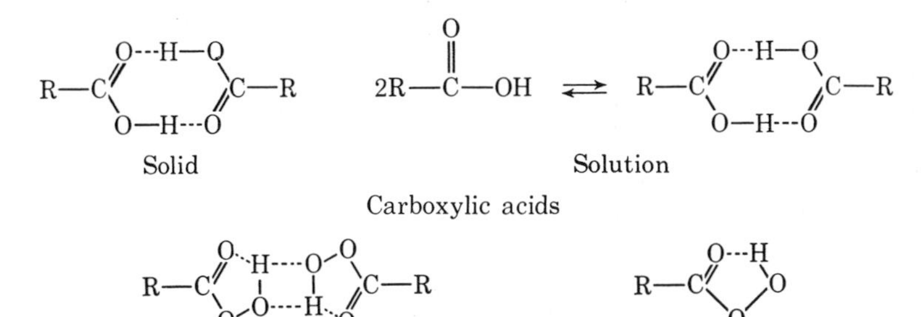

Solid Solution

Peroxy acids

Dipole moment data show the hydrogen-bonded peroxy acid ring to be puckered rather than planar (190). The C—$\overset{O}{\overset{\|}{C}}$—O—O grouping is essentially coplanar, with the —OH group projecting out of plane not quite vertically (dihedral angle between C—O—O and O—O—H planes is 72°, O—O—H angle is 105°).

As would be expected from their bonded hydroxyl group, peroxy acids have considerably weaker acid strengths than have the corresponding carboxylic acids. Short-chain peroxy acids in aqueous solutions (53,212) and long-chain ones in alcoholic systems (191) are both only about 1/1000 as strong as the corresponding carboxylic acids. On the other hand, in reactions typical of the —O—O— group, peroxy acids are the most reactive of all peroxides.

2. Diacyl Peroxides

These peroxides are analogs of anhydrides. In solution, they have a fixed skew structure about the —O—O— group with carbonyl oxygens facing inward. Their dihedral angle between the two essentially planar

TABLE I

Structural Classification of Organic Peroxides

Main class	Subclasses	Nonperoxy analogs
1. Peroxy acids, $R—C(=O)—OOH$	Aliphatic, aromatic	Carboxylic acids, $R—C(=O)—OH$
2. Diacyl peroxides, $R—C(=O)—OO—C(=O)—R'$	Aliphatic, aromatic, mixed	Carboxylic anhydrides, $R—C(=O)—O—C(=O)—R'$
3. Alkyl hydroperoxides, $R—OOH$	Primary, secondary, tertiary; aralkyl, cycloparaffin	Alcohols, $R—OH$
4. Peroxy esters, $R—C(=O)—OO—R'$	Primary, secondary, tertiary	Esters, $R—C(=O)—O—R'$
5. Dialkyl peroxides, $R—OO—R'$	Bis and mixed peroxides, where R and R′ are alkyl or aralkyl, and primary, secondary, or tertiary	Ethers, $R—O—R'$
6. Polymeric peroxides (copolymers of oxygen and substituted olefins): 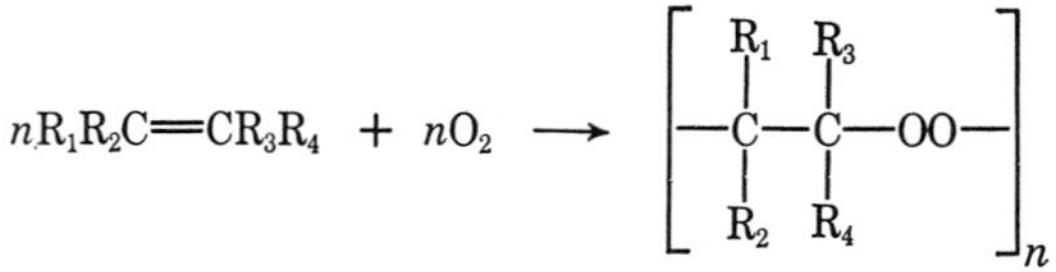 $nR_1R_2C{=}CR_3R_4 + nO_2 \longrightarrow [—C(R_1)(R_2)—C(R_3)(R_4)—OO—]_n$	Up to three of the four groups can be H; others can be alkyl, aryl, cyano, or alkoxyl	Polyalkylene oxides, 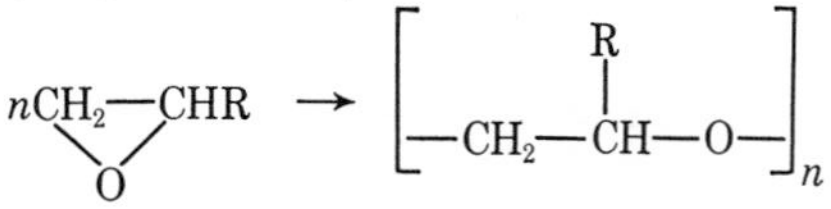 $nCH_2—CHR$ (epoxide, bridged by O) $\longrightarrow [—CH_2—CH(R)—O—]_n$
7 1,4-Epiperoxides, 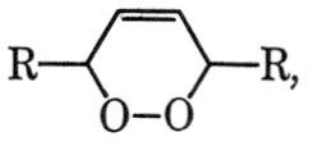 R—(cyclic, O–O)—R, and endoperoxides (transannular), 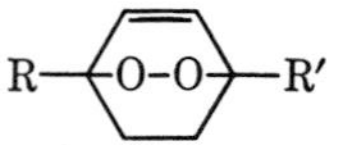 R—(O-O bridged ring)—R′	Terpene peroxides (ascaridole), polynuclear aromatic (anthracene peroxides), steroid peroxides	1,4-Epoxides, 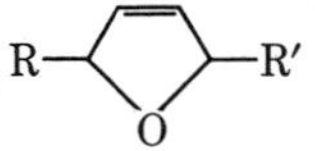 R—(cyclic, O)—R′ and R—(O bridged ring)—R′ 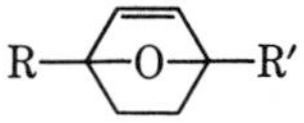

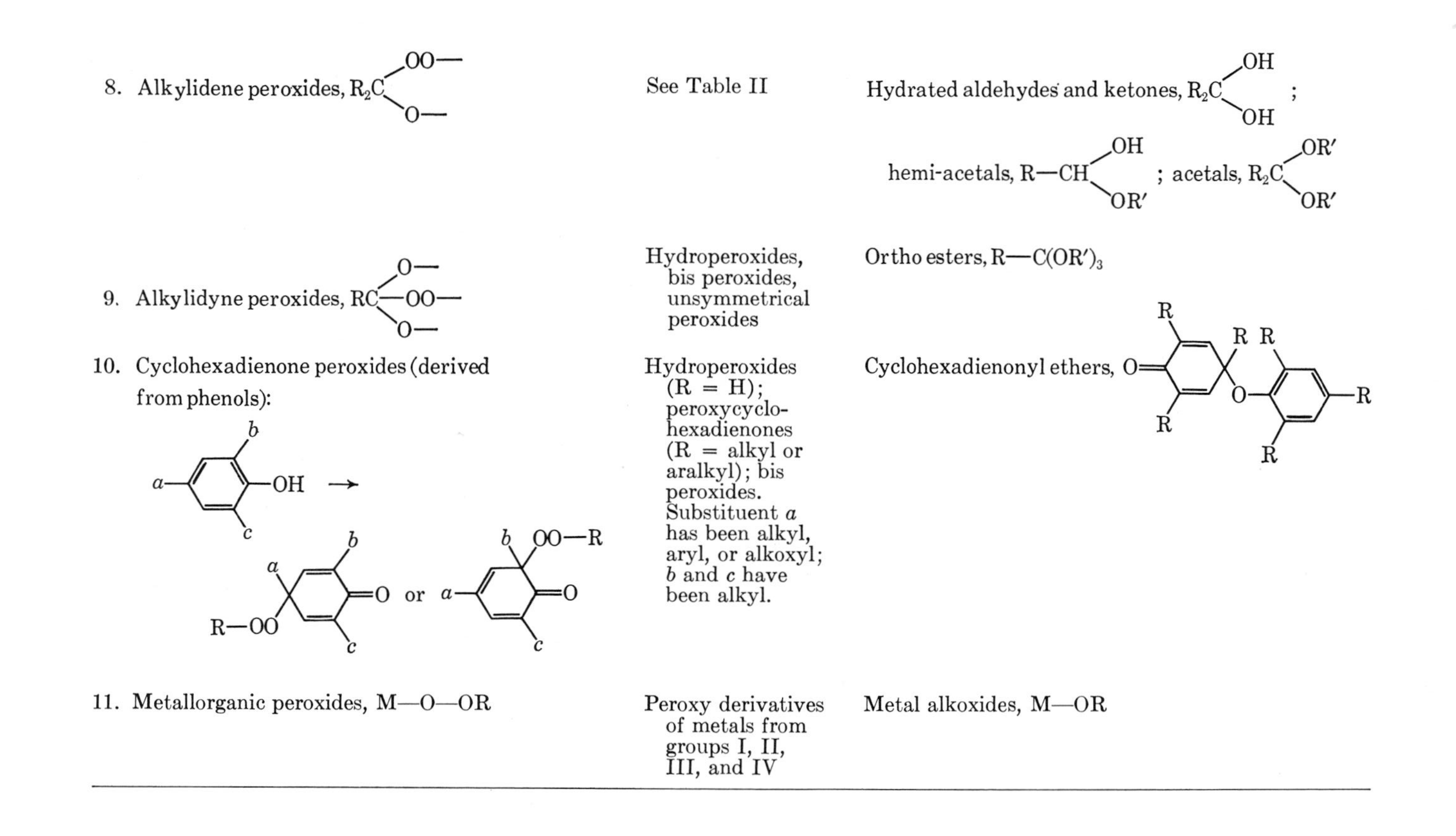

8. Alkylidene peroxides, $R_2C(OO—)(O—)$	See Table II	Hydrated aldehydes and ketones, $R_2C(OH)_2$; hemi-acetals, $R—CH(OH)(OR')$; acetals, $R_2C(OR')_2$
9. Alkylidyne peroxides, $RC(O—)(OO—)(O—)$	Hydroperoxides, bis peroxides, unsymmetrical peroxides	Ortho esters, $R—C(OR')_3$
10. Cyclohexadienone peroxides (derived from phenols):	Hydroperoxides (R = H); peroxycyclohexadienones (R = alkyl or aralkyl); bis peroxides. Substituent *a* has been alkyl, aryl, or alkoxyl; *b* and *c* have been alkyl.	Cyclohexadienonyl ethers
11. Metallorganic peroxides, M—O—OR	Peroxy derivatives of metals from groups I, II, III, and IV	Metal alkoxides, M—OR

$$\mathrm{C{-}\overset{\overset{\Large O}{\|}}{C}{-}O{-}O}$$ structures is slightly more than 100° (110). X-ray diffraction studies indicate a skewed structure in the crystalline solid also (174), an observed foreshortening of the —O—O— spacing being consistent with a dihedral angle of about 125°.

In their chemical reactivity diacyl peroxides approach that of the highly active peroxy acids. For both classes this property stems from their facile decomposition into carboxyl radicals, which are resonance-stabilized.

$$\mathrm{R{-}\overset{\overset{O}{\|}}{C}{-}O{-}O{-}\overset{\overset{O}{\|}}{C}{-}R \longrightarrow 2R{-}C{\genfrac{}{}{0pt}{}{\nearrow O}{\searrow O\cdot}} \rightleftarrows R{-}C{\genfrac{}{}{0pt}{}{\nearrow O\cdot}{\searrow O}}} \tag{1}$$

3. Alkyl Hydroperoxides

These compounds are encountered broadly, since they are the initial product in most autoxidations. The —O—O—H group can appear on almost any organic molecule—saturated or unsaturated; linear, branched, cyclic, or heterocyclic; monomeric or polymeric. (A significant exception is the benzene ring, which never carries a directly linked hydroperoxide group (171).) These compounds are alkylated hydrogen peroxide and their structures are similarly skewed, with C—O—O—H dihedral angles of about 100° (110,158).

Primary, secondary, and tertiary hydroperoxides are principal subdivisions, and have notable differences in reactivity and chemical stability. Commercially important classes include aralkyl (benzyl), *tert*-alkyl, and cycloalkyl hydroperoxides.

Alkyl hydroperoxides characteristically have greater acid strength than the analogous alcohols. This permits the formation of salts with alkali, a reaction useful for separation and purification. The following range of acid strengths is instructive (53):

Compound	pK_a
Carboxylic acids	3.6– 4.9
Peroxy acids	7.1– 8.2
Alkyl hydroperoxides	11.6–12.8
Hydrogen peroxide	11.6

[Although hydrogen peroxide is a weak acid in aqueous solution, it becomes a very strong one when concentrated (132).]

4. Peroxy Esters

Members of this class are more correctly considered derivatives of hydroperoxides than derivatives of peroxy acids, since they are formed by acylating hydroperoxides and upon hydrolysis yield acid and hydroperoxide rather than peroxy acid and alcohol. The members most significant industrially are the *tert*-butylperoxy esters, a relatively stable group of compounds (129).

There is little information about relative rates of hydrolysis or saponification for peroxy and nonperoxy esters; in one case Bartlett and Pincock studied the hydrolysis of the half-peroxy ester, ethyl-*tert*-butylperoxy oxalate (14), and found the peroxy ester group to react 2.3 times as fast as the ester group. Peroxy esters respond to most analytical reagents for hydroperoxides; they do so relatively slowly, however, and hydrolysis is usually the rate-controlling step.

5. Dialkyl Peroxides

These peroxides are often considered stable and unreactive, with di-*tert*-butyl peroxide regularly cited as the ultimate among peroxides in inertness toward analytical reagents. Nevertheless, dialkyl peroxides as a class show a broad range of reactivities. Whereas *tert*-alkyl peroxy groups are typically inert or sluggish, *n*-alkyl peroxy groups are much more reactive. Dimethyl and diethyl peroxides, for example, are explosive at their boiling points; and methyl cumyl peroxide reacts at an appreciable rate with iodide reagents that are without effect on dicumyl peroxide. Steric factors have a primary influence on dialkyl peroxide reactivity.

6. Olefin-Oxygen Copolymers

These compose a less familiar class that has been reviewed by Davies (35, Ch. 5). An olefinic bond that carries one or more substituents such as aryl, vinyl, or cyano, can form a 1:1 alternating copolymer having a —C—C—O—O—C—C—O—O— backbone. A great many such compounds are known, analogs of the nonperoxidic polyalkylene oxides. These copolymers are syrupy liquids or amorphous solids containing up to about 10 repeating units, and frequently explode on heating. (They differ, however, from the extremely explosive polymeric ethylidene peroxides, which have a —C—O—O—C—O—O—C—O—O— backbone.) The peroxide copolymers are more difficult to reduce than hydroperoxides, but somewhat less difficult than most dialkyl peroxides.

7. 1,4-Epiperoxides

These cyclic peroxides are formed in much the same way as the peroxide copolymers and are similar to them in reducibility. They differ fundamentally, however, in requiring synthesis by photocatalysis, frequently with the aid of a photosensitizer. These peroxides result from addition of oxygen to 1,3-dienes, to compounds having vinyl groups conjugated with aryl, and to polynuclear aromatics. The name endoperoxide signifies a compound in which this peroxide group is transannular.

Many of the 1,4-epiperoxides of substituted anthracene (acene epiperoxides) form reversibly and give up as much as 95% of their combined oxygen, often with luminescence, when heated to 150–180°C. The case of α-terpinene, cited by Davies (35, Ch. 5), exemplifies the close relation between peroxide copolymers and 1,4-epiperoxides. In the absence of light this terpene reacts with oxygen to form a polymeric peroxide, but when irradiated in the presence of a photosensitizer, it gives the endoperoxide, ascaridol.

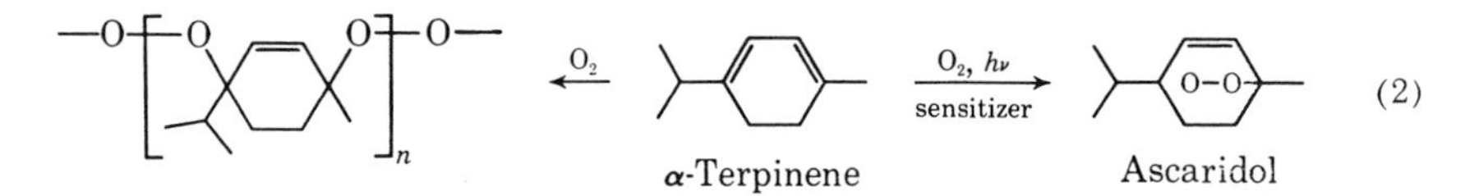

α-Terpinene   Ascaridol

8. Alkylidene Peroxides

This is the name used by Davies (35, Ch. 4) for a class of peroxides and hydroperoxides that is both large and complex; witness the array of subclasses in Table II. These compounds are formed by reactions of hydroperoxides or hydrogen peroxide with aldehydes and ketones, or with dialkyl, vinyl, and α-substituted ethers; autoxidation of ethers or alcohols; and ozonation of olefins. Their nonperoxidic analogs are acetals and hemiacetals.

It is characteristic of this class that the primary reaction products, e.g., of a ketone with hydrogen peroxide, are bifunctional and highly reactive. Consequently, they can react further and rearrange to produce mixtures of truly formidable complexity. The advent of separations by paper and thin-layer chromatography has made effective study of these reactions possible.

It is also characteristic of this class that its members tend to be explosive, often violently so. This is not a rigorous rule, as practical conditions for the safe utilization of a number of these products have been established.

Ozonation of olefins has been studied extensively by Criegee (27) and reviewed by Bailey (7). The initial ozone attack forms a complex (molo-

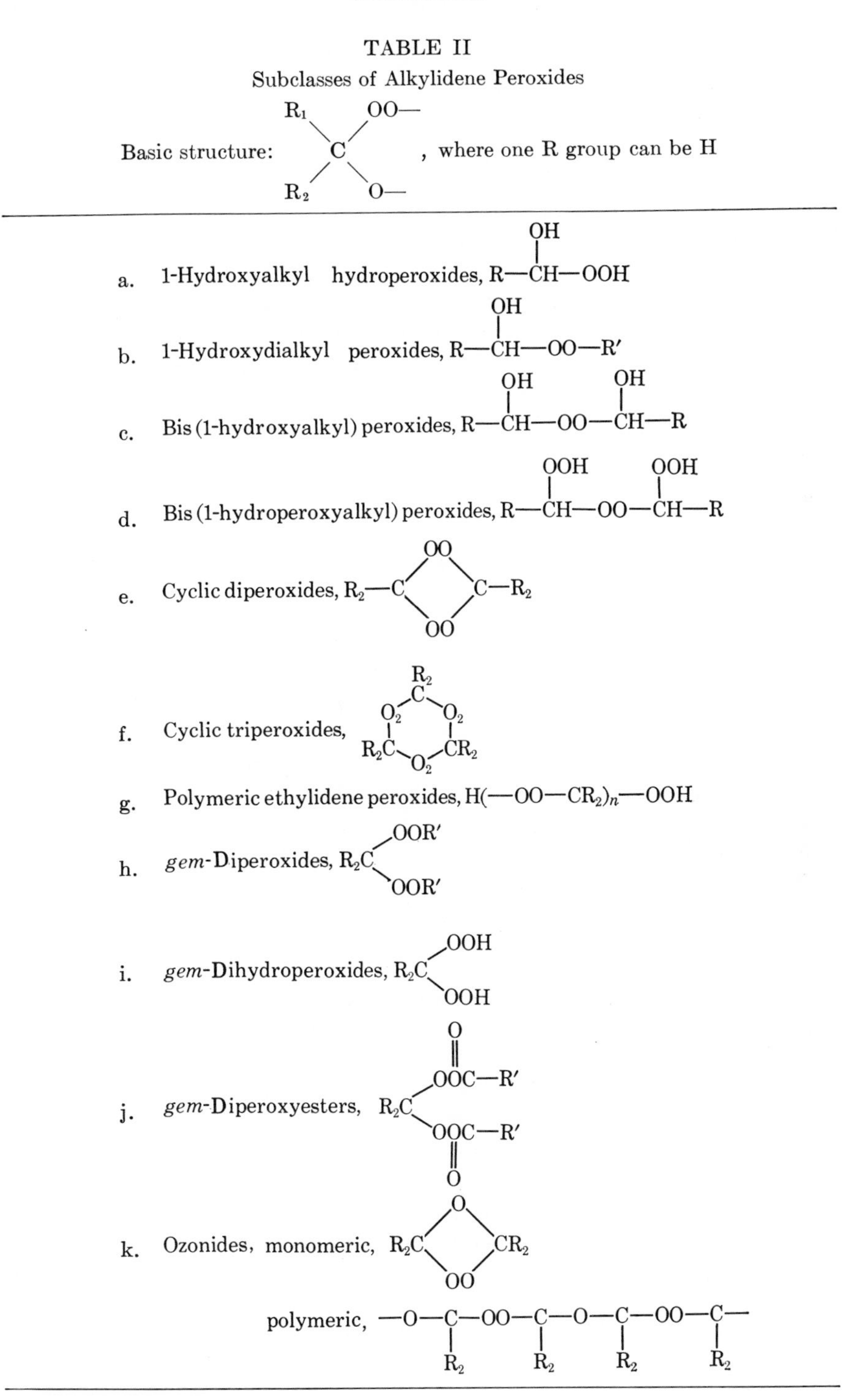

TABLE II

Subclasses of Alkylidene Peroxides

Basic structure: $(R_1)(R_2)C(OO-)(O-)$, where one R group can be H

a. 1-Hydroxyalkyl hydroperoxides, R—CH(OH)—OOH

b. 1-Hydroxydialkyl peroxides, R—CH(OH)—OO—R′

c. Bis (1-hydroxyalkyl) peroxides, R—CH(OH)—OO—CH(OH)—R

d. Bis (1-hydroperoxyalkyl) peroxides, R—CH(OOH)—OO—CH(OOH)—R

e. Cyclic diperoxides, R_2—C(OO)(OO)C—R_2

f. Cyclic triperoxides, $(CR_2O_2)_3$

g. Polymeric ethylidene peroxides, $H(—OO—CR_2)_n—OOH$

h. *gem*-Diperoxides, $R_2C(OOR')_2$

i. *gem*-Dihydroperoxides, $R_2C(OOH)_2$

j. *gem*-Diperoxyesters, $R_2C(OOC(=O)—R')_2$

k. Ozonides, monomeric, $R_2C(O)(OO)CR_2$

polymeric, —O—C(R_2)—OO—C(R_2)—O—C(R_2)—OO—C(R_2)—

zonide) of somewhat indefinite structure; this breaks down to give a carbonyl compound and a zwitterion [reaction (3)]. Ultimate products

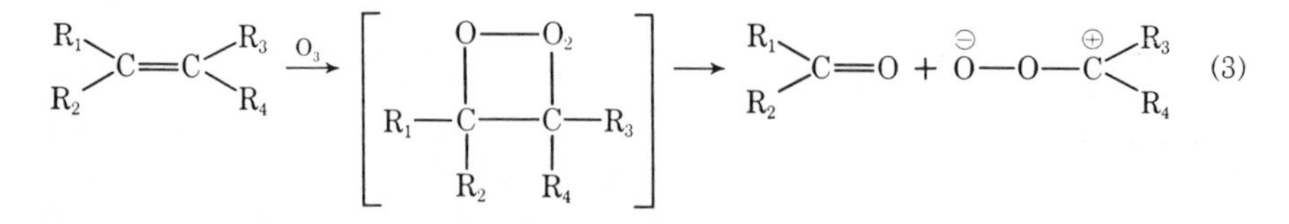

depend upon the behavior of the zwitterion, i.e., whether it (*a*) reacts with solvents containing active hydrogens to form alkylidene hydroperoxides; (*b*) dimerizes in an inert solvent to give a cyclic diperoxide; or (*c*) adds to the carbonyl product or another carbonyl reagent to produce the ozonide:

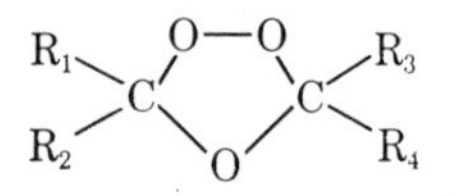

9. Other Peroxide Classes

The next two peroxide classes in Table I are of limited interest, analytically. The *alkylidyne peroxides* (Davies' nomenclature) are analogous to orthoesters, from which they can be obtained by reaction with hydroperoxides or hydrogen peroxide; they are also formed by air oxidation of ethylene acetals. *Cyclohexadienone peroxides* result from autoxidation of substituted phenols or their reaction with alkylperoxy radicals. (These peroxides are involved in the antioxidant behavior of such phenols.)

At one time diaryl peroxides, Ar—OO—Ar, were believed to be formed through dimerization of the aryloxy radicals produced from phenols:

$$\text{Ar—OH} \rightarrow \text{ArO}\cdot \rightleftharpoons \text{Ar—OO—Ar} \qquad (4)$$

It is now known, however (35, Ch. 7), that the supposed dimers are actually nonperoxidic cyclohexadienonyl phenyl ethers. It would appear that aryl peroxides, like the aryl hydroperoxides mentioned above, are unknown.

Organoperoxymetallic compounds make up the final class in Table I. They have received much study in recent years; these studies have been summarized by Davies (35, Ch. 8) and by Silbert (171). The organometallic peroxides, R—O—O—M, are derivatives of, and hydrolyze readily to, hydroperoxides. They are formed either from the reaction of hydroperoxides with metal compounds or from autoxidation of metal alkyl bonds. Typically, metal peroxides result from the reaction of hydroperoxides: with excess alkali metal hydroxide (periodic group I); with alkaline earth oxides or metal dialkyl compounds (group II); with boron halides and aluminum halides or alkyls (group III); and with similar intermediates of silicon, germanium, tin, and lead (group IV).

Regarding the autoxidation of metal alkyls from groups I, II, and III, two general reactions are pertinent: (*a*) The initial product is a peroxide ($RM + O_2 \rightarrow R—O—O—M$), and (*b*) when the oxygen supply is deficient, the peroxide is reduced to an alkoxide by excess alkyl ($R—O—O—M + RM \rightarrow 2R—O—M$). With aluminum alkyls, we have obtained greater than 95% of exact stoichiometry for this reaction pair by measuring oxygen absorption under conditions of starved supply.

B. RELATIVE ANALYTICAL REACTIVITIES

Analysts tend to classify peroxides by reactivity toward analytical methods or reagents rather than by structure. From this viewpoint, we have arranged the peroxide classes in order of apparent decreasing ease of reduction. The sequence, based on peroxide analyses by the polarograph (99,171,190), by catalytic hydrogenation (29), and by reaction with iodide (29,116), is as follows:

1. Peroxy acids
2. Diacyl peroxides
3. Hydroperoxides
4. Ozonides
5. Peroxy esters
6. Di-*n*-alkyl, di-*sec*-alkyl, and mixed peroxides; olefin-oxygen copolymers; 1,4-epiperoxides; dimeric ketone peroxides
7. Di-*tert*-aralkyl peroxides; endoperoxides
8. Trimeric ketone peroxides; di-*tert*-alkyl peroxides

Our laboratory has recently proposed another simple classification (116) based on response to the widely used boiling sodium iodide–isopropyl alcohol reagent: *easily reduced* (stoichiometric iodine liberation within 5 min); *moderately stable* (partial reaction in 5 min); and *difficulty reduced* (inert to this reagent). This classification will be the basis of our later discussion of iodine-liberation methods (Section IV-B-3-d).

C. OCCURRENCE

Hydroperoxides and products of their further reaction occur spontaneously through the autoxidations that cause gum formation in gasolines, rancidity in edible oils, drying of paints, bleaching of palm oils, spontaneous combustions, and the formation of explosive residues in many solvents, especially ethers. Peroxyacyl nitrates occur in the air over many of our cities, where they are formed from olefins, oxygen, and nitrogen oxides and cause eye irritation and plant damage (124,184). Some organic peroxides are produced in living plants by photosynthesis; e.g., ascaridole is the active ingredient of wormseed oil.

As industrial chemicals, peroxides are manufactured in increasing total volume. The list of peroxides that are commercially available is quite variable, however, reflecting vigorous competition for markets. A recent compilation showed more than 40 different peroxides offered by 19 manufacturers or suppliers under more than 140 trade names (54). Principal uses of these chemicals are as polymerization catalysts and polymer crosslinking agents. Growth in the latter area is rapid but selective, with emphasis on those peroxides having attractive balances of crosslinking efficiency and price. Most of the currently available commercial peroxides are shown in Table III.

TABLE III

Commercially Available Organic Peroxides[a]

Hydroperoxides

- *tert*-Butyl hydroperoxide (with 25% di-*tert*-butyl peroxide)
- Cumene hydroperoxide
- Diisopropylbenzene hydroperoxide
- *p*-Menthane hydroperoxide

Alkyl Peroxides

- Di-*tert*-butyl peroxide
- Di-*tert*-amyl peroxide
- 2,5-Dimethyl-2,5-di(*tert*-butylperoxy) hexane
- 2,5-Dimethyl-2,5-di(*tert*-butylperoxy) hexyne-3
- 2,2-Bis(*tert*-butylperoxy)butane
- Dicumyl peroxide
- *tert*-Butyl cumyl peroxide
- α,α'-Di(*tert*-butylperoxy)diisopropylbenzene

Alkyl Peroxy Esters

- *tert*-Butyl peracetate
- *tert*-Butyl perisobutyrate
- *tert*-Butyl perlaurate
- Mono-*tert*-butyl permaleate
- *tert*-Butyl peroctoate
- *tert*-Butyl perpivalate
- *tert*-Butyl peroxyisopropyl carbonate
- Diisopropyl peroxydicarbonate
- *tert*-Butyl perbenzoate
- 2,5-Dimethyl-2,5-diperbenzoxyhexane
- Di-*tert*-butyl perphthalate

Acyl Peroxides

- Peracetic acid
- Diacetyl peroxide
- Dipropionyl peroxide
- Dioctanoyl peroxide
- Didecanoyl peroxide
- Dilauroyl peroxide
- Peroxydisuccinic acid
- Dibenzoyl peroxide
- Bis(*p*-chlorobenzoyl) peroxide
- Bis(2,4-dichlorobenzoyl) peroxide

Ketone Peroxides

- Methyl ethyl ketone peroxide
- Methyl isobutyl ketone peroxide
- Mesityl oxide peroxide
- Cyclohexanone peroxide

Aldehyde Peroxides

- Dibenzal diperoxide
- Di(α-hydroxyheptyl) peroxide

Other

- Acetyl cyclohexansulfonyl peroxide

[a]Listed in a Bayer trade publication (54); some may be development products.

The industrial peroxide made in the largest tonnage is cumene hydroperoxide, which cleaves into phenol and acetone under acid conditions. World-wide acceptance of the cumene oxidation–cleavage process during the past decade testifies to the economic soundness of this new route to these two industrial chemicals. *p*-Cresol is produced similarly from the analogous *p*-cymene hydroperoxide.

D. STABILITY

Although for naturally occurring peroxides stability is of some importance, for the commercial peroxides it plays two key roles: The stability of a peroxide determines its individual safety hazard rating and defines the areas of its technological utility.

The study of hazards from peroxides and of techniques for the safe handling of these materials in plant and laboratory has kept pace with their commercial development. Several publications have recently reviewed the available information (25,67,136,137,162,169). Siemens categorized the hazardous properties of peroxides as flammability, thermal sensitivity, mechanical sensitivity, and explosiveness (169), a list to which Noller and Bolton have added physiological effects (136). Safe handling and storage of organic peroxides in the laboratory has been discussed by these authors (136) and by Sambale (162). Noller and associates have proposed a relative hazard classification for organic peroxides (137), based on specific test procedures; this summarizes the hazard potential for some 40 forms of the 26 peroxides (from seven classes or subclasses) that are in widest commercial use. Especially valuable in this subject area is a recent publication of ASTM Committee E-15, by Castrantas, Banerjee, and Noller (25).

Chemical stability is important in determining areas of utility for hydroperoxides, and to some extent for peroxy acids and peroxy esters. The property of overriding significance to most commercial organic peroxides, however, is thermal stability, because end uses depend on their ability to dissociate into active free radicals at a technologically desired temperature. Such thermal decompositions are first order in peroxide concentration and so give linear graphs of log half-life vs $1/T$ (°K). Over limited temperature ranges, a plot of log half-life vs T (°C) is also substantially linear. We have used this arrangement in Fig. 1 to illustrate the broad range of activity that is available.

It may seem surprising to observe in this figure that the alkyl hydroperoxides are the thermally most stable class, since they are chemically much less stable than dialkyl peroxides or peroxy esters. It is interesting also that the thermal dissociation of hydroperoxides into radicals is greatly

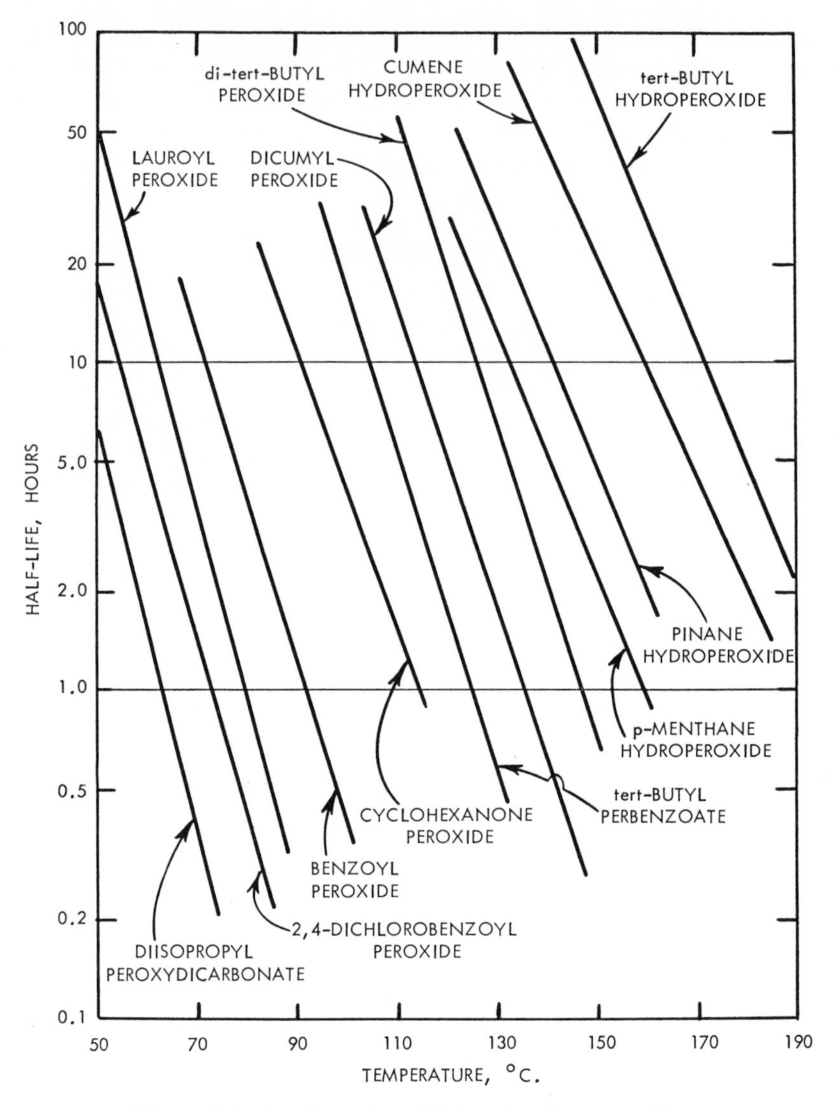

Fig. 1. Relative thermal stabilities of various peroxides.

enhanced by the presence of di-*tert*-butyl peroxide (115). For instance, a 1:1 mixture of the latter with cumene or *tert*-butyl hydroperoxide decreases the half-times of the hydroperoxides by 60- and 450-fold, respectively.

Doehnert and Mageli list activation energies and frequency factors for many peroxides (44), and Guillet and co-workers have shown how these parameters of the Arrhenius equation vary with molecular structure for the diacyl peroxides (68).

Molyneux has recently published a careful evaluation of the literature data on Arrhenius parameters for dialkyl or diaralkyl peroxides (135), based on the assumed homolytic, unimolecular scission:

$$ROOR' \rightarrow RO\cdot + R'O\cdot \quad (5)$$

Theoretically, the activation energy, E, for this relatively weak bond should vary much less within this class than is indicated by the observed range of 31–38 kcal/mole between diethyl and di-*tert*-butyl peroxides. Likewise, the frequency factors, A, should all be close to 10^{13} sec^{-1}, whereas they are found to vary between 10^{13} and 10^{16} sec^{-1}. Furthermore, there is a linear "isokinetic" relationship between log A and E that implies a temperature (210°C) at which all RO—OR′ peroxides should have identical rate constants, $k = 0.1$ sec^{-1}.

Molyneux explored many potential avenues for rationalizing the existing data, without success. He recommends, ". . . a suspension of judgment on what is the correct value for this bond dissociation energy," pending further attempts to resolve the impasse.

III. SEPARATION AND CONVERSION OF ORGANIC PEROXIDES

The judicious and imaginative use of a variety of separation techniques can greatly simplify the determination of organic peroxides in complex systems. The simpler fractions thus produced may make an ambiguous situation clear-cut, ready for straightforward analysis; or the separation procedure by itself may provide adequate analytical information. Alternatively, converting the peroxide or peroxides into other more readily determined compounds may achieve the desired analysis more readily. The intent of this section is not to present complete specific separation or conversion procedures, but rather to stimulate awareness of such possibilities by discussing a number of them briefly. It is better left to the individual chemist to tailor the appropriate technique to his particular system.

In analysis, situations requiring the separation of peroxides are widely varied and may include: (*a*) separation at concentrations ranging from trace (e.g., removal of peroxidic contaminants from analytical solvents) to very high (e.g., removal of impurities from organic peroxides to be used as analytical standards); (*b*) separation of a specific class of peroxides from other peroxides and nonperoxides (e.g., isolation of peroxy acids from acid mixtures by selective extraction, or separation of dialkyl peroxides from hydroperoxides in a variety of ways); or (*c*) separation between members of the same class (e.g., resolution of the hydroperoxide mixtures produced in hydrocarbon autoxidation, or the alkylidene peroxide mixtures produced by the reaction of hydrogen peroxide with aldehydes and ketones). In such endeavors, one should utilize the chemical and physical properties of both the wanted and the unwanted components. One should be alert also to the possibility that two or more different approaches may be feasible, and should consciously try to select that best suited to the situation.

All of the usual laboratory separation techniques can come into play in organic peroxide analysis. Extraction (liquid–liquid partition) in separatory funnels may suffice, by virtue of its simplicity and predictability (even though the separations be less than optimum), when only limited effort can be justified. Laboratories possessing appropriate countercurrent distribution apparatus, such as the Craig-Post equipment or a Scheibel column, can obtain very effective separations of complex mixtures, including commercial peroxide preparations, for both analytical (referee) and preparative purposes. Precipitation–filtration will apply in the separation of *tert*-hydroperoxides with concentrated alkali, or the purification of a peroxide by recrystallization. Distillation is not widely employed except as a step in preparative work, having been generally replaced in analysis by gas chromatography. Chromatography is applicable in all its forms, and we shall consider it separately in Section IV-A-1-a.

A. REMOVAL OF PEROXIDES FROM SOLVENTS

Probably the earliest separation problem with peroxides was their removal from organic solvents, especially ethers, in which peroxides, formed by autoxidation, are contaminants and constitute a well-recognized and ever-present hazard. Chemical methods, summarized by Davies (35, Ch. 14) and Hawkins (74, Ch. 11), have included treatment of the solvents with alkalies, sodium sulfite, acidic ferrous sulfate, stannous chloride, lead dioxide, zinc and acid, sodium and alcohol, lithium aluminum hydride, and triphenyl phosphine. Treatment with solid cerous hydroxide, as described by Ramsey and Aldridge, appears particularly effective for ethers (150). Shaking with copper-treated zinc dust (56) has also been recommended (203). Percolation through an active alumina column, introduced by Dasler and Bauer (32), is undoubtedly among the simplest, most effective, and most generally applicable of solvent purification techniques; for example, it removes from hydrocarbons not only peroxides but most other oxygenated impurities. Column purification on strongly basic ion-exchange resins (55) also has been successful for ethers, and peroxide reduction on resin columns in the sulfite form has been suggested (35, Ch. 14).

B. SEPARATION OF PEROXIDES FROM MIXTURES BY MEANS OF ALKALI

A widely used method for purifying hydroperoxides is to precipitate and isolate their sodium salts and then regenerate the hydroperoxide with a weak acid; this approach has been reviewed by Davies (35, Ch. 8). Primary and secondary hydroperoxides require low temperature and prompt regeneration in order to minimize the base-catalyzed decomposition to carbonyl compounds (35, Ch. 13; 96; 166). Tertiary hydroperoxides are more stable but require the use of substantial excess alkali to avoid formation of alcohol through slow elimination of oxygen (35, Chs. 8 and 13).

A variation in behavior is shown by *tert*-butyl hydroperoxide. Although this compound dissolves without decomposition, neither its sodium nor potassium salt precipitates from concentrated aqueous alkali solutions. An alcoholic medium produces opposite results, however: Cooling hot solutions of *tert*-butyl hydroperoxide and sodium or potassium butylate in *tert*-butyl alcohol (128), or diluting ice cold solutions in concentrated ethanolic alkali with a large volume of cold acetone (97), causes precipitation of stable salts (containing alcohol of crystallization).

We have found that the concentration of the aqueous alkali has an effect on the consistency of the precipitated salts: With cumene hydroperoxide, concentrations below 20% sodium hydroxide produce slimy precipitates; with *p*-cymene hydroperoxides, to produce precipitates of good consistency we recommend concentration of 40–50% sodium hydroxide at ice temperatures. Our preferred procedure is to add a few milliters of 40% sodium hydroxide with vigorous stirring to a cold, dilute petroleum ether solution of the oxidate. Sintered stainless steel funnels are better than sintered glass for filtrations involving such concentrated alkali.

The alkali precipitation technique is a convenient way to isolate the impurities from concentrated hydroperoxides for further analysis. For instance, dicumyl peroxide and α,α-dimethylbenzyl alcohol can be separated almost quantitatively from, say, 70% cumene hydroperoxide in this way, and each determined differentially in the presence of residual hydroperoxide by using a series of three different iodine-liberation procedures (116).

Partition between more dilute aqueous alkali and an immiscible organic solvent is useful in the analysis of autoxidized hydrocarbons, particularly aralkyl types such as the di- and triisopropylbenzenes. Typical distribution data on some pure components obtained in our laboratories are shown in Table IV. With these compounds, we have observed no decomposition of hydroperoxide groups in sodium hydroxide/hydrocarbon or sodium hydroxide/ether partition systems. Hydroperoxide loss occurred with sodium hydroxide/methyl isobutyl ketone, however.

C. CONVERSION OF PEROXIDES TO DERIVATIVES

Peroxides may be converted into derivatives that are subsequently identified and determined in lieu of the parent peroxides. The most common approach, much used for proof of structure in studies of oxidation processes, is chemical reduction. There are many effective reagents (34, 36,81,97,118,119,209), including thiophenol, sodium sulfite, lithium aluminum hydride, sodium borohydride, stannous chloride, zinc and acetic acid, zinc and hydrochloric acid, and sodium toluene-*p*-sulfinate and pyridine, as well as catalytic hydrogenation over nickel, platinum, or palladium. Trialkyl phosphines are especially useful, since they can be relied upon to convert virtually every peroxide into the analog that corresponds to the plucking out of one of the oxygen atoms from its —O—O— bond.

Formation of derivatives from which the peroxide can be regenerated is frequently useful. In addition to the salt formation with alkali discussed above, this technique includes isolation of alkyl hydroperoxides as xanthhydryl or triphenylmethyl derivatives (37) and as derivatives of substituted cyclic immonium pseudo bases, salts, or ethers (154); for example, derivatives useful for hydroperoxide separation and characterization were formed with 1-methyl-6,8-dinitro-2-ethoxy-1,2-dihydroquinoline.

IV. DETERMINATION OF ORGANIC PEROXIDES

Dominant in the detection, identification, and determination of organic peroxides are their physical and chemical properties. Each analytical method is based on the use of one or several of these attributes. We have indicated the wide ranges in physical and chemical properties of these compounds by the material in Sections II-A, II-B, and II-D. This breadth has also been implied in the discussion of separations and conversions in Section III.

The many established or promising peroxide methods have been listed in Table V, in three groups. One consists of physical or instrumental methods, based principally on physical properties, while the other two are chemical methods, based on reaction with reducing agents or colorimetric reagents. The table indicates the scope of each method, and lists some principal references. We will systematically review these groups of per-

TABLE IV

Distribution Coefficients for Some Autoxidized *p*-Diisopropylbenzene Components

Solvent pair[a]	Distribution coefficients[b] for the oxidate components				
	Monoalcohol	Monohydroperoxide	Dialcohol	Alcohol hydroperoxide	Dihydroperoxide
1. 7% aq. NaOH/benzene	0.0084	0.042	0.85	7.7	145
2. 7% aq. NaOH/methyl ethyl ketone	—	—	—	0.20	3.5
3. 10% aq. NaOH/ether	—	—	—	0.6	15.1
4. 20% aq. NaOH/ether	—	—	—	0.3	16.6
5. 20% aq. NaOH/methyl isobutyl ketone	—	—	—	<0.1	7.7

[a] Lower phase/upper phase.

[b] Ratio of concentration in lower (alkali) phase to concentration in upper phase.

TABLE V

Established or Promising Methods for Determining Organic Peroxides

Method or technique	Applicable peroxide types	Applicable concentration ranges	Useful features	Principal ref.
A. Determination by Physical or Instrumental Means				
Gas chromatography	Relatively volatile and thermally stable	From trace to major component	Simple and reliable when applicable	1, 20
Liquid column chromatography	General	From trace to major component	Can purify standards, analyze mixtures	51
Thin-layer chromatography	All except relatively volatile	From trace to major component	Able to resolve mixtures of polyfunctional compounds	21a, 21b, 74a, 91, 156
Paper chromatography	All except relatively volatile	From trace to major component	Able to resolve mixtures of polyfunctional compounds	1, 22, 43, 126, 127, 155
Pyrolysis-GC	Broad potential	From low to high	Quick assay for unreactive peroxides, and relatively simple mixtures	84
Mass spectrometry	General; need slight volatility	From low to high	Where applicable, can analyze mixtures directly	—

Infrared and near-IR absorption	General, but accent on hydroperoxides	From moderately low to high	Can distinguish some ROOH's from analogous ROH's	15, 40, 61, 66, 80, 98, 131, 171, 207
Ultraviolet absorption	Aromatic peroxides	From low to high	Combines well with silicic acid column chromatography	89
Nuclear magnetic resonance	Broad potential	Above 10 percent	Structure determination; quick method for assay in crude autoxidation mixtures	38, 39, 59
Electron paramagnetic resonance	Accent on hydroperoxides	From trace to moderate	Quick determination, possible selectivity	90a
Differential thermal analysis	All types	From moderately low to high	Quantitative analysis, sometimes of mixtures, convenient for determining thermal decomposition characteristics	201
Polarography (reduction with electrons)	All except di-*tert*-alkyl and cyclic trimeric peroxides	From trace to major component	Convenient to follow kinetics of peroxide formation and decomposition in chemical processing and autoxidation	86, 99, 100, 117, 141, 159, 164a, 174, 210

(*continued*)

TABLE V (*continued*)

Reaction or reagent	Applicable peroxide types	Applicable concentration ranges	Useful features	Principal ref.
	B. Determination by Chemical Reaction with Reducing Agents			
Reduction with iodide	General	From trace to major component	Variety of procedures to fit many peroxide types	74, Ch. 11; 116; 117
Reduction with diphenyl sulfide	Peroxy acids only	From low to high	Specific reaction	82, 83
Reduction with triethyl arsine	Peroxy acids, diacyl peroxides and alkyl hydroperoxides	From low to high	Differentiates dialkyl peroxides	82
Reduction with tertiary phosphines	General	From low to high	Derivative prep. for gas chromatography	48, 81
Reduction with stannous ion	Peroxy acids(?), diacyl peroxides and alkyl hydroperoxides	From low to high	Avoids interference from organic sulfur compounds	12, 50
Reduction with arsenite ion	Peroxy acids(?), diacyl peroxide and alkyl hydroperoxides	From trace to high	Avoids interference from organic sulfur compounds	170, 198
Oxidation with $KMnO_4$	Cyclohexanone hydroperoxides	From low to high	Active catalyst content	191a

Reduction with hydrogen (catalytic)	General	From low to high	Derivative Prep.	8; 29; 35, Ch. 14
Reduction with $LiAlH_4$	General	From low to high	Derivative Prep.	4, 119, 188
Reduction with ferrous ion[a]	—	—	—	—
C. Determination by Reaction with Colorimetric Reagents				
Ferrous thiocyanate	Easily reduced	Trace to low	Rapid, precise, sensitive	93, 102, 180
Leuco chlorophenolindophenol[a]	Easily reduced	Trace to low	—	72, 73, 102
N,N'-Diphenyl-*p*-phenylene-diamine	Easily reduced	Trace to low	Rapid, precise, sensitive	161
Leuco methylene blue	Easily reduced	Trace to low	Constant molar response, precise, very sensitive	9, 47, 181, 192
N-Benzoyl leuco methylene blue	Easily reduced	Trace to low	Reagent unaffected by air	52
sym-Diphenylcarbohydrazide	Hydroperoxides and diacyl peroxides	Trace to low	Rapid, precise, sensitive	69b
Iodide	General	Trace to low	Constant molar reponse, precise, sensitive, widely applicable	10, 11, 45, 75
Titanium(IV)[a]	Hydroperoxides and peroxy esters	Trace to low	—	147, 163, 213
Triphenyl phosphine	Hydroperoxides	Trace to low	Constant molar response, reagent unaffected by air	184a

[a] Discussed but not recommended.

oxide methods in Section IV-A. In Section IV-B, we will present a more extensive critical evaluation of those reactions and instrumental methods that appear most useful and broadly applicable.

A. SURVEY OF ESTABLISHED AND PROMISING PEROXIDE METHODS

1. Physical or Instrumental Methods (Table V(A))

a. Chromatographic Techniques

(1) Gas Chromatography

Because instrumentation for this technique is so highly developed and widely available today, gas chromatography contends for the position of preferred peroxide method wherever it is applicable, that is, for the analysis of those peroxides that are sufficiently volatile and stable. Some of the successful applications are listed in Table VI; in many instances these have included isolation and purification on a preparative scale. Bukata and co-workers have recently summarized the earlier literature (20); Hahto and Beaulieu list a more extensive bibliography (69).

TABLE VI

Some Peroxides Determined by Gas Chromatography

Compound	Ref.
Diacetyl peroxide	69
Peroxyacetyl nitrate	31
Peroxypropionyl nitrate	31
Peroxybutyryl nitrate	31
Ethyl hydroperoxide	90
Isopropyl hydroperoxide	90
Allyl hydroperoxide	49
n-Butyl hydroperoxide	214
Isobutyl hydroperoxide	38, 214
tert-Butyl hydroperoxide	1, 20, 90
tert-Pentyl hydroperoxide	1, 20
trans-2-Methyl-3-pentenyl-2-hydroperoxide	18
4-Methyl-3-pentenyl-2-hydroperoxide	18
tert-Butyl peroxyacetate	20
tert-Butyl peroxyisobutyrate	20
Diisopropyl peroxide	121
Di-*tert*-butyl peroxide	1, 20
Di-*tert*-pentyl peroxide	1
2,5-Dimethyl-2,5-di(*tert*-butylperoxy)hexane	20
2,5-Dimethyl-2,5-di(*tert*-butylperoxy)-3-hexyne	20

Most separations have been on *tert*-alkyl hydroperoxides and di-*tert*-alkyl peroxides (1,20), although primary and secondary hydroperoxides have been both analyzed and purified (28,90,214), and the dangerous diacetyl peroxide has been purified and characterized (69). The last two compounds in Table VI probably approach the practical molecular weight limit for gas chromatography of peroxides. Bukata and co-workers used a column temperature of 138°C to separate this pair, and reported relative standard deviations of 1%. The more volatile di-*tert*-butyl peroxide had better precision, with a relative standard deviation of 0.3%.

A remarkably sensitive gas chromatographic peroxide method is the determination of the peroxyacyl nitrates (along with many other nitrogen oxide derivatives) in air (31). These peroxides have been identified by Stephens and co-workers as primary eye irritants and phytotoxicants occurring in smog (124,184). Because of the extreme sensitivity of nitrate groups to the electron-capture mode of gas chromatographic detection, both peroxyacetyl nitrate and the even more damaging peroxypropionyl nitrate can be determined in a 2-ml sample of air at concentrations below 5 parts per billion.

Brill used preparative and analytical gas chromatography as the basis for an important study (18) in which he showed that isomerization of hydroperoxides can occur readily in dilute nonpolar solutions. The compounds in equilibrium (6) were readily interconverted. Rate was inversely

$$CH_3—C(CH_3)(OOH)—CH{=}CH—CH_3 \rightleftarrows CH_3—C(CH_3){=}CH—CH(OOH)—CH_3 \qquad (6)$$

trans-2-Methyl-3-pentenyl-2-hydroperoxide   4-Methyl-3-pentenyl-2-hydroperoxide

proportional to concentration, an apparent effect of hydrogen bonding. Brill's results cast doubt on the interpretation of many studies of the mechanism of autoxidation.

Hyden has taken another approach to the use of gas chromatography in the analysis of peroxides, one that has many intriguing possibilities (84). This is to inject a peroxide sample solution directly into the *hot* injection port of a gas chromatograph where the peroxide decomposes instantly; separation of its volatile decomposition products can then serve both to qualitatively identify and quantitatively determine it. Hyden reports excellent results for di-*tert*-butyl peroxide, which decomposed into ethane and acetone at 310°C.

Apparently related is the method of Courtier (26b), who used columns as hot as 200°C and injection ports as hot as 250°C for identifying both the peroxides and the diluting solvents in samples of commercial catalyst preparations for polyester polymerization.

Pyrolysis methods are not restricted to peroxides, of course, but are applicable to analysis in general. Characteristic features are the strong tendency of pyrogram patterns to vary with pyrolysis conditions, and the consequent need for a pyrolysis technique that is both flexible and precise. The recent approach of Simon and co-workers (176), which employs high frequency induction heating of samples as thin films on ferromagnetic conductors, is a long step toward such control; in their apparatus, the Curie temperature of the conductor automatically provides a precisely reproducible pyrolysis temperature. Analytical chemists should consider broad development of the pyrolysis–gas chromatography approach, possibly coupled to liquid column or thin-layer chromatography, and monitored when necessary by mass spectrometry, as has recently been described (194).

(2) Liquid Column Chromatography

Both adsorption and liquid–liquid partition chromatography are effective in the separation and analysis of peroxides. In adsorption chromatography, separations are due to differences in the strength of adsorptive forces that hold the various components to the adsorbing substrate. Eggersglüss has studied such factors extensively for peroxides (51). Based on 37 pure compounds from six peroxide classes, run on columns of seven individual adsorbents, and eluted with three solvents, he concluded that adsorption was a function of the number and polarity of functional groups, and that relative strength of adsorption increased throughout the following series:

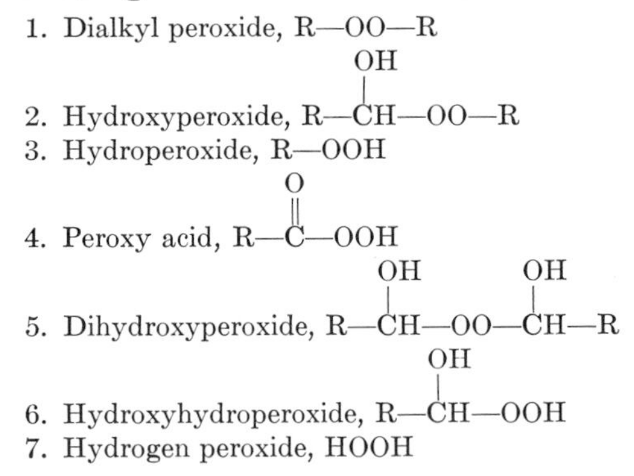

The effect of moisture in the eluting solvent should be considered in peroxide separations on adsorption columns; too much water may decrease the adsorptive power of the substrate for specific compounds. Decomposition of peroxides on adsorbing substrates is also a drawback in this type of separation. Alkaline alumina is a prime offender, although it is said to be improved by pretreatment with acetic acid (35, Ch. 14). Davies has suggested (35, Ch. 14) that ion-exchange resins used in ether purification (96) might separate hydroperoxide components of mixtures; gradient elution by a buffer of continuously varying pH should be considered in this approach.

In our laboratories, an effective scheme for the analysis of autoxidized isopropylbenzenes combined elution from silicic acid adsorption columns with identification and determination of the individual components (including nonperoxides) by ultraviolet absorption (140).

The distribution coefficients in Table IV suggest that liquid–liquid partition chromatography in aqueous alkali/organic systems should be effective in resolving peroxide mixtures. This has proved to be the case. Nonalkaline systems should also be operable, based on distribution coefficients in water/ether and water/petroleum ether systems tabulated by Eggersglüss for all six peroxide classes used in his adsorption study (51).

We have used a 20% aqueous sodium hydroxide/Celite column successfully in the analysis of isopropylbenzene oxidates (64). To prepare a typical column we mixed 10 ml of 20 wt % aqueous sodium hydroxide into 10 g of Celite, with a mortar and pestle, and packed it into a No. 2 chromatographic column; elution with hexane, diethyl ether, and ethyl acetate–acetic acid in succession gave quantitative recoveries of monohydroperoxide, alcohol hydroperoxide, and dihydroperoxide from both *m*- and *p*-diisopropylbenzene oxidates.

In our experience with such ultraviolet-absorbing samples, the silicic acid adsorption column coupled with ultraviolet analysis is more convenient and useful than partition systems. Particularly effective has been the UV-Scanalyzer, an automatic gradient elution column chromatograph developed in our laboratories by Kenyon and co-workers (89) and used here steadily for more than 10 years. Recently introduced detectors for liquid column chromatography based on differential refractometry (200), differential heats of adsorption (85), differential vapor osmometry (175), and continuous residue pyrolysis (186), along with rapidly developing improvements in techniques and in commercially available hardware, promise broader utility for liquid column separations in the analysis of peroxides and nonperoxides.

(3) Paper Chromatography

A number of workers have developed applications of paper chromatography to peroxide analysis. The principal contributions are summarized

TABLE VII

Summary of the Principal Systems for Paper Chromatographic Separations of Organic Peroxides

Authors	Type	Stationary phase	Moving solvent	Spray reagent	Ref.
1. Abraham, Davies Llewellyn, and Thain	Reversed phase	Whatman No. 4 Paper, soaked in 5% silicone in cyclohexane, then dried	H_2O–EtOH–$CHCl_3$ (20:17:2)	Ferrous thiocyanate (freshly prepared): to soln. of NH_4SCN (5% w/v) and H_2SO_4 (1% v/v) add and dissolve $FeSO_4 \cdot (NH_4)_2SO_4 \cdot 6H_2O$ (7% w/v)	1
	Applications (and R_f ranges): H_2O_2 (0.84); aralkyl hydroperoxides (0.75–0.85); cycloalkyl hydroperoxides (0.5–0.6); trityl hydroperoxides and *tert*-butyl peroxy esters (0.15–0.2); benzoyl peroxide (0.0)				
2. Rieche and Schulz	Reversed phase	Partially acetylated Schleicher and Schüll 2043b paper	EtOAc-dioxane–H_2O (2:4.5:4.6)	(*1*) 3% *p*-amino dimethyl-aniline hydrochloride and 2% AcOH in 1:1 MeOH–H_2O (*2*) 3% KI in 1:1 MeOH–H_2O	155
	Application (and R_f ranges): H_2O_2 (0.83); aralkyl cycloalkyl and cyclic ether hydroperoxides (0.35–0.75); succinyl peroxide (0.8); benzoyl peroxide (0.2); lauroyl and stearoyl peroxide (0.0)				
3. Milas and Belic	Normal	Whatman No. 1 paper, (*1*) soaked in 50% dimethyl formamide (*2*) soaked in 50% *N*-methyl-*N*-formamide (dried in each case) (*3*) Untreated	 Decalin Decalin *n*-BuOH–EtOH–H_2O (45:5:50)	(*1*) 0.1% *p*-aminodimethyl-aniline hydrochloride (*2*) AcOH–sat. aq. KI–starch solution (3:2:5) (*3*) HI(56%)–AcOH(10:90)	126

	Applications (*and R_f ranges*): Some 22 pure peroxides of many types characterized by R_f's from 0.0 to 1.0, which varied widely among the three solvent systems. Also used to identify six complex peroxides produced by reacting diethyl ketone with H_2O_2.				
4. Cartlidge and Tipper	(*1*) Normal	Whatman No. 3 paper, untreated, between glass plates	Ether; or *n*-BuOH–ether–H_2O (10:10:1);	(*1*) Ferrous thiocyanate as in 1, above (*2*) *p*-Phenylene diamine and acetaldehyde in aq. AcOH (*3*) HI(56%)–AcOH (10:90)	22
	(*2*) Normal	Whatman No. 3 paper, treated with (*a*) 5% and (*b*) 20% ethylene glycol in acetone, between silicone-treated glass plates	For (*a*), *n*-BuOH–pet. ether (10:90); for (*b*), either $CHCl_3$–pet. ether (1:1) or ether–pet. ether (1:20)		
	(*3*) Reversed phase	Whatman No. 3 paper, treated with (*a*) 5% and (*b*) 20% Silicone oil in pet. ether, between silicone-treated glass plates	For (*a*), H_2O–EtOH–$CHCl_3$ (20:17:2); for (*b*), H_2O–MeOH (1:4)		
	Applications (*and R_f ranges*): (*1*) H_2O_2 separated from hydroperoxides (R_f's 0.1–0.5, and 0.95–1.0) (*2*) Excellent separations among many members of many peroxide classes; see original reference for example. (*3*) Dialkyl peroxides (0–0.5; H_2O_2, 0.7); hydroxyperoxides, cumene hydroperoxide (0.8–0.9), alkyl hydroperoxides (1.0).				
5. Dobson and Hughes	Reversed phase	Whatman No. 3 paper, treated with 5% silicone oil solution, between silicone-treated glass plates	(*a*) 10% H_2O in MeOH (*b*) 15% H_2O in MeOH	Ferrous thiocyanate as in 1, above	43
	Applications (*and R_f ranges*): Dialkyl peroxides (0.2–0.7); alkylhydroperoxides (1.0)				

in Table VII. Normal systems have employed untreated paper (126) and paper coated with dimethyl formamide (126) or ethylene glycol (22). In reversed-phase applications, both silicone oil-loaded (1,22,43) and acetylated (155) papers have been used. The success of these investigators in varying R_f values widely by modifying the composition of stationary and mobile phases is a clear invitation to other peroxide analysts to try similar innovations.

Milas and Golubovic described a practical extension of this technique (127), wherein they scaled up a paper chromatographic separation (126) to preparative scale on a cellulose powder–dimethyl formamide column, eluted with pentane. In this way, they separated and determined the structures of six peroxides formed in the acid-catalyzed reaction of diethyl ketone and hydrogen peroxide, viz.,

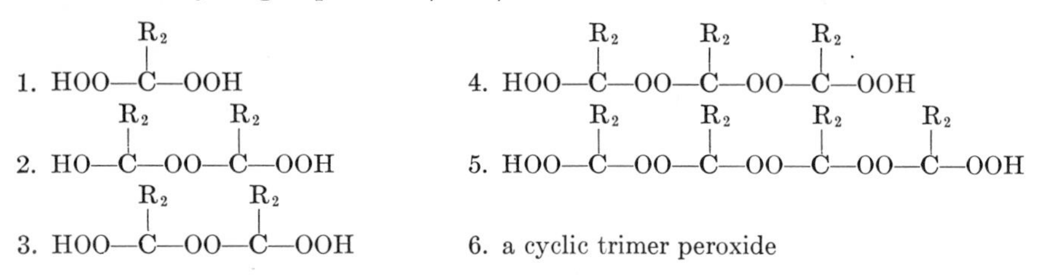

This approach deserves consideration, particularly for complex peroxidic systems.

(4) Thin-Layer Chromatography

Stemming mainly from the work of Stahl (182), thin-layer chromatography (TLC) increasingly competes with paper chromatography for a leading role in the separation of many compounds, including organic peroxides. The first significant work in this area was by Knappe and Peteri, who applied TLC to five important peroxide classes (91). They studied the separating ability of many mobile phases on silica gel and selected a pair that in combination usually gave unequivocal identification of a particular peroxide (confirmed by running a known). Still a third solvent mixture could resolve peroxides that had identical R_f values with both the standard mobile phases. These TLC systems and R_f values obtained by means of them for 18 peroxides are listed in Table VIIIA. Except for diacyl peroxides, which moved relatively rapidly, the peroxides did not separate systematically by classes. Commercial ketone peroxides proved to be mixtures of several components.

Another pertinent study has been that of Hayano and co-workers (74a), who studied a series of seven compounds representing four classes of

TABLE VIIIA

R_f Values for Some Organic Peroxides by Thin-Layer Chromatography—Solvent Systems of Knappe and Peteri (91)

Compound	Average R_f value (on silica gel)		
	I Toluene–CCl_4 (2:1)	II Toluene–HOAc (19:1)	III Pet. ether–EtOAc (49:1)
Lauroyl peroxide	0.85 (0.71)[a]	0.95 (0.92)[a]	
2,4-Dichlorobenzoyl peroxide	0.81 (0.66)[a]	0.88 (0.92)[a]	
4-Chlorobenzoyl peroxide	0.74	0.94	
Benzoyl peroxide	0.55 (0.41)[a]	0.70 (0.59)[a]	
tert-Butyl peroctoate	0.28	0.55	
Methyl isobutyl ketone peroxide, major component A	0.25	0.55	
tert-Butyl perbenzoate	0.24	0.47	
Cyclohexanone peroxide, major component A	0.21	0.38	
Methyl ethyl ketone peroxide, major component A	0.16	0.42	
tert-Butyl peracetate	0.12	0.32	0.18
Cumene hydroperoxide	0.11 (0.14)[a]	0.33 (0.34)[a]	0.09
2,2-Bis(*tert*-butyl peroxy)-butene	0.10	0.35	
Methyl ethyl ketone peroxide, major component B	0.10	0.10	
tert-Butyl hydroperoxide	0.05 (0.06)[a]	0.30 (0.34)[a]	
Methyl isobutyl ketone peroxide, major component B	0.00	0.12	
Cyclohexanone peroxide, major component B	0.00	0.12	
Cyclohexanone peroxide, minor component C	0.00	0.10	
Hydrogen peroxide	0.00	0.00	0.00

[a] Data of Hayano, Ota, and Fukushima (74a).

organic peroxides. These workers compared the performance of Knappe and Peteri's solvent systems with that of their own trio of three-component solvents, on silica gel. The latter systems and the R_f values obtained by using them are presented in Table VIIIB, while R_f values obtained with the Knappe and Peteri solvents have been inserted in Table VIIIA. Note in that case that agreement between the two groups of workers is good for a pair of hydroperoxides, but that discrepancies for three diacyl peroxides are sizable. Hayano and co-workers reported that the R_f values in Table VIIIB were quite precise, having standard deviations in general below 0.05 R_f units.

TABLE VIIIB

R_f Values for Some Organic Peroxides by Thin-Layer Chromatography—Solvent Systems of Hayano, Ota, and Fukushima (74a)

	Average R_f value (on silica gel)[a]		
Compound	I Benzene–HOAc–MeOH (10:1:1)	I Chloroform–HOAc–water (8:1:2)	III Benzene–HOAc–water (8:1:2)
Lauroyl peroxide	0.90	0.97	0.82
2,4-Dichlorobenzoyl peroxide	0.87	0.90	0.80
Benzoyl peroxide	0.80	0.87	0.63
Dicumyl peroxide	0.85	0.93	0.73
Di-*tert*-butyl perphthalate	0.73	0.66[b]	0.24
Cumene hydroperoxide	0.54	0.59[b]	0.27[b]
tert-Butyl hydroperoxide	0.34	0.37	0.13

[a] Standard deviation of the R_f values is 0.01–0.04, except as noted.
[b] Standard deviation is 0.12.

The most important contributions to the TLC of organic peroxides appear to be the comprehensive studies of Buzlanova and co-workers (21a,21b); they reported on a total of 39 peroxides in five classes. Table VIIIC summarizes the solvent systems used by these workers, 15:1 heptane–ether and 20:3 toluene–methanol, and the R_f values found on silica gel for some 26 peroxides, representing four major classes (21a).

Data are presented also that were obtained with the toluene–methanol solvent on alumina; these results are similar to the data from silica gel, although the spots were more diffuse. Buzlanova and co-workers report partial decomposition on alumina for diacyl peroxides (21a) and for ketone peroxides (21b).

The peroxide spots were located by means of a spray reagent of *N*,*N*-dimethyl-*p*-phenylenediamine hydrochloride in methanol (128 ml of methanol, 25 ml of water, 1 ml of glacial acetic acid, and 1.5 g of the amine hydrochloride). This reagent was said to be selective for detecting organic peroxides, giving a red-violet spot against a light background, and to have a sensitivity of 1 μg peroxide. Color was said to develop at different rates depending on the type of peroxide, with hydrogen peroxide and hydroperoxides showing up immediately, and peresters requiring heating at 70°C for 10 min. Detection of dialkyl peroxides required that the chromatogram be first sprayed with concentrated hydrochloric acid and then heated at 70°C for 10–15 min, before developing with the peroxide spray reagent. Diaralkyl peroxides showed up much more readily than

TABLE VIIIC

R_f Values for Some Organic Peroxides by Thin-Layer Chromatography—Solvent Systems of Buzlanova, Stepanovskaya, and Antonovskii (21a)

Compound	R_f value		
	Silica gel		Alumina
	I Heptane–Et_2O (15:1)	II Toluene–MeOH (20:3)	II Toluene–MeOH (20:3)
Dialkyl Peroxides			
Di-*tert*-butyl peroxide	—	—	—
tert-Butyl pinyl peroxide	0.76	0.90	—
tert-Butyl-*tert*-amyl peroxide	0.74	0.95	—
tert-Butyl cumyl peroxide	0.64	0.95	0.74
Dicumyl peroxide	0.55	0.95	—
Diacyl Peroxides			
Stearoyl peroxide	0.41	0.98	0.98
Lauroyl peroxide	0.37	0.98	0.98
p-Chlorobenzoyl peroxide	0.32	0.91	—
Benzoyl peroxide	0.21	0.74	0.95
Peroxy Esters			
tert-Butyl per(*o*-benzoyl)-benzoate	0.11	0.94	—
tert-Butyl per(*m*-nitro)-benzoate	0.08	0.90	—
tert-Butyl percaproate	—	0.88	—
tert-Butyl perpelargonate	—	0.87	—
1,1′-Diperbenzoyldicyclohexyl peroxide	—	0.83	—
tert-Amyl peracetate	0.14	0.80	0.87
tert-Butyl perbenzoate	0.18	0.78	0.93
1,1-Diperbenzoylcyclohexane	—	0.78	—
tert-Butyl peracetate	0.13	0.68	0.86
Hydroperoxides			
1,1-Diphenylethane hydroperoxide	0.04	0.80	—
p-Methane hydroperoxide[a]	0.07	0.66	0.51
		0.50	0.39
		0.25	—
		0.18	—
Cumene hydroperoxide	0.05	0.54	0.37
Pinane hydroperoxide[b]	0.20	0.45	—
	0.09	0.35	—
		0.23	—
Tetralin hydroperoxide	0.05	0.43	0.32
tert-Amyl hydroperoxide	0.04	0.42	—
tert-Butyl hydroperoxide	0.05	0.37	0.24
m-Diisopropylbenzene dihydroperoxide	0.00	0.30	0.11
Hydrogen peroxide	0.00	0.05	0.10

[a] A mixture of the oxidation products of *p*-methane.
[b] A mixture of the oxidation products of pinane.

dialkyl peroxides such as *tert*-butyl-*tert*-amyl peroxide. Di-*tert*-butyl peroxide was too inert and too volatile to be detected. It was pointed out (21a) that Knappe and Peteri had apparently detected (91) not di-*tert*-butyl peroxide but rather *tert*-butyl hydroperoxide present as an impurity.

It can be seen in Table VIIIC that all peroxides have lower mobilities in the heptane–ether solvent. This system is very effective for separating mixtures containing dialkyl and diacyl peroxides, which are relatively mobile, but has almost no effect on the more strongly adsorbed peresters and hydroperoxides. On the other hand, the toluene–methanol system is effective in separating mixtures containing peresters and hydroperoxides; in it all dialkyl and diacyl peroxides have R_f values of about 1, and differ little from each other.

The increased mobility in toluene–methanol was attributed both to the greater polarity of the system and to the ability of methanol to displace peroxides from the adsorbent and form hydrogen bonds to the active sites (21a). Increased concentrations of methanol in toluene produced linear increases in R_f; at 35% methanol, in fact, hydrogen peroxide, which is notoriously difficult to move on a TLC plate, has an R_f value above 0.5.

In a separate study of cyclic ketone peroxides (21b), Buzlanova and co-workers chromatographed a group of 12 cyclohexanone and methylcyclohexanone peroxides on silica gel with 20:3 toluene–methanol as solvent. They found the interesting result that there were only three characteristic R_f values (which were essentially independent of the presence or location of methyl substitution): (*1*) All 1,1′-dihydroxy peroxides had an R_f of 0.05, identical to that of hydrogen peroxide, confirming the dissociation of these compounds into hydrogen peroxide and the original ketone; (*2*) the 1-hydroxy-1′-hydroperoxy peroxides had R_f's of 0.12–0.14, less than those of the alkyl hydroperoxides, at 0.30–0.50; and (*3*) the 1,1′-dihydroperoxy peroxides all had R_f values of 0.74–0.75, which is consistent with intermolecular hydrogen bonding of their hydroperoxy group.

Both Buzlanova and co-workers and Hayano and co-workers found R_f's to be identical for a compound whether alone or in a mixture.

Like paper chromatography, TLC is more frequently used as a qualitative than as a quantitative tool. With a running time of about 15 min and the ability to process many samples simultaneously on a single plate, TLC is valuable for preliminary surveys of new or complex peroxide samples, and for monitoring column separations and conversion reactions. Studies of the autoxidation of fats by Rieche and co-workers (156) exemplify the practical value of TLC in such applications. TLC separations of dicumyl peroxide for quantitative estimation has been reported (17). Silica gel with pore sizes of 20–50 Å has been recommended for effective peroxide separations (165).

b. Infrared Absorption

The infrared literature on organic peroxides in quite well summarized by Davies (35, Ch. 14) and Hawkins (74, Ch. 11), discussed by Bellamy (15), and reviewed by Karyakin (88).

(1) General Considerations

An overemphasis of the limitations of infrared absorption in the analysis of organic peroxides has been based on the following two concepts: (*a*) that the —O—O— stretching mode is relatively symmetrical, with only slight vibrational changes in dipole moment, so that its absorption band can have only low intensity, and (*b*) that the force constants in the —C—C—, —C—O—, and —O—O— bonds must all be similar, in view of their similar masses, so that the —O—O— group should contribute principally to the structurally sensitive skeletal modes, without displaying a characteristic set of frequencies (167).

These theoretical predictions do not consider the fact that organic peroxides possess a skewed, modified hydrogen peroxide structure, with a characteristic dihedral angle of about 90° between the two molecular planes through the —O—O— bond. Such a structure should have substantial dipole moment changes during vibrations, and, provided its electronic interaction with the rest of the molecule is slight, could well have characteristic peroxide vibrational modes (perhaps more complex in form than a simple —O—O— stretch), able to generate infrared absorptions at characteristic frequencies and with moderate intensities; this turns out to be the case for the hydroperoxide group, as will be seen in Section IV-B-1.

It is the present consensus of infrared analysts (15,66,98,130,131,171) that a characteristic —O—O— stretching frequency does exist in the region of 850–890 cm^{-1} for most peroxides, and that vibration of the dihedral angle in hydroperoxides produces another usually strong absorption in the 820–840 cm^{-1} region. Because of accidental overlap by other (particularly skeletal) vibrations in the molecule, and by characteristic vibrations of alcohols, ethers, and oxiranes, however, analytical chemists must use these peroxide bands with caution in the qualitative examination of unknown systems, and should always obtain independent confirmation.

(2) Empirical Correlations

Characteristic spectral features involving structural components other than the —O—O— bond have been established for some classes of peroxides. These empirical correlations have been made by examining the groups of peroxide spectra that have appeared from time to time, often in the wake of advances in synthesis techniques. Principal contributors have included Minkoff, who presented spectra of more than 30 peroxides from eight classes (131); Davison, who reported on the carbonyl absorptions of 24 symmetrical and unsymmetrical diacyl peroxides and five peroxy esters (40); Swern and co-workers, who studied eight long-chain peroxy acids (191); and Williams and Mosher, who gave infrared spectral data for 17 primary and secondary alkyl hydroperoxides and the corresponding alcohols (207), and with Welch tabulated spectra for 10 primary and secondary dialkyl peroxides. Hoffman has reported spectral data for several allyl hydroperoxides (78); Criegee and Paulig have shown spectra for three cyclic peroxides (28).

Probably the most useful infrared spectra–structure correlations among peroxides are those of the peroxidic carbonyls, based mainly on Davison's work (40). For diacyl peroxides, carbonyl absorption produces doublets split by about 25 cm^{-1}; they are centered in the range of 1769–1794 cm^{-1} (5.65–5.57 μ) for aromatic members of the class, and in the slightly higher range of 1798–1808 cm^{-1} (5.56–5.53 μ) for aliphatic members. These bands are readily differentiated from the carbonyl doublets of anhydrides by the wider 65 cm^{-1} frequency splitting of the latter. Peroxy esters have a single band in the range of 1758–1783 cm^{-1} (5.69–5.61 μ), higher in frequency than the singlet of its corresponding simple ester by an average of 36 cm^{-1} (range of 1715–1765 cm^{-1} or 5.83–5.67 μ). Aliphatic peroxy acids have a similar band at 1747–1748 cm^{-1} (5.72–5.71 μ) (191) due to hydrogen-bonded peroxidic carbonyl compared to 1710 cm^{-1} (5.85 μ) for hydrogen-bonded carboxylic acid carbonyl; one can positively identify peroxy acids by the failure of dilution to produce any unbonded carbonyl or hydroxyl absorption.

In other empirical correlations, Williams and Mosher (207) noted that primary alkyl hydroperoxides have three bands at 1433 cm^{-1} (6.98 μ), 1470 cm^{-1} (6.80 μ), and 1488 cm^{-1} (6.72 μ), whereas the secondary hydroperoxides and all the corresponding primary and secondary alcohols have only a single absorption near 1470 cm^{-1} (6.80 μ). They found that secondary hydroperoxides have a double band with peaks at about 1340 cm^{-1} (7.46 μ) and 1380 cm^{-1} (7.25 μ), whereas primary hydroperoxides and the alcohols show the 1380 cm^{-1} (7.25 μ) absorption only. Among *n*-alkyl hydroperoxides, they observed a difference in behavior between homologs with even and odd numbers of carbon atoms: Even-numbered compounds have a doublet at 807 cm^{-1} (12.39 μ) and 826 cm^{-1} (12.10 μ), whereas the odd-numbered members have a single band at about 813 cm^{-1} (12.30 μ).

(3) Near-Infrared Analysis

The utility of near-infrared absorption for hydroperoxide analysis deserves wide recognition. The value of this technique is due to the combination of high resolution by modern instruments in the 3300–3700 cm^{-1} (3–2.7 μ) region of interest and the high molar absorptivities of hydroperoxide bands occurring there. In our laboratories, Goddu has studied the near-infrared analysis of hydroperoxide and derived alcohols, and has summarized the pertinent information in a general treatment of near-infrared spectrophotometry (61). He found that the hydroxyl bands of hydroperoxides at 3515–3550 cm^{-1} (2.845–2.817 μ), are well separated from those of alcohols at 3590–3640 cm^{-1} (2.785–2.747 μ), and phenols at 3610 cm^{-1} (2.77 μ), so that mixtures of such components can be analyzed. In carbon tetrachloride, alkyl and alicyclic hydroperoxides absorb at about 3550 cm^{-1} (2.817 μ) with molar absorptivities of 80–90 liters/mole-cm. For aralkyl hydroperoxides this absorption is split into a doublet, typically at 3545 cm^{-1} (2.821 μ) and 3520 cm^{-1} (2.841 μ) each having an absorptivity value in the 30–40 liters/mole-cm range. Electron-attracting substituents present in variants such as *p*-chlorocumyl, 3,4-dichlorocumyl, and *p*-nitrocumyl hydroperoxides withdraw electrons from the benzene ring and eliminate the splitting. The use of methylene chloride as solvent rather than carbon tetrachloride also removes this splitting from the unsubstituted aralkyl hydroperoxides and, in addition, lowers the resulting single frequency by about 20 cm^{-1}.

Near-infrared should also be excellent for determining peroxy acids, in view of the data of Swern and co-workers (191): For eight long-chain peroxy acids the absorption frequency of the hydroxyl stretching mode was constant, at 3280 cm^{-1} (3.048 μ), over a concentration range in carbon tetrachloride from 0.006*M* to at least 0.3*M*, and the molar absorptivities varied but little. The average for all peracids over all concentrations was 51.6 liters/mole-cm, with a maximum variation at any one concentration of only 3%. This behavior is due, of course, to the intramolecularly hydrogen-bonded monomeric structure of peracids in solution. Carboxylic acids, in contrast, show extremely variable absorptivities (61) due to their well-known equilibrium between free monomer and hydrogen-bonded dimer.

Near-infrared absorption is an excellent technique for following the decomposition and formation of hydroperoxides. It should also be valuable for studying the hydrolysis of peroxy esters, and for identifying and determining the alcohol reduction products of many peroxides. Other workers have described such uses (191c).

c. Ultraviolet Absorption

Although absorption coefficients are usually much higher than in the infrared region, ultraviolet spectra are much simpler and less specific. We have found this spectral region useful in studying autoxidation of isopropyl benzenes (140), particularly when coupled with automatic liquid column chromatography on silica gel (89). Ultraviolet absorption is also used to determine excess reagent in a method based on chemical reduction by triphenyl phosphine (48); this is discussed below.

d. Nuclear Magnetic Resonance

Nuclear magnetic resonance (NMR), well established as a potent instrumental technique for elucidating molecular structures, is applicable to both qualitative and quantitative analysis of organic peroxides, although it has been relatively little used in this area. Pioneering studies on alkyl hydroperoxides by Fujiwara and co-workers (59) showed large differences in chemical shift among hydroperoxides, alcohols, and carboxylic acids not only for their oxygenated proton but also for the α-methylene protons. Davies and co-workers studied the proton magnetic resonance spectra of *tert*-butyl and isobutyl groups in boron compounds (38), and have used NMR to follow the autoxidation of several isomeric tributyl borons (39); they could readily differentiate *tert*-butyl, *tert*-butoxy, *tert*-butylperoxy, isobutyl, isobutoxy, and isobutylperoxy groups attached to boron. Brill also used NMR to identify isomeric hydroperoxides in his isomerization studies (18) mentioned earlier (Section IV-A-1-a-(1)). These four are the only references we have found on the analysis of organic peroxides by NMR.

In the analysis of peroxides by proton magnetic resonance we are concerned with hydroperoxy protons and protons attached to carbons that are alpha (α-CH) and beta (β-CH) to the functional group. The shielding of the hydroperoxy proton is appreciably less than that of the hydroxyl proton (59), as indicated by the greater downfield shift of its resonance ($\delta = -9.1$ and -4.4 ppm, respectively, for —OOH and —OH), and is similar to that for the carboxyl proton ($\delta = -9.8$ ppm). Available data comparing chemical shifts of α-CH and of β-CH for alcohols and hydroperoxides are summarized in Table IX; shielding is again less in the hydroperoxides than in the alcohols, for both classes of protons. Resolution of α-CH resonances in mixtures of hydroperoxides and alcohols should be straightforward, since there is a separation of about 0.4 ppm.

Recent data from our laboratory (199), in Table IXB and in Fig. 2, show that the chemical shifts for β-CH in alcohols and the corresponding hydroperoxides differ by only 0.02–0.07 ppm. Such small differences can be resolved by currently available instruments such as the Varian A-60A spectrometer. The NMR spectra in Fig. 2 show the resolution of β-CH

TABLE IX

Chemical Shift of Protons α and β to Hydroxyl and Hydroperoxide Groups, in Proton Magnetic Resonance Spectra

A. Alpha Protons

R	$(\delta_{CH_3} - \delta_{CH_2})_{ROH}$,[a,b] ppm	$(\delta_{CH_3} - \delta_{CH_2})_{ROOH}$,[a,c] ppm
n-Propyl	2.53	2.93
n-Butyl	2.58	2.93
n-Octyl	–	2.93
n-Nonyl	2.58	–
Isobutyl	2.43	2.81

B. Beta Protons

R	$(\delta_{CH_3})_{ROH}$,[a,d] ppm	$(\delta_{CH_3})_{ROOH}$,[a,d] ppm
$(CH_3)_3C$	−1.22	−1.24
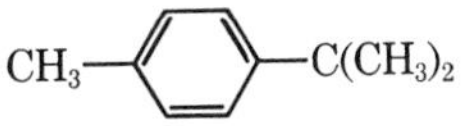 $CH_3-C_6H_4-C(CH_3)_2$	−1.46	−1.53
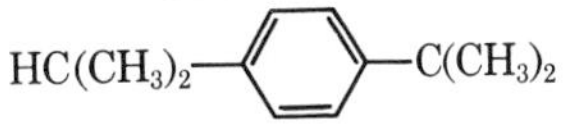 $HC(CH_3)_2-C_6H_4-C(CH_3)_2$	−1.53	−1.57
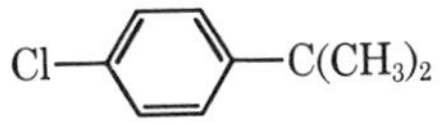 $Cl-C_6H_4-C(CH_3)_2$	−1.47	−1.50
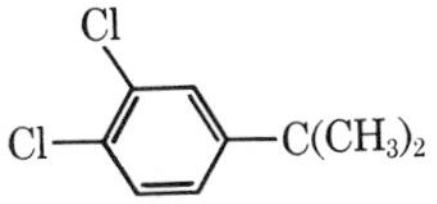 $Cl_2C_6H_3-C(CH_3)_2$ (3,4-dichloro)	−1.50	−1.52

[a] $\delta_{R'}$ ≡ chemical shift of proton resonance in group R′ downfield from the proton resonance of tetramethylsilane (TMS).

[b] Data taken from A.P.I. NMR Spectral Data Catalog (A.P.I. Project No. 44).

[c] Calculated from data of Fujiwara and co-workers (59).

[d] Unpublished data from current studies in our laboratory by Ward (199); samples were dissolved in deuterochloroform and run on a Varian A-60A spectrometer.

peaks for two three-component mixtures, each containing a dialkyl peroxide and the corresponding hydroperoxide and alcohol. The three peaks in Fig. 2*A* represent, from left to right, *tert*-butyl hydroperoxide, *tert*-butyl alcohol, and di-*tert*-butyl peroxide, in deuterochloroform; the first two are separated by 0.017 ppm. Figure 2*B* shows similar peaks for (from left to right) dicumyl peroxide, cumene hydroperoxide, and α,α-dimethylbenzyl alcohol in deuteroacetone; here the first pair are separated by only 0.014 ppm. (Concentrations were of the same order of magnitude, but unknown.)

Note that the dialkyl peroxides in Fig. 2 are at opposite ends of the triads. This variation illustrates the fact that resonance position for the

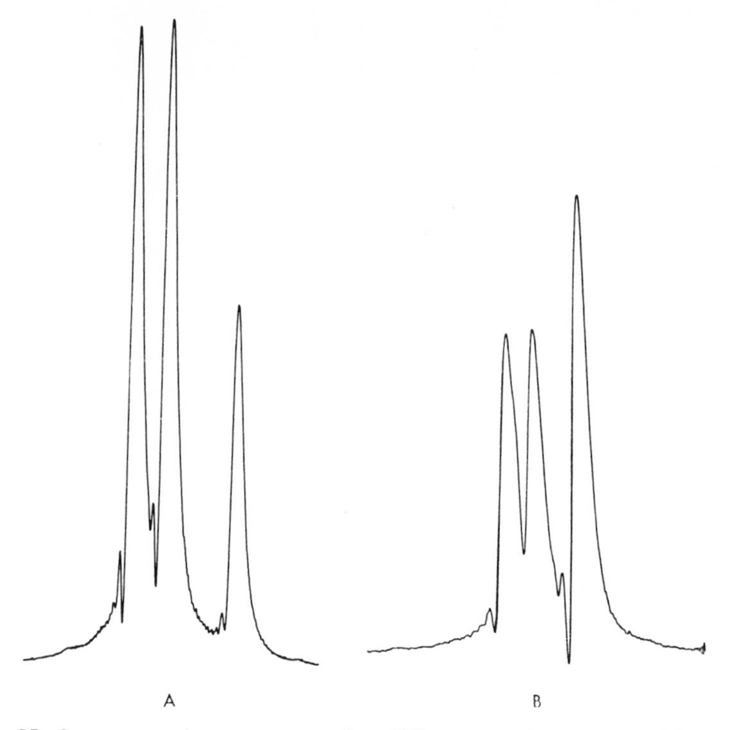

Fig. 2. Nuclear magnetic resonances of β-CH protons in two peroxide systems, obtained with a Varian A-60A spectrometer. (*A*) From left to right, with chemical shifts from tetramethylsilane (TMS) in ppm: *tert*-butyl hydroperoxide, −1.248; *tert*-butyl alcohol, −1.231; and di-*tert*-butyl peroxide, −1.196. Solvent: deuterochloroform. (*B*) From left to right, with chemical shifts from TMS in ppm: dicumyl peroxide, −1.534; cumene hydroperoxide, −1.520; and cumyl (α,α-dimethylbenzyl) alcohol, −1.493. Solvent: deuteroacetone.

several components is frequently solvent dependent; in other runs with the three cumene oxidate components, for example, there was superposition of alcohol and peroxide bands in deuterochloroform, while in the complexing solvents dimethyl sulfoxide and pyridine, superposition of alcohol and hydroperoxide occurred. In some cases such variations are an asset rather than a liability in analytical studies. Resonance position can also be concentration dependent (59).

In conjunction with NMR spectroscopy, conversion to nonperoxides by reaction with triphenyl phosphine, as mentioned in Section III-5, should be particularly valuable in the analysis of peroxides, although use of such a combination has not been reported. The reaction is said to be quantitative, unequivocal, and rapid for almost every peroxide in chlorinated solvents and benzene (excellent for NMR work), and neither the reagent nor its oxidation product will interfere. Changes in band areas before and after treatment with the triphenyl phosphine reagent can be expected to give effective qualitative and quantitative analytical data.

e. Electron Paramagnetic Resonance

Electron paramagnetic resonance (EPR) has been studied by Klyueva and co-workers for the quantitative determination of hydroperoxides (90a). Their approach has two steps, the first of which is decomposition of hydroperoxide by means of variable-valence metal ions according to the mechanism:

$$ROOH + Me^{n} \rightarrow RO\cdot + Me^{n+1} + HO^{-} \quad (7a)$$
$$ROOH + Me^{n+1} \rightarrow ROO\cdot + Me^{n} + H^{+} \quad (7b)$$

The second is interaction of the short-lived RO· and ROO· free radicals with a special reagent, a so-called reference compound, which is thereby converted into stable radicals that are quantitatively estimated in the EPR instrument. In order of decreasing activity, metal salts useful for hydroperoxide decomposition are Co, Cu, Pd, Mn, Pb, Ag, Cr, Ni, and Fe naphthenates and stearates.

Initial work on benzene solutions of mono- and dihydroperoxides of *m*- and *p*-diisopropylbenzenes employed diphenylamine (DPA) as the reference material and cobalt stearate as catalyst. Small amounts (0.07 ml) of benzene solutions of hydroperoxide containing a twofold excess of DPA were added to ampules containing some solid cobalt stearate on the bottom. These tubes were sealed, placed in hot water (85–90°C) for 1 min, and then quenched in cold water.

Using *m*-diisopropylbenzene monohydroperoxide, Klyueva and co-workers studied the quantitative relationship between hydroperoxide concentration and EPR signal intensity; for the latter, they used the maximum span of the EPR spectrum (three approximately equal lines), since the shape of the spectrum did not change with hydroperoxide concentration. Response was linear over the hydroperoxide concentration range 0.01–0.2 wt %, and then increased less rapidly at higher concentrations, probably because of an increase in the recombination rate of the hydroperoxide radicals. Sensitivity was 0.01 wt % or $6 \times 10^{-4}M$, corresponding to 8 μg of hydroperoxide per ampule. The relative error was about 5%.

On trying to determine cumene hydroperoxide in a reaction mixture containing 30–35% acetone, 55–60% phenol, and 10% condensation products, they found that DPA was an unsuitable reference. A new reference, phenothiazine (thiodiphenylamine), proved acceptable. In fact, stable free radicals readily formed during reaction of cumene hydroperoxide and phenothiazine at room temperature without need for a catalyst. Again, response (a four-line signal) was linear to about 0.2 wt % hydroperoxide and then fell off rapidly.

The use of EPR for estimating organic peroxides is an interesting approach whose eventual utility remains to be seen. There would perhaps be possibilities for selectively estimating certain peroxides in mixtures.

f. Thermal Analysis

Differential thermal analysis (DTA) can be of auxiliary service in the analysis and characterization of organic peroxides and compositions containing them. Melting and freezing peaks give information on identity, purity, and quantity for crystalline peroxides. Decomposition exotherms indicate relative thermal stabilities, and can help define the role of specific peroxides in the curing of resin compositions.

A relatively new DTA technique known as differential scanning calorimetry (DSC) records the instantaneous rate of physical or chemical thermal processes as a function of temperature or time (201), thus giving peak areas directly in energy units. This development increases the applicability of thermal analysis to many types of compounds, including peroxides. Mathematical analysis of DSC peak shapes and areas can efficiently yield much information; from a single melting thermogram, for example, mole percent purity can be determined, and the kinetic parameters of thermal decompositions can be calculated from decomposition exotherms.

g. Polarography

This method has been used in determining organic peroxides for more than a quarter century (42,63,185). It is of great value in specific areas because of its high sensitivity, its rapidity, and its potential for qualitatively identifying and differentiating the components of mixtures. The status of this technique in peroxide analysis has been reviewed recently (86,117,164a).

The polarographic characterization of peroxides has involved a good deal of research. Several groups have studied water-soluble peroxides in predominantly aqueous media (19,42,179,185), an interest that arose mainly in researches on combustion chemistry. Other groups with a primary concern for autoxidation processes in fats and oils have devised and explored a general-purpose, nonaqueous polarographic system (99,

107,141,210) and have applied it broadly to studies of interest (87,100,106, 153,189,211). Still others have compared aqueous and nonaqueous media and have examined the effects of such parameters as solvent polarity, nature of the supporting electrolyte, peroxide structure, molecular weight, and adsorption phenomena at the electrode on half-wave potential, diffusion current constant, and polarographic irreversibility (16,159,164,179). A firm basis has recently been established, also, for the analysis of oxidized oils by oscillographic polarography (112a). These matters are discussed further in Section IV-B-2.

In spite of the substantial body of information on the polarography of organic peroxides, much remains to be done. In the opinion of Romantsev and Levin (159), this area ". . . is still in its initial stage of development; the effect of structure and other substituents in the molecule on the polarographic characteristics has not been examined, and there is almost no information on the effect of the nature of the solvent and the supporting electrolyte. The analytical methods suggested are accordingly not sufficently well founded and are hardly the best ones." For the most part, we must concur.

A novel related technique for peroxide analysis due to Kober (91a) is of interest. He has described a galvanic cell for the continuous and discontinuous quantitative determination of oxygen and inorganic and organic peroxides and hydroperoxides in liquids. The cell consists of a corrodible electrode made of lead, zinc, iron, tin, or cadmium tubing 1–2 m long and 3–9 mm in diameter, coated on the outside with an inert plastic; this tube is separated by a concentric membrane sheath from an inert coaxial inner electrode of gold, platinum, silver, or copper wire; the cell is connected to a current-measuring apparatus and an integrator. The electrolyte is a 10–30% potassium hydroxide solution, which may contain auxiliary solvents. Oxygen and peroxides act as depolarizers in this system, and their content is calculated according to Faraday's law by the equation

$$m = (M/Fz) \int_0^t I \, dt \tag{8}$$

where m is the mass of the depolarizer, in grams, M its molecular weight, F the Faraday number, z the number of electrons transmitted in the reduction, and $\int_0^t I \, dt$ the integral of the time changes of current strength, in ampere-seconds. Sensitivities down to $10^{-7}\%$ and an accuracy of $\pm 2\%$ were claimed. As examples, Kober cited the continuous and discontinuous analysis for cumene hydroperoxide at about 1% concentration, using lead and silver electrodes.

2. Chemical Reduction Methods

Methods depending on quantitative reduction of peroxides by chemical reagents are summarized in Table VB. These are primarily wet-chemical volumetric procedures.

a. Iodide Methods

Reduction with iodide is the most widely used and the most generally useful of all the reactions for peroxide analysis. A confusing variety of iodide procedures have accumulated, however, as the result of a procession of innovations intended to remedy earlier deficiencies. This situation has been realistically depicted in Hawkin's review (74, Ch. 11).

It is worth attempting to select the better ones of these methods, as Martin has done in giving detailed instructions for a half-dozen recommended procedures (117). It is also worth trying to identify the relevant factors in the performance of iodide reagents on various peroxides and in specific sample systems, as has been done in a paper from our laboratory (116), since such knowledge enables analytical chemists to tailor the reducing power of their reagents to particular sample compositions.

Basically, one can prescribe iodide reagent systems with several levels of reactivity: relatively mild reagents, capable of determining easily reduced peroxides; reagents of intermediate strength, able to reduce more stable peroxides also, yet without interference from related nonperoxides; and those with very strong reducing power, able rapidly to reduce the most stable peroxide and many nonperoxides as well. We shall defer our main discussion of these important iodide reagents until Section IV-B-3.

b. Organic Sulfide and Arsine Methods

Diphenyl sulfide rapidly and quantitatively reduces peroxy acids, with no significant attack on other peroxides, including diacyl peroxides and alkyl hydroperoxides (82). Another thioether, *p*-nitrothioanisole, has the same selectivity. It should be recognized, however, that this behavior of these two reagents is not the rule for thioethers but rather a useful exception; only these two among the organic sulfides react so slowly with diacyl peroxides and hydroperoxides. Thiophane and diethyl sulfide (83), in contrast, reduce dibenzoyl peroxide quantitatively within 15 min at room temperature. Table X shows how reactivity toward a typical diacyl peroxide varies with structure of the thioether.

Triethyl arsine quantitatively reduces peroxy acids, diacyl peroxides, and alkyl hydroperoxides, producing 1 and 2 moles of carboxylic acid and

TABLE X

Relative Effectiveness of Thioethers in Decomposing Dibenzoyl Peroxide[a]

Thioether	Reaction half-time
Thiophane	4 min
Diethyl sulfide	5 min
Dibutyl sulfide	24 min
p-Methoxythioanisole	50 min
Ethyl-*p*-tolyl sulfide	85 min
p-Methylthioanisole	122 min
Dibenzyl sulfide	122 min
Diisopropyl sulfide	150 min
Thioanisole	185 min
p-Nitrothioanisole	7 days
Diphenyl sulfide	7 days

[a] Data of Horner and Jürgens (83).

1 mole of alcohol, respectively (82). Reduction with triphenyl arsine is equally quantitative, but slower.

c. Iodide, Alkali, Sulfide, and Arsine Methods: A Composite Scheme

Horner and Jürgens have coupled reductions with diphenyl sulfide and triethyl arsine to two iodide reduction procedures (one mild and one more vigorous) and an acidimetric titration to obtain a flexible scheme for quantitatively resolving mixtures of any two or all of the four peroxide classes: peroxy acids, diacyl peroxides, alkyl hydroperoxides, and dialkyl peroxides, along with organic acids (82). Such discrimination should make this approach a powerful tool for studying peroxide syntheses and reactions, two particularly promising applications being the alkylidene peroxide mixtures formed by the reaction of hydrogen peroxide or alkyl hydroperoxides with aldehydes or ketones and certain other compounds, and the mixtures resulting from extensive autoxidation of fats and oils.

Horner and Jürgens' analytical scheme consists of a group of 11 procedures; these are summarized in Table XI. Each procedure is designed to analyze one specific peroxide mixture out of all possible combinations of the four classes; nonperoxidic organic acids are determined as well. Specificity is obtained primarily by the use of two reagents: diphenyl sulfide, which rapidly and specifically converts peroxy acids to carboxylic acids while itself being converted to diphenyl sulfoxide; and triethyl arsine, which is inert to dialkyl peroxides but reacts quantitatively with the

TABLE XI

Resolution of Peroxide Classes—Quantitative Determination of Mixtures Containing up to Four Classes: Peroxy Acids, Diacyl Peroxides, Alkyl Hydroperoxides, and Dialkyl Peroxides (82)

Components present	Procedure outline	Calculations[a]
A. Organic acids, peroxy acids, diacyl peroxides	1. *Determine total peroxide content with KI:* React an aliquot part of the sample mixture in 2–5 ml benzene or chloroform with 1 ml saturated aqueous KI in 30 ml glacial acetic acid under CO_2 for 10–15 min at room temperature; dilute with 30–50 ml water and titrate liberated iodine with 0.1*N* sodium thiosulfate. Consumption: *a* ml 0.1*N* thiosulfate 2. *Determine diacyl peroxide with KI after destroying peroxy acid with Ph_2S:* Add a second aliquot of the sample mixture to diphenyl sulfide slightly in excess of its total peroxide content, allow to react for 10 min and then determine the diacyl peroxide iodimetrically as above. Consumption: *b* ml 0.1*N* thiosulfate 3. *Determine total acid content with NaOH:* Titrate a third aliquot of the sample mixture with 0.1*N* NaOH to a phenolphthalein endpoint. Consumption: *c* ml 0.1*N* alkali	$a \times 0.1 = \text{meq}_{ox}$, peroxy acid + diacyl peroxide $b \times 0.1 = \text{meq}_{ox}$, diacyl peroxide $c \times 0.1 = \text{meq}_{ac}$, acid + peroxy acid $(a - b) \times 0.1 = \text{meq}_{ox}$, peroxy acid $(c - (a - b)/2) \times 0.1 = \text{meq}_{ac}$, organic acid
B. Organic acids, peroxy acids, alkyl hydroperoxides	1. *Determine total peroxide content with KI:* as in A1. Consumption: *a* ml 0.1*N* thiosulfate	As in A, simply inserting alkyl hydroperoxide for diacyl peroxide

	2. *Determine alkyl hydroperoxide with KI after destroying peroxy acid with* Ph_2S: As in A2. Consumption: b ml 0.1N thiosulfate	
	3. *Determine total acid content with NaOH:* As in A3. Consumption: c ml 0.1N alkali	
C. Diacyl peroxides, alkyl hydroperoxide	1. *Determine total peroxide content with KI:* As in A1. Consumption: a ml 0.1N thiosulfate	$a \times 0.1 = \text{meq}_{ox}$, diacyl peroxide + alkyl hydroperoxide $b \times 0.1 = \text{meq}_{ac} = \text{meq}_{ox}$, diacyl peroxide $(a - b) \times 0.1 = \text{meq}_{ox}$, alkyl hydroperoxide
	2. *Determine diacyl peroxide with NaOH after conversion to acid with* Et_3As: To a second aliquot part of the sample mixture under nitrogen add 300–400 mg triethyl arsine and let stand 30 min; dilute with 30 ml water and titrate the resulting acids with 0.1N NaOH to a phenolphthalein endpoint; determine an acidity blank on the triethyl arsine reagent and subtract from the sample titration. Consumption (net): b ml 0.1N alkali	
D. Organic acids, peroxy acids, dialkyl peroxides	1. *Determine total peroxide content with HI:* React an aliquot part of the sample mixture in 2–5 ml benzene or chloroform with 1 ml saturated aqueous KI and 5 ml concentrated HCl in 30 ml glacial acetic acid under CO_2 or N_2 by boiling for 30 min; after cooling, dilute with 30–50 ml water and titrate liberated iodine with 0.1N sodium thiosulfate. Consumption: a ml 0.1N thiosulfate	$a \times 0.1 = \text{meq}_{ox}$, peroxy acid + dialkyl peroxide $b \times 0.1 = \text{meq}_{ox}$, peroxy acid $(a - b) \times 0.1 = \text{meq}_{ox}$, dialkyl peroxide $(c - {}^1/_2\, b) \times 0.1 = \text{meq}_{ac}$, organic acid

(continued)

TABLE XI (*continued*)

Components present	Procedure outline	Calculations[a]
	2. *Determine peroxy acid with standard Et_3As:* To a second aliquot part of the sample mixture under CO_2 add an excess of 0.1*N* (0.05*M*) solution of triethyl arsine in benzene and allow to stand for 15 min; then back-titrate the unconsumed triethyl arsine with an 0.1*N* iodine solution. Consumption: *b* ml 0.1*N* triethyl arsine	
	3. *Determine total acid content with NaOH:* As in A3. Consumption: *c* ml. 0.1*N* alkali	
E. Diacyl peroxides, dialkyl peroxides	1. *Determine total peroxide content with HI:* As in D1. Consumption: *a* ml 0.1*N* thiosulfate	$a \times 0.1 = meq_{ox}$, diacyl peroxide + dialkyl peroxide $b \times 0.1 = meq_{ox}$, diacyl peroxide $(a - b) \times 0.1 = meq_{ox}$, dialkyl peroxide
	2. *Determine diacyl peroxide with standard Et_3As:* As in D2, but with the reaction period extended to 45 min. Consumption: *b* ml 0.1*N* triethyl arsine	
F. Alkyl hydroperoxides, dialkyl peroxides	1. *Determine the total peroxide content with HI:* As in D1. Consumption: *a* ml 0.1*N* thiosulfate	As in E, simply substituting alkyl hydroperoxide for diacyl peroxide
	2. *Determine alkyl hydroperoxide with standard Et_3As:* As in E2. Consumption: *b* ml 0.1*N* triethyl arsine	
G. Organic acids, peroxy acids, diacyl peroxides, alkyl hydroperoxides	1. *Determine the total peroxide content with KI:* As in A1. Consumption: *a* ml 0.1*N* thiosulfate: Σ peroxy acid, diacyl peroxide, alkyl hydroperoxide	$(a - b) \times 0.1 = meq_{ox}$, peroxy acid $(d - c) \times 0.1 = meq_{ac} = meq_{ox}$, diacyl peroxide $(c - (a - b)/2) \times 0.1 = meq_{ac}$, organic acid $[b - (d - c)] \times 0.1 = meq_{ox}$, alkyl hydroperoxide

	2. *Determine diacyl peroxide + alkyl hydroperoxide with KI, after destroying peroxy acid with Ph_2S:* React a second aliquot portion of the sample mixture with ca. 300 mg diphenyl sulfide for 10 min. Determine remaining peroxides iodimetrically as in A1. Consumption: b ml $0.1N$ thiosulfate: Σ diacyl peroxide, alkyl hydroperoxide	
	3. *Determine initial total acid content with NaOH:* As in A3. Consumption: c ml $0.1N$ alkali: Σ acid, peroxyacid	
	4. *Determine total acid content with NaOH after converting diacyl peroxide to acid with Et_3As:* As in C2 (30 min reaction); include blank correction. Consumption (net): d ml $0.1N$ alkali: Σ acid, peroxy acid, diacyl peroxide.	
H. Organic acids, peroxy acids, diacyl peroxides, dialkyl peroxides	1. *Determine total peroxide content with HI:* As in D1. Consumption: a ml $0.1N$ thiosulfate: Σ peroxy acid, diacyl peroxide, dialkyl peroxide.	$(a - b) \times 0.1 = \text{meq}_{ox}$, dialkyl peroxide $c \times 0.1 = \text{meq}_{ox}$, diacyl peroxide $(b - c) \times 0.1 = \text{meq}_{ox}$, peroxy acid $(d - (b - c)/2) \times 0.1 = \text{meq}_{ac}$, organic acid
	2. *Determine peroxyacid + diacyl peroxide with standard Et_3As:* As in E2 (45 min reaction period). Consumption: b ml $0.1N$ triethyl arsine: Σ peroxy acid, diacyl peroxide.	
	3. *Determine diacyl peroxide with standard Et_3As, after destroying peroxy acid with Ph_2S:* React a third aliquot portion of the sample mixture with about 200–300	

(continued)

TABLE XI (*continued*)

Components present	Procedure outline	Calculations[a]
	mg diphenyl sulfide for 10 min; then determine diacyl peroxide on this reaction mixture according to H2 (or E2). Consumption: c ml 0.1N triethyl arsine: diacyl peroxide	
	4. *Determine organic acid + peroxy acid with NaOH:* As in A3. Consumption: d ml 0.1N alkali: Σ acid, peroxy acid.	
I. Organic acids, peroxy acids, alkyl hydroperoxides, dialkyl peroxides	1. *Apply steps H1, H2, H3 and H4:* substitute alkyl hydroperoxide for diacyl peroxide whenever the latter occurs in method H.	Calculate as for method H, substituting alkyl hydroperoxide for diacyl peroxide
J. Organic acids, diacyl peroxides, alkyl hydroperoxides, dialkyl peroxides	1. *Determine total peroxide content with HI:* As in D1. Consumption: a ml 0.1N thiosulfate: Σ diacyl peroxide, alkyl hydroperoxide, dialkyl peroxide	$(a - b) \times 0.1 = meq_{ox}$, dialkyl peroxide
	2. *Determine diacyl peroxide + alkyl hydroperoxide with standard Et_3As.* Consumption: b ml 0.1N triethyl arsine: Σ diacyl peroxide, alkyl hydroperoxide	$(d - c) \times 0.1 = meq_{ac} = meq_{ox}$, diacyl peroxide
	3. *Determine organic acid with NaOH:* As in A3. Consumption: c ml 0.1N alkali: acid	$c \times 0.1 = meq_{ac}$, organic acid
	4. *Determine total acid content with NaOH, after converting diacyl peroxide to acid with Et_3As:* As in G4. Consumption: d ml 0.1N alkali: Σ acid, diacyl peroxide	$[b - (d - c)] \times 0.1 = meq_{ox}$, alkyl hydroperoxide

K. Organic acids, peroxy acids, diacyl peroxides, alkyl hydroperoxides, dialkyl peroxides	1. *Determine total peroxide content with HI:* As in D1. Consumption: a ml 0.1N thiosulfate: Σ peroxy acid, diacyl peroxide, alkyl hydroperoxide, dialkyl peroxide 2. *Determine peroxy acid + diacyl peroxide + alkyl hydroperoxide with standard* Et_3As: As in E2. Consumption: b ml 0.1N triethyl arsine: Σ peroxy acid, diacyl peroxide, alkyl hydroperoxide 3. *Determine diacyl peroxide + alkyl hydroperoxide with standard* Et_3As*, after destroying peroxy acid with* Ph_2S: As in H3. Consumption: c ml 0.1N triethyl arsine: Σ diacyl peroxide, alkyl hydroperoxide 4. *Determine organic acid + peroxy acid with NaOH:* As in A3. Consumption: d ml 0.1N alkali: Σ acid, peroxy acid 5. *Determine total acid content with NaOH, after converting diacyl peroxide to acid with* Et_3As: As in G4. Consumption: e ml 0.1N alkali: Σ acid, peroxy acid, diacyl peroxide	$(a - b) \times 0.1 = meq_{ox}$, dialkyl peroxide $(b - c) \times 0.1 = meq_{ox}$, peroxy acid $(e - d) \times 0.1 = meq_{ox}$, diacyl peroxide $[c - (e - d)] \times 0.1 = meq_{ox}$, alkyl hydroperoxide $(d - (b - c)/2) \times 0.1 = meq_{ac}$, organic acid

[a] In the calculations, a milliequivalent of peroxide based on reduction of the —O—O— bond (meq_{ox}) equals $^1/_2$ millimole, whereas a milliequivalent of an organic acid (meq_{ac}) equals 1 millimole.

three other peroxide types according to the following equations.

$$RC(=O)-OOH + Et_3As \xrightarrow{H_2O} RC(=O)-OH + Et_3AsO \cdot H_2O \quad (9)$$

$$RC(=O)-OO-C(=O)R + Et_3As \xrightarrow{H_2O} 2RC(=O)-OH + Et_3AsO \cdot H_2O \quad (10)$$

$$R-OOH + Et_3As \xrightarrow{H_2O} R-OH + Et_3AsO \cdot H_2O \quad (11)$$

Triethyl arsine is used in two modes: one as a standard solution whose consumption is a measure of the total easily reduced peroxides, and the other as a conversion agent that specifically reveals diacyl peroxides through the additional acidity (2 moles) produced by their reaction.

A potential interference should be pointed out to those who might be contemplating variations of the reaction schemes in Table XI. Organic sulfoxides, e.g., diphenyl sulfoxide produced by reduction of peroxy acids in method A2, although inert to iodide in the potassium iodide reagent of method A1, liberate iodine quantitatively from stronger hydriodic acid reagents (83) such as that in method D1.

The methods in Table XI appear to be aimed at the analysis of mixtures whose qualitative compositions are known, at least as to peroxide classes. In some cases this will represent the analytical situation realistically. In others, however, there will be no positive information about the nature of the sample. It will then be necessary to use method K, the final one in the table, which covers the possible presence of all four peroxide classes.

The general utility of this scheme for the resolution of peroxide mixtures into their component classes would appear to be greatly enhanced if there were a way to include peroxy esters. We have found no indication in the literature that this feat has been achieved, and we have no personal experience in this particular area. We encourage a study of the response of peroxy esters to organic sulfides and arsines.

d. Organic Phosphine Methods

According to Horner and Jurgeleit, virtually every type of peroxide reacts with tertiary phosphines (81) to give the corresponding tertiary phosphine oxide and the analogous nonperoxy product corresponding to the reaction:

$$R_1-O-O-R_2 + PR_3 \rightarrow R_1-O-R_2 + OPR_3 \quad (12)$$

They cite numerous examples of the following conversions:

Alkyl hydroperoxides	→	alcohols
Unsaturated hydroperoxides	→	unsaturated alcohols
Dialkyl peroxides	→	ethers
Hydroxyalkyl hydroperoxides	→	aldehydes
Dihydroxyalkyl peroxides	→	ketones
Peroxy acids	→	acids
Peroxy esters	→	esters
Diacyl peroxides	→	anhydrides
Endoperoxides	→	endooxides
Ozonides	→	ketones (or aldehydes)

Reactions were rapid at room temperature except for dialkyl peroxides, which required heating. They obtained high yields of the specific products indicated by the above equation, in which R_1 and R_2 can be H, acyl, alkyl, or aralkyl. The only exception noted was with diacyl peroxides, which in the presence of alcohol yield some ester as well as anhydride; this is the circumstance that led Horner and Jürgens to select tertiary arsines as their preferred reagent (82).

Phosphine conversion of the readily available α-hydroperoxy olefins and diacyl peroxides (173) into α-hydroxy olefins and unsymmetrical acid anhydrides has been suggested as a useful organic preparative reaction.

This study of phosphines made clear the possibility of differentiating alkyl hydroperoxides from dialkyl peroxides through the much more rapid reaction of the hydroperoxides (81). Dulog and Burg developed this idea into what appears to be a sound and effective method (48) for determining organic peroxides of every class (except a few that are outstandingly unreactive), and for differentiating quantitatively between the easily reduced peroxides (hydroperoxides, peroxy acids, diacyl peroxides, and ketone peroxides) and the more sluggish dialkyl peroxides.

In all cases a weighed (excess) quantity of the stable and easily purified triphenyl phosphine is reacted with the peroxide in an aromatic hydrocarbon medium in an inert atmosphere. The easily reduced peroxides react quantitatively within 10 min at room temperature, whereas the dialkyl peroxides require refluxing in xylene under nitrogen for 2 hr (some peroxides of intermediate reactivity, such as *tert*-butyl perbenzoate, are refluxed in benzene for 1 hr).

Three alternatives have been provided for estimating the amount of excess phosphine reagent, which is the final step in the quantitative procedure. These are a gravimetric, a spectrophotometric, and a titrimetric procedure. The first two are based on conversion of triphenyl phosphine into the water-soluble α-hydroxymethyltriphenylphosphonium chloride by reaction with formaldehyde and hydrochloric acid, as described by Hoffmann (79):

$$P(C_6H_5)_3 + HCHO \underset{NaOH}{\overset{HCl}{\rightleftharpoons}} [(C_6H_5)_3PCH_2OH]^+Cl^- \quad (13)$$

In the gravimetric method the excess phosphine is separated by extraction into water as the hydroxymethyl complex, then converted back to phosphine with alkali, and extracted directly into peroxide-free ether for evaporation and final weighing. This finish-up procedure seems particularly suited for occasional analyses; although it requires more manipulation than is desirable in routine determinations, it requires no special reagent preparation or standardization.

In the spectrophotometric approach excess phosphine is estimated from absorption at 275 mμ by the aqueous solution of the hydroxymethyl complex after an appropriate dilution; absorbance is independent of both formaldehyde and hydrochloric acid concentration. We think the procedure could be made quicker and easier by simply extracting the reaction mixture with a measured volume of acidified formaldehyde solution and then withdrawing an aliquot from the resulting aqueous layer for an appropriate dilution and photometry; this would avoid the washing steps necessary for transfer of the aqueous layer, as specified in the published procedure (48).

For the titrimetric finish a different approach is taken, in which the excess phosphine is reacted with a known (excess) amount of hydrogen peroxide, and the excess of the latter then titrated with standard ceric sulfate. This is a straightforward, simple procedure, whose only apparent disadvantage appears to be the additional 30 min required for the phosphine–hydrogen peroxide reaction.

Dulog and Burg report consistently good agreement between these triphenyl phosphine methods and iodimetric methods. In addition, they quantitatively resolved mixtures of cumene hydroperoxide and dicumyl peroxide, and of methyl linoleate hydroperoxide and 2,2-bis(*tert*-butyl peroxy) butane, both in methyl linoleate.

The importance of an inert atmosphere should be emphasized. In current studies we have found that results can be high and variable unless oxygen is rigorously excluded from dissolved triphenyl phosphine. In this regard it appears to be more demanding of good technique than iodide–acetic acid reagents. There may be a simple solution to this problem, however, which we have recently noted: Stein and Slawson (184a) found that the presence of an anti-oxidant (di-*tert*-butyl-*p*-cresol) made the autoxidation of triphenyl phosphine in isopropyl alcohol negligible—no spectrophotometric change in a reagent blank was detected after as long as three hours in the presence of air. The use of anti-oxidants to render triphenyl phosphine a more tractable reagent in the above set of methods should be evaluated.

The quantitative determination of ozonides by reaction with triphenyl phosphine was recently described by Lorenz (112), who found that this peroxide class will liberate iodine stoichiometrically from iodide only when the reduction yields ketones rather than aldehydes. By reaction with a weighed excess of the phosphine reagent in ethanol or in heptane–phenol solution nine out of ten purified monomeric ozonides prepared from a group of olefins were estimated to be more than 95% pure. Excess reagent was titrated with standard aqueous iodine solution. Reaction at room temperature required up to three days for completion, but probably could be safely speeded up by using a somewhat higher temperature or a more reactive reagent such as triethyl phosphine (81). Air was excluded to prevent direct oxidation of reagent or autoxidation of aldehydes, both of which could cause high results.

We propose yet another approach to the use of tertiary phosphines in the resolution of peroxide mixtures, suggested by their ability to achieve clean-cut, usually quantitative, and broadly general conversions of organic peroxides into specific nonperoxide analogs. In this scheme total peroxidic functionality would be determined from the consumption of phosphine reagent, and individual peroxides would be estimated from their analogs via established chemical or instrumental methods. This approach should be particularly valuable for studying the formation and reactions of organic peroxides in labile systems.

e. Stannous Methods

Reduction with stannous ion avoids interference in the iodide methods by sulfur compounds (12) that consume liberated iodine. Barnard and Hargrave have described earlier work with the stannous chloride reagent and studied the effects of many parameters (12). They devised a method that is highly precise and accurate for hydroperoxides, although rather long and complex; diacyl peroxides and, presumably, peroxy acids are also determined. Egerton and co-workers (50) found reaction to be much more rapid in alkaline media, without apparent hydroperoxide loss despite the known instability of primary and secondary hydroperoxides to alkali (35, Ch. 13; 96; 166).

f. Arsenite Methods

Reduction with arsenite ion is another reaction that has been found useful where there are interferences with iodide methods. Walker and Conway have modified Siggia's original proposal (170) into a remarkably accurate, precise, and sensitive method for determining hydroperoxide in petroleum products down to concentration levels of less than 0.1 meq/liter (peroxide number 0.1) (198). Gasolines and fuel oils, but not refined mineral oil, were found to contain components that interfere seriously in iodide reduction methods (92,196) at the typically low hydroperoxide concentrations, by consuming liberated iodine.

This interference is probably due to the sulfur compounds present in most petroleum products. It might be argued that iodine absorption by olefins in the sample is perhaps also a contributing factor. We tend to discount this possibility, being convinced that the triiodide ion is unable in general to react with olefins. (See the discussion in Section IV-B-3-c.) In view of the extremely high ratio of olefin to peroxide (or triiodide) that must have existed in certain of Walker and Conway's samples, however, we cannot with certainty rule out the chance that some of the iodine was absorbed by the olefins.

The Walker and Conway method appears quite practical for product control testing in a petroleum refinery laboratory, the application for which it was devised. We do not recommend it for general use, however, because it is slow.

Apparently no one has described any attempt to apply either stannous or arsenite reagents to peroxy esters or dialkyl peroxides. We might expect peroxy esters to reduce readily in an alkaline medium, and primary and secondary dialkyl peroxides might also respond. The scope of these two reagents should be surveyed more broadly, since useful specificity may exist.

g. Potassium Permanganate

Titration with potassium permanganate oxidizes hydrogen peroxide to oxygen and water, and serves to determine it quantitatively. Szobor and Back have studied the possibility of utilizing this titration, along with differences in rates of conversion to hydrogen peroxide by hydrolysis, to resolve the mixtures of peroxy compounds that constitute commercial cyclohexanone peroxide (191a). They came up with some unexpected and useful results.

The principal cyclohexanone peroxide components active as polymerization catalysts are 1-hydroxy-1′-hydroperoxycyclohexyl peroxide and the 1,1′-dihydroperoxy derivative. Szobor and Back found that when these compounds were dissolved in glacial acetic acid, diluted with three volumes of water, acidified with sulfuric acid, and allowed to stand for 24 hr at 25°C, titration of the reaction mixture with $0.1N$ potassium permanganate was accompanied by vigorous evolution of oxygen gas and consumed titrant equivalent to the total active oxygen content of the starting peroxides; this was the expected result of quantitative hydrolysis to hydrogen peroxide. On the other hand, when the same diluted solutions were titrated with permanganate immediately, there was no evolution of oxygen, and the consumption of titrant corresponded to the active oxygen from the hydroperoxy groups alone. Evidence was presented that a Mn(II)–peroxide complex had been formed. Szobor and Back propose that such a titration be used to determine the total active hydroperoxy content of cyclohexanone peroxide mixtures.

The 1,1′-dihydroxycyclohexyl peroxide component behaved differently (191a): On immediate titration, it consumed titrant equivalent to its active oxygen content, and simultaneously evolved oxygen gas. This behavior indicates the rapid hydrolysis of the dihydroxy derivative, which was noted also by Buzlanova and co-workers in a TLC evaluation of the peroxide mixture (21b), discussed earlier, in Section IV-A-1a-(4).

The dihydroxycyclohexyl peroxide will interfere with the proposed titration, if present. Such a problem will be signalled by oxygen evolution, however. Researchers studying this application would be well advised to use TLC also (21b).

h. Catalytic Hydrogenation

This technique has been relatively little used, although recommended by Davies (35, Ch. 14). Its chief utility in peroxide analysis has been for derivative preparations in proofs of structure. Kinetic studies of the hydrogenation of various peroxide and nonperoxide groups on several types of catalyst (8) showed numerous examples of selectivity, although the order of reactivity for various peroxides varied considerably with the type of catalyst employed, e.g., nickel, palladium, or platinum.

Criegee and co-workers used rates of hydrogenation on palladium black in acetic acid and rates of iodine liberation from sodium iodide–acetic acid reagent to differentiate between and characterize a group of peroxides (29). The two reduction methods gave closely similar results: Hydroperoxides reacted quantitatively in a few minutes, whereas peroxy esters and dialkyl peroxides were substantially less reactive, and dimeric and trimeric acetone peroxides still less so.

i. Lithium Aluminum Hydride

This is another reagent that has been relatively little used with peroxides. There is the drawback that lithium aluminum hydride is not specific in its action, reacting generally with all compounds containing an active hydrogen, and quantitatively reducing ketones and esters (76). It may be a useful conversion agent, however, whenever peroxide reduction should produce characteristic alcohols or when there is other pertinent information about a sample. Studies of *p*-cymene oxidation (166) provided an excellent example of such an application.

Lithium aluminum hydride reduces organic hydroperoxides smoothly at room temperature to the corresponding alcohols (188) with the evolution of exactly 2 moles of hydrogen per mole of hydroperoxide. Other peroxide classes form 1 mole of hydrogen per mole, dibenzoyl peroxide reacting with almost explosive vigor, methyl α-tetralyl peroxide and ascaridole at moderate rates, and di-*tert*-butyl peroxide very slowly. The stoichiometry corresponds to fission of the —O—O— bond and direct formation of lithium aluminum alkoxide. This reaction has been used to determine methyl linoleate hydroperoxide quantitatively (48), and to determine hydroperoxide impurities in isopropyl- and isopropenylacetylene (191b).

Use of excess hydride reagent converts peroxides to the corresponding alcohols quantitatively. With excess peroxide, however, it is clear that only two of the four hydrogens in lithium aluminum hydride are reactive (119) and that the resulting dialkoxide will not reduce peroxides:

$$\text{ROOR} + \text{LiAlH}_4 \rightarrow \text{LiAl(OR)}_2\text{H}_2 + \text{H}_2 \quad (14)$$

$$\text{ROOR} + \text{LiAl(OR)}_2\text{H}_2 \rightarrow \text{no reaction} \quad (15)$$

There is parallel behavior in the reduction of nitriles by this reagent at 35°C or lower (4).

An analogous but milder reagent, sodium borohydride, is particularly useful for the reduction of triglyceride peroxides, since it does not reduce ester groups and so leaves the basic fat structure intact (118).

j. Ferrous Methods

In the past, chemists have made extensive use of reduction by ferrous ion for the quantitative determination of easily reduced organic peroxides, particularly the alkyl hydroperoxides formed in the development of rancidity and in other cases of autoxidation. These methods have long been controversial. On the one hand, results under a given set of conditions are consistently precise for a given peroxide or autoxidized substrate, and hence proportional to results by reliable iodometric procedures. On the other hand, the proportionality factor varies with the nature of the reaction solvent, has unsystematically different values for different peroxidized materials, and for a given peroxide and a given solvent can vary by a factor of as much as 5 or 6 with the extent of aeration of the reaction mixture—results being higher than iodometric values in the presence of oxygen and lower than iodometric in its absence.

Many careful studies have been undertaken in efforts to understand these phenomena (12,26,102,103,195,197), but have failed in varying degrees because the chemistry that is involved was not comprehended. A substantial understanding of the ferrous methods has come about, however, through the definitive work of Kolthoff and Medalia (93,94), who drew on basic studies of the reaction mechanisms to elucidate the analytical difficulties.

Their treatment recognizes both positive and negative error factors. Positive errors occur only when air is present during peroxide reduction; this situation induces the formation of additional hydroperoxide from the dissolved oxygen by a free-radical chain reaction, tending to produce excess ferric ion in proportion to the original peroxide concentration and the oxygen content. Negative errors arise from several chain reactions in which RO· radicals can be consumed without a corresponding oxidation of ferrous ions and, therefore, without production of the theoretical 2 moles of ferric ion per mole of peroxide. The nature of the solvent has a very large effect on these negative factors, as various types of solvent molecules participate in the several chain reactions to differing degrees. Alcohols are doubly harmful because they can convert a portion of the ferric ion product back to ferrous, by a reaction in which alcohol free radicals are oxidized to carbonyl compounds. Other workers have confirmed obtaining lower results with alcohol than with acetone-based reagents (26,102).

As both positive and negative errors can occur in the presence of air, it is apparent that these may sometimes offset one another to yield approximately correct results fortuitously (195). When air has been rigorously excluded, however, only negative errors are possible, and unless they are eliminated, results by anaerobic ferrous methods must be low. Kolthoff and Medalia's procedures (93) based on deaerated aqueous acetone are the soundest attempt in this direction of which we are aware. Chapman and Mackay have used a similar anaerobic acetone reagent (26).

In view of the extremely complex reaction system in ferrous methods, with so many significant variables to be controlled (93), we recommend that volumetric ferrous *macro* methods be rejected in favor of others based on more tractable reagents. The use of ferrous methods continues to be widespread in trace colorimetric analyses (26,93,102,180) because of speed, simplicity, and sensitivity and because it may be more important to obtain relative peroxide levels than absolute concentrations. We discuss such applications in the following section.

3. Colorimetric and Photometric Methods

Although one may choose to view colorimetric peroxide methods as either instrumental or chemical in nature, it is justifiable to treat them as a fully independent class because of the characteristic way in which they are used. Analytical chemists usually turn to a colorimetric method in the special situation where they must estimate traces of peroxides in groups of samples, and frequently in limited portions. Such a problem typically involves the estimation of autoxidation products or of peroxide catalysts and crosslinking agents.

Methods for determining peroxides by colorimetric and photometric procedures are summarized in Table VC.

a. Colorimetric Ferrous Methods

About colorimetric ferrous methods there are two opposing schools of thought. One advocates striving to eliminate errors as thoroughly as possible. As we have seen in the section immediately above, gross errors of opposite sign can occur with these methods simultaneously. Kolthoff and Medalia feel that this is inherently undesirable, and that such errors should be systematically eliminated (93): positive errors by using only anaerobic systems, and negative errors by finding better solvents than acetone and trying additives such as bromide ion or maleic and fumaric acids that are known to suppress free-radical oxidations (123). There would appear to be distinct possibilities of improving stoichiometry to the point where more than a few types of autoxidized substrates would yield precise and fairly accurate peroxide analyses.

The other position, taken by Lea (102), is that there is "... little prospect of so complicated a reaction being made truly stoichiometric particularly when applied to fats of variable fatty acid composition oxidized under a variety of conditions." He feels that nothing is to be gained by excluding atmospheric oxygen, because the method is then less sensitive and less simple. Furthermore, the characteristically good precision of the reactions under many empirically fixed situations, including the presence of air, should make it possible to get accurate results when necessary, for a particular substrate, by calibrating the selected colorimetric ferrous procedure with an iodometric or other reliable method.

A procedure representing each position is detailed in Section V. That selected for use in the presence of oxygen is Smith's adaptation (180) of a method by Loftus-Hills and Thiel (111). We should mention also another adaptation of the latter method, due to Driver, Koch, and Salwin (44a). By changing from a (30:70) methanol–benzene to a (80:20) ethanol–

benzene solvent system, these workers found that color developed satisfactorily at room temperature, so that heating was unnecessary. This circumstance permitted a considerably simplified procedure, with color development directly in a colorimeter cuvet. The method has been evaluated by Hamm and Hammond (69a).

b. Leuco Chlorophenolindophenol

Two careful evaluations (72,102) of Hartmann and Glavind's method (73) based on 3,5-dichloro-4,4′-dihydroxydiphenylamine (leuco dichlorophenolindophenol) have shown it to behave similarly to the ferrous method, with results severely high in the presence and low in the absence of oxygen. Precision was much inferior to that of preferred aerobic ferrous methods (102), and sensitivity only one-fourth as great, with respective molar absorptivities of 7300 vs 28,200 liters/mole-cm (72,180). The leuco chlorophenolindophenol method accordingly is not recommended for general use.

c. *N*,*N*′-Diphenyl-*p*-phenylenediamine

The best of the aromatic diamine reagents appears to be *N*,*N*′-diphenyl-*p*-phenylenediamine, which has been recommended by Ryland (161). The first thorough evaluation of this type of reagent was by Ueberreiter and Sorge (192), who developed a procedure based on the oxidation of 4,4′-diaminodiphenylamine to the indamine dye, phenylene blue:

$$NH{=}C_6H_4{=}N{-}C_6H_4{-}NH_2$$

They found this method inferior to their well-known leuco methylene blue method (discussed below in Sections IV-A-3-d and IV-B-6), because rate of formation of indamine and fading due to solvolysis both depended strongly on the effective acid strength of the solvent medium. Molar absorptivity (at 640 mμ) varied similarly; with fluorenone peroxide, for example, it increased from 28,000 to 42,000 as the acetic acid was modified with 5% trichloroacetic acid, and to 54,000 with the further addition of 1% water.

With *N*,*N*′-diphenyl-*p*-phenylenediamine, Ryland (161) found good sensitivity, good reagent stability, and low blanks under her preferred conditions (given in Table XII). Color development varied with solvent composition, however; for example, *tert*-butyl hydroperoxide gave the following molar absorptivities in media containing 65 vol % of the indicated solvents:

Auxiliary solvent	Molar absorptivity (liters/mole-cm)
Cyclohexane	25,000
Methyl alcohol	21,000
Acetone	10,500
None (all benzene)	24,000

In benzene all the peroxides studied obeyed Beer's law.

Table XII summarizes other data for the benzene system. It is clear that the stoichiometry of the color development, a one-electron oxidation of the diamine to the acid-stable semiquinone free radical (109,161), is inexact in this system and variable among the peroxides, although constant for a given peroxide. In particular, three of the peroxides in the table formed more than the theoretical 2 moles of semiquinone per mole of peroxide, thus giving apparent molar absorptivities above the nominal limit of 27,400 liters/mole-cm. This behavior is similar to that observed with ferrous thiocyanate (93,102) and leuco dichlorophenolindophenol (72,102) in the presence of air, and probably has a similar cause, namely, induced formation of hydroperoxide by dissolved oxygen during the free-radical chain reactions that are part of the peroxide reduction process. (See further discussion in Section IV-B-4.)

TABLE XII

Colorimetric Determination of Peroxides with *N,N'*-Diphenyl-*p*-phenylenediamine[a]

Peroxide	Rate of color development[b]	Molar absorptivity[c,d] (liters/mole-cm)
Succinyl peroxide	Rapid[e]	17,600
Benzoyl peroxide	Rapid[e]	24,400
tert-Butyl hydroperoxide	Rapid[e]	24,400
Cumene hydroperoxide	Slower[f]	24,400
Lauroyl peroxide	Slower[f]	32,600
tert-Butyl perbenzoate	Slower[f]	35,500
tert-Butyl peracetate	Rapid[e]	54,000
Di-*tert*-butyl peroxide	No reaction	—

[a] Data and procedure of A. L. Ryland (161).

[b] 2 ml of 0.05*M* diamine and 2 ml of 0.18*M* trichloroacetic acid in benzene, plus the peroxide sample in benzene, mixed at ambient temperature in a 25-ml volumetric flask (wrapped in aluminum foil) and diluted to the mark with benzene.

[c] At 700 mμ; there is another band at 385 mμ which has some 50% greater sensitivity.

[d] Theoretical value is 27,400 based on the figure of 13,700 for one-electron oxidation of the diamine to the semiquinone (109).

[e] Almost instantaneous, color fully developed in 5 min.

[f] A little slower, but full color developed within 30 min.

In order to appraise realistically the role of aromatic diamine reagents in trace peroxide analysis, analytical chemists need to study the effects of deaeration, overlooked previously (46,161,192). If color development were found grossly low in the absence of oxygen, paralleling the behavior of ferrous thiocyanate and leuco dichlorophenolindophenol reagents, then one should rate *N,N'*-diphenyl-*p*-phenylenediamine (the best of the aromatic diamines) as approximately equivalent to but not significantly better than the best ferrous thiocyanate procedure (102,180), whether in speed, simplicity, convenience, precision, or sensitivity.

Ryland evaluated several other diamine reagents (161). *N,N'*-Dimethyl- and *N,N,N',N'*-tetramethyl-*p*-phenylenediamines were relatively rapidly oxidized by atmospheric oxygen, and the latter was much slower in color development than the *N,N'*-diphenyl derivative; neither offered any advantage. Methylene-4,4'-bis(*N,N*-dimethylaniline), also known as tetramethyl base and described as a spot test reagent for peroxides (58), showed no color development in solution under conditions similar to those used with the other diamines. If the reaction mixture was evaporated on filter paper, however, the characteristic blue color developed very rapidly, as in the original spot test.

d. Leuco Methylene Blue

The colorimetric determination of trace concentrations of organic peroxides by reduction with leuco methylene blue, developed by Ueberreiter and Sorge (181,192), has two characteristics highly to be prized in such methods. One is a relatively high sensitivity, about threefold better than that of the ferrous thiocyanate (180), aromatic diamine (161,192), and iodide colorimetric methods (45,75). The second is an essentially constant molar response for all reactive peroxide types—a property documented by statistical analysis of a mass of data (181) obtained over about a tenfold range of concentrations on seven peroxides that included four ketone peroxides, two diacyl peroxides, and an alkyl hydroperoxide. We calculated that the average molar absorptivity in this group was 77,200, with a 95% confidence interval of $\pm$3900 liters/mole-cm. Among other important colorimetric peroxide methods, only the iodometric procedures (to be discussed in the following section) have a similarly constant absolute response. This behavior is valuable because it allows a general calibration to be obtained from any conveniently available known peroxide.

Along with this sensitivity and absolute accuracy is a precision quite adequate for a trace method. The pooled relative standard deviation of Ueberreiter and Sorge's data (181), with 43 degrees of freedom, is 1.85%.

This indicates at the 95% confidence level that duplicate determinations should differ by not more than 5.2% for peroxide numbers down to 0.025. In addition, the procedure is rapid and simple, and can be carried out in the presence of air. It is applicable to benzene-soluble samples, including polymers, and color development for slower reacting peroxides can be accelerated safely by heating briefly (47,181).

An excellent description of this method by Martin (117) includes an improved procedure for the phenylhydrazine reduction of methylene blue (101) (by which the reagent is prepared) and a careful presentation of Ueberreiter and Sorge's original procedure.

Despite all the foregoing advantages of this method, many analytical chemists in need of reliable trace peroxide analyses have turned away from it because of difficulties in preparing suitably pure leuco methylene blue reagent and in storing its dilute benzene solution. For instance, although the presence of oxygen seems harmless during color development, both oxygen and water vapor cause deterioration of the dilute reagent solution during storage and must be rigorously excluded; the leuco methylene blue must be prepared, recrystallized, and dried under a similarly inert atmosphere (181).

It is our view that this method has too high a potential value to warrant being dismissed because of the cited difficulties. It would seem a better course, rather, to try to develop preparation and handling techniques that would render these difficulties tolerable. Steps taken in this direction, as well as an interesting comparison with the benzoyl leuco methylene blue method of Eiss and Giesecke (52), are discussed in Section IV-B-4.

e. *sym*-Diphenylcarbohydrazide

Earlier work on the use of this reagent in the colorimetric detection and estimation of peroxides has been summarized by Hamm and co-workers (69b). Stamm originally proposed the use of diphenylcarbohydrazide (182a) for detecting rancidity in oils, suggesting that the red color resulted from reaction with acids, aldehydes, and ketones. Hartmann and Glavind proposed, however, that the basis of the Stamm test was the oxidation of diphenylcarbohydrazide to diphenylcarbazone by hydroperoxide:

$$\text{Ph}\cdot\text{NH}\cdot\text{NH}\cdot\text{CO}\cdot\text{NH}\cdot\text{NH}\cdot\text{Ph} + \text{ROOH} \rightarrow \text{Ph}\cdot\text{N{:}N}\cdot\text{CO}\cdot\text{NH}\cdot\text{NH}\cdot\text{Ph} + \text{ROH} + \text{H}_2\text{O} \quad (16)$$

since its results paralleled their indophenol test (73).

Hamm and co-workers have come up with a sensitive and apparently soundly based quantitative method for determining peroxides in fats and oils by use of the diphenylcarbohydrazide reagent (69b); their method should be seriously considered by anyone trying to select a suitable colorimetric peroxide method. The reaction is carried out in (80:20) tetrachloroethane–acetic acid containing 0.1% of purified 1,5-diphenylcarbohydrazide; a 3-min heating period in boiling water is specified, to develop the color. Absorbance is measured at 565 mμ.

It has been shown that diphenylcarbazone is the reaction product in this modified Stamm test, and that in the absence of oxygen each mole of peroxide (lauroyl peroxide) produced 1 mole of this derivative. When oxygen was not excluded, each mole of peroxide oxidized 2.631 moles of diphenylcarbohydrazide. When this factor was included in the calculation, agreement was excellent between analyses obtained by the colorimetric method and by an iodometric method, on oxidized soybean oil (69b).

The absorbance is quite temperature sensitive, decreasing about 1.5% for a temperature increase of 1°C. Color intensity should be read within 30 min of heating.

Hamm and Hammond have made a comparison (69a) of an iodometric procedure, the modified Stamm method, and the colorimetric iron method of Driver and co-workers (44a), calibrating the latter two with lauroyl peroxide (these produced, respectively, 2.63 and 1.46 moles of oxidized reagent per mole of peroxide). For the analysis of milk fat, corrected values were in excellent agreement by the three methods. The iron method could determine peroxide numbers down to 0.009, while the modified Stamm test could go down to 0.003.

f. Colorimetric Iodide Methods

The range of peroxide analyses via the well-established iodide reduction methods can be extended to concentration levels well below the practical limits of volumetric procedures by estimating the liberated iodine colorimetrically, either directly (5,10,75,168) or indirectly (45). Most widely used (10,75) has been reaction at room temperature with potassium iodide in 2:1 acetic acid–chloroform, a reagent system similar to that commonly employed for the volumetric determination of peroxides in fats and oils (104,145), and with the same limitations as to types of peroxides that may be determined; for example, there was no trace of reaction by dicumyl or di-*tert*-butyl peroxide (10).

The absorbing species in these procedures is the triiodide ion (75), the form in which iodine exists quantitatively at iodide/iodine ratios of 5 and larger (75). The absorption has a molar absorptivity of 22,700 liters/mole-cm in the acetic acid–chloroform solvent, 0.05M in potassium iodide, with a maximum at 362 mμ.

In practice, none of the colorimetric iodine methods use measurements at the absorption peak, partly because borosilicate glass has a slight absorption that would necessitate careful cell matching (a difficulty readily eliminated by the use of fused quartz or other high silica cells) and partly because absorbances would usually be too high. Heaton and Uri's procedure (75) was to carry out their photometry at a wavelength between 380 and 500 mμ where the triiodide solutions from their samples had absorbances in a desirable range, and then convert these values by means of the absorptivities listed in Table XIII into equivalent absorbances at

TABLE XIII

Molar Absorptivities of Triiodide in Acetic Acid–Chloroform (2:1) (75)

Wavelength, mμ	Molar absorptivity, liter/mole-cm	Wavelength, mμ	Molar absorptivity, liter/mole-cm
380	17,900	450	1610
390	12,300	460	1150
400	7,900	470	814
410	5,300	480	553
420	3,800	490	379
430	2,800	500	269
440	2,160	—	—

400 mμ, where the calibration had been established. Calibrations based on standards of either iodine or linoleic acid hydroperoxide were equivalent, and Beer's law was followed up to iodine concentrations of $5 \times 10^{-4}M$. To eliminate detectable blank values, a special cell was used in which the operations of deaeration, sample reduction, and triiodide spectrophotometry were carried out.

Banerjee and Budke, in modifying this method, chose to set up two procedures covering the 0–1 and 1–10 ranges of peroxide number, the first calibrated with triiodide absorption at 410 mμ, the second at 470 mμ (10). For the lower range they used a specially constructed cell related to that of Heaton and Uri, and maintained positive nitrogen pressure throughout the reaction, but for the higher range they used ordinary volumetric flasks and measured the absorbance (rapidly) in regular 1-cm cells. Evaluation of the method showed satisfactory recoveries for 17 commercial peroxides representing four easily reduced types.

Heaton and Uri's careful work demonstrated in both the volumetric assay and trace colorimetric ranges that scrupulous elimination of atmospheric oxygen can reduce iodometric blank values to the vanishing point, at least in methods for easily reduced peroxides. To bubble nitrogen through a solution for an hour in order to effect deaeration, as these workers did, appears excessively cautious. Banerjee and Budke obtained acceptable results even at the lowest concentration levels using a total deaeration period of only 6 min. Their blanks corresponded to $<2\ \mu g$ active oxygen, or a peroxide number of <0.05 on a 5-ml sample (10).

As with the volumetric methods, it is evident that analytical chemists using colorimetric iodide reduction procedures for determining peroxides have much leeway in devising variants that best fit their requirements of speed, simplicity, precision, accuracy, and type of sample. An interesting example is Anton's adaptation for determining hydroperoxides in nylon yarn (5); sample and potassium iodide–acetic acid reagent were reacted in tetrafluoropropanol at room temperature, and triiodide absorbance measured at 400 mμ—relative standard deviation was $\pm 5\%$ and the sensitivity 1 μmole/g.

There have been no published colorimetric versions of the methods for difficultly reduced peroxides. If such methods are needed, however, their development should be straightforward. It would be important to eliminate the traces of reducing impurities in acetic acid that are believed to be responsible for most of the iodine liberation in blanks with adequately vigorous reagents (116); refluxing over powdered dichromate, as described by Lea (102), might be suitable. If there should be significant interference from highly colored sample substrates, this might be eliminated by an appropriate extraction with an immiscible solvent, after the peroxide reduction.

Dubouloz and co-workers, in determining lipid peroxides, have used extensively a colorimetric method that employs indirect estimation of the liberated iodine. In its modern version (45) this method appears quite practical. The reaction medium, two parts of 10% benzoic acid in chloroform and one part of methanol, contains a standard (excess) amount of an intensely colored compound, thiofluorescein, that is quantitatively decolorized by iodine; in a 5-min reaction at 70°C in nitrogen-purged centrifuge tubes, peroxides are reduced by potassium iodide, a proportionate amount of thiofluorescein is consumed, and the remaining color then extracted into 10.0 ml of 10% aqueous sodium carbonate containing cyanide or Complexon III and estimated by absorbance at 590 mμ after centrifuging. Peroxide numbers are obtained from the absorbance decrease relative to a blank run.

This indirect determination, although effective, has no real advantage over direct triiodide spectrophotometry. Sensitivities are about equal, with thiofluorescein decoloration by iodine corresponding to a molar absorptivity of about 25,000 and triiodide absorption at 362 mμ to about

23,000 liters/mole-cm. Supposed advantages, based on relative freedom from iodine losses due to either volatility or reabsorption by olefinic substrates, can be discounted because of the excellent quantitative record of the direct methods (10,75) and the results of a study by Banerjee and Budke (11) of trace peroxide analyses in olefins and other unsaturated solvents.

In this connection we wish to emphasize our conviction (discussed in Section IV-B-3-c) that loss of iodine through volatility or reabsorption by olefins does not occur when an adequate excess of iodide is present, although many have considered such errors to be potential hazards (33,45, 48,192,196).

g. Other Colorimetric Methods

A spectrophotometric method for determining hydroperoxides in autoxidized fats and oils, based on the quantitative and selective conversion of triphenyl phosphine to triphenyl phosphine oxide has only recently come to our attention. Stein and Slawson (184a) have developed a simple, apparently widely applicable procedure in which the reagent consists of 0.1% triphenyl phosphine and 0.01% di-*tert*-p-cresol (an antioxidant) in isopropyl alcohol. Reaction is quantitative in 20 min at room temperature. Hydroperoxide is estimated from the *decrease* in absorbance at 260 mμ, as triphenyl phosphine (molar absorptivity 11,000 liters/mole-cm.) is converted to triphenyl phosphine oxide (molar absorptivity 1,520 liters/mole-cm.). Air is not excluded; the antioxidant prevents detectable air oxidation of the reagent for as long as 3 hrs. The method was applied successfully at from 0.5 to over 1500 meq peroxide per kg oil; results were only slightly lower than those by an iodometric procedure.

Formation of a yellow complex between titanium(IV) and hydrogen peroxide has long been known as a moderately sensitive method for the colorimetric estimation of either the metal ion or the peroxide (95). It can also determine those organic peroxides that first decompose to hydrogen peroxide by hydrolysis (147,163,213), principally hydroperoxides and peroxy esters.

Martin has discussed a number of other colorimetric methods (117), but these have relatively little utility. They include the oxidation of phenolphthalin to phenolphthalein by hydrogen peroxide (138), and color development from a reagent consisting of vanillin in 70% sulfuric acid (6).

A recent method for determining hydroperoxide groups in oxidized polyethylene, by Mitchell and Perkins (133), differs from conventional colorimetric methods in using absorption of infrared rather than visible radiation as the final quantitative measuring step. Reaction of oxidized polyethylene films with gaseous sulfur dioxide for as little as 10 min at room temperature gave apparently quantitative conversion of hydroperoxide groups (which absorb at 3520 cm^{-1}) into dialkyl sulfate groups (with a principal absorption at 1195 cm^{-1}). Presumably these groups arise through formation of an alkyl hydrogen sulfate intermediate by the reaction (134):

$$ROOH + SO_2 \rightarrow ROSO_3H \tag{17}$$

followed by esterification with hydroxyl groups present in the oxidized polymer film. The method was calibrated with dilauryl sulfate. A sensitivity of 2 ppm —OOH was observed, with a potential limit of 0.1 ppm by use of thicker specimens. This is due to the relatively high absorptivity of the 1195 cm^{-1} sulfate band, which was consistently 14-fold stronger than that of the 3520 cm^{-1} hydroperoxide band. Extension to other polyolefins was suggested (133).

Purcell and Cohen have studied the microdetermination of peroxides by kinetic colorimetry (148a), a technique wherein the rate of color development aids in identifying the peroxides present in a sample. They were concerned particularly with the classes of oxidants found in polluted atmospheres typical of smog conditions. From a group of procedures evaluated for applicability to atmospheric oxidants (26a), they selected three reagents for their kinetic studies: ferric thiocyanate, neutral potassium iodide, and molybdate-catalyzed potassium iodide. All three reagents gave immediate maximum color development with ozone and peracetic acid, and more gradual color development with "slow oxidants"—acetyl peroxide, nitrogen dioxide, alkyl hydroperoxides, and peroxyacyl nitrates. These slow oxidants were evaluated by plotting the undeveloped absorbance ($A_\infty - A$) vs time; single-component systems gave the straight line of a pseudo-first-order reaction, with a characteristic half-time. Hydrogen peroxide gave slow color development only with neutral potassium iodide; addition of the molybdate catalyst caused immediate full reaction.

B. FURTHER DISCUSSION OF SELECTED TOPICS

1. The Infrared Spectrum of the Hydroperoxide Group

Hydroperoxides possess a molecular structure property that permits a relatively sophisticated treatment of their spectral properties: The —C—O—O—H group is electronically isolated from the rest of the molecule to a degree sufficient to give it a characteristic set of vibrational frequencies and infrared absorption bands.

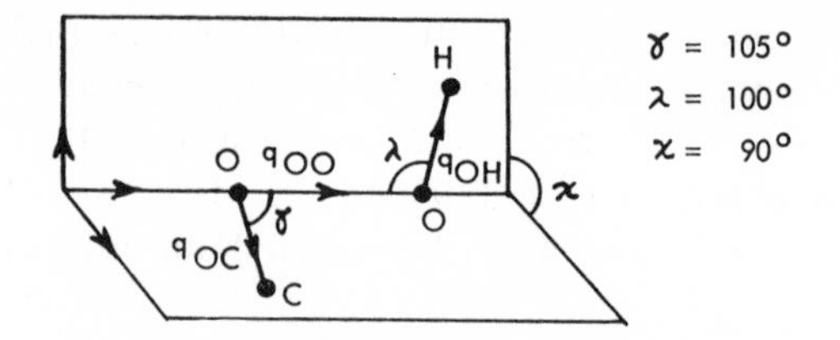

Fig. 3. Vibrational coordinates for the model molecule, COOH.

Coordinate	Assigned stretching (ν) and bending (δ) vibrations
q_{CO}	ν_{CO}
γ	δ_{COO}
q_{OO}	ν_{OO}
λ	δ_{OOH}
q_{OH}	ν_{OH}
χ	$\delta_{\text{dihedral angle}}$

Kovner, Karyakin, and Efimov (98) and Gribov and Karyakin (66) have presented the clearest work on this subject. In both studies the hydroperoxide group was treated as a separate four-atom entity, C—O—O—H, whose vibrational motions were described in terms of the coordinates depicted in Fig. 3; such an assemblage should have six fundamental frequencies (i.e., $3n - 6$ for an n-atom molecule). Bond distances and angles were assumed to be those known for performic acid, and a value of 90° was used for the dihedral angle. Six frequencies from the spectrum of cumene hydroperoxide, including a newly observed intense band at 585 cm^{-1} (17.09 μ), and five from deuterated cumene hydroperoxide were used as fundamental frequencies of the hypothetical COOH and COOD molecules; the requisite set of 21 force constants was then devised from these frequencies, from analogous force constants in methyl alcohol, from certain reasonable zero-value assignments, and from determination by the variational method (for the diagonal force constants). From these parameters, Kovner and associates (98) calculated vibrational frequencies for the four-atom models that were in excellent agreement with those observed for cumene hydroperoxide and deuterohydroperoxide, as can be seen in Table XIV.

Gribov and Karyakin (66) extended this work by refining the calculation of frequencies for the COOH model and computing the configurations of its normal vibrations, along with the relative position shifts of all atoms during the vibrations. These results are summarized in Table XV. Of particular interest is the finding that the 880 cm^{-1} (11.36 μ) band previously assigned to a simple —O—O— stretching vibration is a more complex

TABLE XIV

Hydroperoxide Vibrational Frequencies, Calculated and Observed (98)

Assigned vibration	Vibrational frequency, cm⁻¹ Calc	Obs[a]	Assigned vibration	Vibrational frequency, cm⁻¹ Calc	Obs[b]
δ_{COO}	586	585	δ_{COO}	549	—
χ^c	834	840	χ^c	680	—
$\nu_{O—O}$	886	880	$\nu_{O—O}$	873	855
$\nu_{C—O}$	1156	1155	$\nu_{C—O}$	1156	1155
δ_{OOH}	1321	1325	δ_{OOD}	997	995
ν_{OH}	3450	3450	ν_{OD}	2539	2550

[a] In the absorption spectrum of cumene hydroperoxide.
[b] In the absorption spectrum of cumene deuterohydroperoxide.
[c] Vibration of the dihedral angle.

mode having important contributions also from COO bending and from flapping of the dihedral planes.

Going another step forward, Gribov and Karyakin made an evaluation of the band intensities, applying an approximation of a general method developed by Gribov (65). They obtained the very satisfactory results shown in Table XVI.

TABLE XVI

Calculated Intensities of the COOH Group Vibrations (66)

Calc frequency, cm⁻¹	Calculated dependence of band intensity[a,b,c]	Relative intensities Calc	Obs[d]	Usual qualitative intensity evaluation
3450	$\lvert\Delta\mu\rvert^2 = 4\left(\frac{\partial\mu_{OH}}{\partial q_{OH}}\right)_0^2$	—	—	very strong
1322	$\lvert\Delta\mu\rvert^2 = 9\mu_{OH}^2$	9	9.5	strong
1155	$\lvert\Delta\mu\rvert^2 = \left(\frac{\partial\mu_{OH}}{\partial q_{OC}}\right)_0^2$	—	—	very strong
883	$\lvert\Delta\mu\rvert^2 = \mu_{OH}^2 + 0.36\mu_{OC}^2$	1.2	1.2	weak
835	$\lvert\Delta\mu\rvert^2 = 9\mu_{OH}^2$	9	9.5	strong
588	$\lvert\Delta\mu\rvert^2 = 4\mu_{OH}^2 + 0.49\mu_{OC}^2$	4.3	11	average

[a] By zero order approximation of method of Gribov (65).
[b] Absorption intensity is given by the square of the change in absolute value of the molecular dipole moment, $\lvert\Delta\mu\rvert^2$.
[c] $\mu_{XY} \equiv$ dipole amount of the bond X—Y.
[d] For cumene hydroperoxide.

TABLE XV

Calculated Frequencies and Normalized Configurations of the COOH Group Vibrations (66)

Frequencies of the normal vibrations			Coordinates significantly involved[b]	Amplitudes of the variations in the vibrational coordinates					
Calculated									
Gribov et al. (66)	Kovner et al. (98)	Observed[a]		q_{CO}	γ	q_{OO}	λ	q_{OH}	χ
3450	3450	3450	q_{OH}	0	0	−0.01	+0.01	1.03	0.01
1322	1321	1325	λ	0.03	−0.01	0.03	−1.14	0.05	−0.06
1155	1156	1155	q_{CO}, γ	0.39	−0.12	−0.09	0.09	0	0.02
883	886	880	γ, q_{OO}, χ	0.03	−0.30	−0.30	−0.03	0	−0.46
835	834	840	χ	0.02	−0.01	0.15	−0.05	0	1.08
588	586	585	γ, χ	0.03	0.30	0.10	−0.04	0	0.18

[a] For cumene hydroperoxide.
[b] Defined in Fig. 3.

For additional support of this concept that the hydroperoxide group is an electronically isolated molecular appendix that has a set of six essentially independent fundamental frequencies, we have examined closely the 17 spectra of alkyl hydroperoxides presented by Williams and Mosher (207). The frequencies in Table XVII are our assignments of the five fundamental COOH absorption bands accessible in these spectra. Karyakin has made a similar comparison (88).

2. Polarographic Analysis

Martin (117) published an excellent review of the polarographic peroxide literature up to 1960. Johnson and Siddiqi (86) summarized the available information through 1964, with an extensive compilation of half-wave potentials. These are key references, along with a paper by Romantsev and Levin (159), who presented new information on the effect of solvent and supporting electrolyte. For details of the general theory and practice of polarography, we refer the reader to Part I of the Treatise, Vol. 4, Chapters 43 and 46, and to Meites's recent textbook (122).

In the following discussion, our aim is to survey the present body of knowledge on peroxide polarography in such a way that one wishing to employ this technique in the solution of a specific problem can either select a suitable set of conditions from among the referenced studies, or plan an appropriate methods development program to work out such conditions for himself.

a. Experimental Factors in the Polarography of Peroxides

It appears to have been as difficult to settle on standard optimum procedures for polarographic analysis as on standard chemical methods, and for some of the same reasons—namely, that the interest of many workers has been limited to peroxides in only a few classes (e.g., water-soluble or water-insoluble members) and that for many applications there are many variations in procedure that work well or reasonably well.

A major factor in selecting the polarographic solvent medium is solubility of the peroxide or the substrate containing it. Water is the usual starting point (19,179) and methanol or ethanol the usual modifying solvent (60,152,159,179), although dioxane has also been used (164) and ethylene glycol monoalkyl ethers undoubtedly would serve (143). Dimethylformamide is a very effective nonaqueous polarographic solvent (57,159,202). The most popular nonaqueous system has been $0.3N$ lithium chloride in 1:1 benzene–methanol; devised by Lewis, Quackenbush, and DeVries (107), this has been an almost exclusive choice for

TABLE XVII

COOH-Group Frequency Assignments for Some Alkyl Hydroperoxides

A. Vibrations of the General COOH Group[a]

Vibrational mode:	ν_{OH}	δ_{OOH}	ν_{CO}	ν_{OO}	χ	δ_{COO}
Significant coordinates:[b]	q_{OH}	λ	q_{CO}, γ	γ, q_{OO}, χ	χ	γ, χ
Calculated frequencies:	3450 cm^{-1}	1322 cm^{-1}	1155 cm^{-1}	883 cm^{-1}	835 cm^{-1}	588 cm^{-1}
Calculated infrared intensities:	very strong	strong	very strong	weak	strong	average

B. Observed Vibrational Frequencies of the Hydroperoxide Group[c]

Hydroperoxide	Infrared absorption band assignments, cm^{-1}					
Cumene	3450	1321	1156	880	840	585
n-propyl	3400	—	—	—	820	—
n-butyl	3380	∽1340	1130	884	821	—
Isoamyl	3390	∽1340	1145	883	822	—
n-amyl	3380	∽1340	1135	890	818	—
n-hexyl	3400	∽1350	1135	893	827	—
n-heptyl	3430	∽1350	1135	905	816	—
n-octyl	3380	∽1350	1130	890	822	—
n-nonyl	3370	∽1350	1125	890	816	—
n-decyl	3360	∽1350	1130	890	821	—
2-butyl	3380	1330	1135	905	857	—
Cyclopentyl	3390	1350	1175	905	827	—
2-pentyl	3380	1350	1140	905	813	—
3-pentyl	3450	1350	1135	—	866	—
2-hexyl	3410	1340	1150	895	833	—
3-hexyl	3390	1360	1140	895	813	—
2-heptyl	3360	1330	1145	895	855	—
2-octyl	3440	1340	1150	—	830	—

[a] After Gribov and Karyakin (66).

[b] Defined in Fig. 3.

[c] Assigned from spectra of Williams and Mosher (207) and of Kovner et al. (98).

polarographic studies of fat and oil peroxidation products (87,100,106,149, 153,189,211), and has been used for characterizing the nonaqueous behavior of many peroxides (99,159,174,210).

Proske took another approach to water-immiscible peroxides and substrates (148) by solubilizing them with relatively high concentrations of polarographically suitable commercial wetting agents (specifically American Cyanamid's Aerosols AY and MA, which were said to be diamyl and dihexyl sodium sulfosuccinate, respectively). With fats, the detergent had to provide all of the supporting electrolyte; benzyl alcohol was used as a solubilizing aid.

Bernard studied the polarographic behavior of more than 25 peroxides of many types (16) and in 14 cases obtained comparative data between aqueous 0.1N sodium chloride, potassium chloride, or potassium sulfate and the nonaqueous system of 0.3N lithium chloride in 1:1 benzene–methanol. When studying water-immiscible substrates with the aqueous system his technique was to extract water-soluble components into 0.1N potassium chloride. In general, his results are compatible with those of other workers, although they have not been included in a recent compilation (86).

One unique feature of Bernard's data was the appearance of a second reduction wave at about −1.1 V for the aqueous extracts of such diverse materials as ammonium persulfate, *tert*-butyl perbenzoate, and peroxidized citral, tetralin, terpinolene, and dicyclopentadiene. In each case he attributed this wave to hydrogen peroxide released by hydrolysis of the corresponding hydroperoxides; he based this identification on observations that the characteristic wave appeared in the extract only after some time and that it was enhanced by additions of hydrogen peroxide (16). It would appear worthwhile to confirm this identification by one of the techniques for removing hydrogen peroxide, either as a titanium (114) or a lanthanum (19) complex.

For the polarography of difficultly reduced compounds, Forman and Wright found dimethylformamide saturated with tetramethylammonium chloride to be a particularly good system (57) because its useful range of reduction potential extended to −2.5 V (vs a mercury pool), well beyond the usual practical limit of −2.0 V. In this system all dialkyl peroxides examined were successfully reduced. These compounds and their half-wave potentials are shown in Table XVIII. It is particularly noteworthy that di-*tert*-butyl peroxide was reduced at an $E_{1/2}$ of −2.10 V with a well-defined wave that was useful for quantitative analysis. We are not aware that the polarographic reduction of this peroxide has been previously reported.

TABLE XVIII

Half-Wave Potentials of Some Dialkyl Peroxides in Dimethylformamide (57)

Compound	$E_{1/2}$ vs Hg pool,[a] V
Di-*tert*-butyl peroxide	−2.10
	−1.93
tert-Butyl-1,1,3,3-tetramethylbutyl peroxide	−1.93
Dipinanyl peroxide	−1.88
Di-*p*-menthyl peroxide	−1.75
Bis(4-isopropylcumyl) peroxide	−1.75
Bis(3,4-dichlorocumyl) peroxide	−1.75
Dicumyl peroxide	−1.60
Bis(2,4,6-tri-*tert*-butylcyclohexadienone) peroxide	−1.37

[a] In dimethylformamide saturated with tetramethylammonium chloride.

Romantsev and Levin (159) found that proton availability in a solvent had a profound effect on the polarographic reduction wave for most types of peroxides. Decreasing protogenic character in the series water (or dilute ethanol), 95% ethanol, 1:1 benzene–methanol, dimethylformamide was paralleled by increasingly irreversible, more elongated wave forms and substantial shifts in half-wave potential toward more negative voltages. An example is the shift in $E_{1/2}$ for dibenzoyl peroxide to −0.4 V in dimethylformamide, the only system in which this peroxide has a reduction potential more negative than −0.1 V.

This effect of solvent composition on irreversibility tends to be offset by adsorption of a peroxide on the mercury electrode, a phenomenon indicated by an increase in drop rate when the peroxide is present. Such adsorption was found for ethyl cumyl peroxide in 20% ethanol and for cumene hydroperoxide in 20% and 95% ethanol (159), and they showed relatively sharp, reversible reduction waves. Bernard (16) and Skoog and Lauwzecha (179) found a similar dependence of wave characteristics on solvent composition.

A variety of salts and acids have been used as supporting electrolytes. Lithium chloride has been a frequent choice because lithium ions allow polarography in a more negative potential range than do sodium or potassium ions. Chloride has been the most widely used anion, although it would appear to be undesirable for aqueous studies of peracids, peresters, and diacyl peroxides because of its ease of oxidation at positive voltages. Both the lithium sulfate electrolyte used by Bruschweiler and Minkoff (19) and an acetate (141) appear to be better choices for reductions of such peroxide classes. Alkaline systems should generally be avoided because they tend to decompose some peroxides. Many of the common buffers have been used (164,179) but with little apparent benefit. If their use is contemplated, however, one should be alert for effects on adsorption equilibria at the electrode in addition to the effects sought from control of proton concentration.

Tetraalkylammonium salts as supporting electrolytes have significant adsorption effects that are well known (144). These quaternary cations exert a screening action that can markedly influence electrode reaction mechanisms involving adsorption and desorption. Romantsev and Levin (159) found $E_{1/2}$ values for peroxides to be consistently shifted toward the negative side whenever these electrolytes replaced alkali salts.

Many peroxides are highly reactive compounds and may react chemically with components of the polarographic system. For example, aliphatic peracids react with the methanol of the widely used 1:1 benzene–methanol solvent (141) (switching to a system of ammonium acetate in glacial acetic acid eliminated this difficulty); peracids (19) and hydroperoxides (16) tend to hydrolyze in aqueous systems to produce hydrogen peroxide; and peroxide decomposition has been attributed to agar agar, gelatin, sintered glass, and mercury (19)—a special cell being designed to minimize such effects (other workers have not reported such difficulties, however).

A pair of small early waves occur in the $0.3N$ lithium chloride, 1:1 benzene–methanol system with fats and oils (87,100,106) and with peracids and diacyl peroxides. Maack and Lück attribute these waves to mercury ions produced when reactive peroxides are reduced by mercury metal (113), reaching this conclusion from cathode ray polarographic studies of appropriate systems. Since these waves have long lacked a definite explanation, this limited examination is a welcome lead that should be followed up. In fact, a study of the reaction of the various classes of peroxides with mercury, as a function of solvent composition and the presence of promotors or inhibitors, would appear long overdue.

To avoid such reactions with mercury, Roberts and Meek (157) developed a vibrating double electrode of bright platinum. It was so arranged that polarities could be reversed at periods in the range of 0.2–5 sec, to help prevent buildup of reaction products. Polarograms were much more stable than with single vibrating electrodes, and were equivalent to those obtained with the dropping mercury electrode. It has been suggested (117) that such electrodes should be of value for continuous process stream analysis.

b. Quantitative Analysis

Although the reduction potential of a peroxide in a well-characterized polarographic system may be useful for qualitative identification, it is the diffusion current constant, I, that is important for absolute quantitative analysis. This is defined by

$$I = \frac{i_d}{Cm^{2/3}t^{1/6}} \tag{18}$$

where i_d is the diffusion current, m is the drop mass, t is the drop period, and C the concentration. The value of I determined for a compound in a particular medium by calibration with a particular capillary can be used to determine the compound quantitatively in the same medium from i_d values obtained on other polarographs with other capillaries. When pure specimens are not available for calibration, standards may be used whose concentration (C) has been determined by appropriate chemical methods.

Diffusion current constants vary both between and within classes of peroxides and are influenced by the composition of the media. Polarography hence must be considered ill-suited for absolute quantitative analysis of the peroxide functional group; calibration with the peroxide to be determined is necessary generally, despite data of Willits and co-workers (210) that show remarkably constant I values for a series of diverse low molecular weight hydroperoxides. Martin suggests (117) that by using values for I of 6.3 for aqueous solutions and 5.3 for 1:1 benzene–methanol, one can obtain order-of-magnitude estimates of peroxide concentration probably accurate to within 30%, without calibration.

These statements do not imply that quantitative peroxide analyses are inherently inaccurate, but rather that calibration is necessary. Indeed, as precision is generally good, even at low concentrations, it needs only calibration with pure samples of the peroxide mixture in question, or with reliable alternative methods of analysis, for polarography to give accurate quantitative analyses at concentrations down to less than $10^{-4}M$.

Most polarographers find it practical when possible to calibrate directly with individual pure peroxides or independently standardized samples (the comparative "standard solution" method described in Section V-B-4 of Chapter 46, in Part 1, Volume 4, of the Treatise) rather than to use absolute diffusion current constants determined independently or obtained from the literature; such direct calibration avoids the additional error from determining $m^{2/3}t^{1/6}$ for the currently used electrode capillary.

c. The Effects of Peroxide Structure

In peroxide analysis there is a close analogy between chemical and polarographic (electrochemical) methods: In the former, reduction of the —O—O— bond is accomplished by chemical reagents, while in the latter the reduction is carried out directly by electrons. In each case, however, it is the strength of the —O—O— bond relative to the chemical or electrical reducing potential of the system that determines the response of a peroxide.

This bond strength is related directly to peroxide structure, that is, to the nature of the one or two substituents by which the H—O—O—H progenitor has been transformed into a particular organic peroxide. In polarographic systems, the strength of the —O—O— bond in the peroxide will be the resultant from structural factors of two types, inductive and environmental.

The inductive effects originate in a substituent group, and should exert their influence independently of environment. Consider the expected behavior of organic peroxides relative to that of hydrogen peroxide. The latter is reduced with a half-wave potential, $E_{1/2}$, of about -1.0 ± 0.1 V (vs. a saturated calomel electrode) in a wide variety of media, as can be seen in Table XIX, where we have summarized the available information on effects of structure in peroxide polarography. Substituents that are electron donors should tend to stabilize the peroxide bond by inducing an increase in its electron density, thereby decreasing its electron affinity (164,174). Alkyl groups are well-known electron donors, with the *tert*-butyl substituent showing the greatest influence; it seems likely, therefore, that this factor is related to the general enhanced chemical stability of dialkyl peroxides and to their more negative half-wave reduction potentials relative to hydrogen peroxide (in anhydrous media). Electron-withdrawing groups such as aromatic and carbonyl substituents should have an opposite effect, and this may be exemplified by the large shift of $E_{1/2}$ for peracids to the positive side of 0.0 V and for diacyl peroxides to the -0.1 to -0.4 V range. In the *tert*-butyl peresters, which reduce with $E_{1/2}$ in the -0.8 to -1.0 V range in anhydrous systems, it would appear that the electron pushing and pulling effects are at a virtual standoff, although the peroxide bond in these compounds should be polarized.

In addition to these inductive effects there are environmental factors that act through the structure of a peroxide to influence its —O—O— bond strength. These are complex and have not been studied extensively. One is the effect of water content (apparent in Table XIX), which suggests that protonation is important at some point in the reduction mechanism (159). Another is the effect of adsorption processes at the dropping mercury electrode (159,164), with competition likely among a number of adsorption equilibria involving peroxides, their reduction products, solvent components, and supporting electrolyte. In many cases, these environmental factors are dominant.

Effects of inductive and environmental factors may be seen in Table XIX for some alkyl hydroperoxides, dialkyl peroxides, and *tert*-butyl peresters.

TABLE XIX

Variations in Polarographic Half-Wave Potential, $E_{1/2}$, with Peroxide Structure and Media Composition, for Several Peroxide Classes

Peroxide class	$E_{1/2}$ (volts vs saturated calomel electrode) in media of decreasing water content				
	Aqueous[a]	20% Ethanol[b]	95% Ethanol[c]	1:1 Benzene–methanol[d]	Ref.
Hydroperoxide					
HOOH	−0.88, −0.93	(−1.07, 50% MeOH)	−1.09	−1.14, (−1.16[e])	16, 19, 159
Methyl	−0.64, −0.79	—	—	—	19, 164
Ethyl	−0.42, −0.44	——	—	—	19, 164
Propyl	−0.35	0.49	—	−0.98	159, 164
n-Butyl	−0.20, −0.21				179
sec-Butyl	−0.28				179
tert-Butyl	−0.35, −0.34, −0.27, −0.28	(−0.74, 50% MeOH)	(−1.04, 90% MeOH)	−0.96, −1.10, (−1.15[e])	16, 19, 99, 179, 210
n-Pentyl	−0.08, −0.19, −0.20	−0.20	—	−0.95	159, 179
n-Hexyl	—	−0.08, −0.12	—	−0.94	159, 179
n-Heptyl	—	−0.03	—	—	179
Heptenyl	—	−0.04	—	−0.88	159
n-Octyl	—	−0.02	—	—	179
n-Nonyl	—	−0.01	—	—	179
Cumyl	−0.10, −0.12	+0.02	−0.54	−0.68, −0.82, (−1.08[e])	99, 159, 210

Dialkyl peroxide					
Diethyl	−0.5, −0.65, −0.63	—	—	−1.63	16, 19, 159
Ethylcumyl	—	−0.22	−1.64	−1.72	159
Dicumyl	—	—	(−1.35, 60% EtOH)	(−1.60, DMF[f])	57
Di-*tert*-butyl	—	—	—	(−2.10, DMF[f])	57
tert-Butyl perester					
Peracetate	−0.3	—	—	−0.97, (−1.02[e])	99, 117
Perbenzoate	positive	(−0.1 and −0.8, 80:20 dioxane–water[g])	—	−0.8, −0.82	16, 117, 164
Perpelargonate	—	—	—	−0.96	174
Percaprate	—	—	—	−0.90	174
Perlaurate	—	—	—	−0.87	174
Permyristate	—	—	—	−0.82	174

[a]Supporting electrolytes: 0.1M $LiSO_4$; 0.01N HCl; 0.1N KCl and 0.04M Britten-Robinson buffer pH 7.5; acetate buffer pH 4; phosphate buffer pH 9.

[b]Supporting electrolytes: 0.1N KCl, 0.1M H_2SO_4, 0.1F H_2SO_4.

[c]Supporting electrolytes: 0.3N LiCl, 0.1N KCl; 0.2N $(CH_3)_4NOH$ in 60% EtOH.

[d]Supporting electrolytes: 0.3N LiCl.

[e]0.03N LiCl + 0.01% ethyl cellulose.

[f]$E_{1/2}$ vs. Hg pool; dimethyl formamide saturated with $(CH_3)_4NCl$; reported irreducible in other systems.

[g]Ten substituted *tert*-butyl perbenzoates show dual waves at −0.1 and −0.8 V in dioxane–water; as water content increases, the first grows while the second diminishes (164).

With hydroperoxides in a nonaqueous medium, the nature of the R group in the R—O—O—H structure apparently has a relatively small effect on $E_{1/2}$ values; these are generally displaced (from that of hydrogen peroxide) 0.2–0.4 V in the positive direction—opposite to the negative shift anticipated from electron induction by the alkyl group. Evidently an overriding environmental factor is operative; moreover, a persistent trend toward more positive reduction potentials with increasing size of the R group shows that this environmental effect is being exerted through the substituent.

In the aqueous or substantially aqueous systems, positive-shifting factors for the hydroperoxides are much stronger, as is also their direct relation to R-group size. In these media, starting from hydrogen peroxide at —0.9 V, there are sharp positive shifts of $E_{1/2}$ with increasing size of the *n*-alkyl substituent up through *n*-hexyl, which has an $E_{1/2}$ of −0.1 V in 20% ethanol. Beyond this point the size effect tapers off quickly, with small regular, positive shifts up through *n*-nonyl hydroperoxide, for which $E_{1/2}$ in 20% ethanol is −0.01 V vs. the saturated calomel electrode.

It will be noted that in this system $E_{1/2}$ is more negative for *tert*-butyl than for *n*-butyl hydroperoxide, which is in line with the greater inductive effect of the *tert*-butyl group.

The effect of water content on half-wave potential is demonstrated throughout Table XIX. Data from a series of three or four solvent compositions are listed for propyl, *tert*-butyl, and cumyl hydroperoxides, ethyl cumyl peroxide, and *tert*-butyl perbenzoate. In every case, the higher the water content, the greater the positive $E_{1/2}$ shifts.

Only in anhydrous media are such environmental effects sometimes small enough not to overwhelm structural factors. In 1:1 benzene–methanol, for instance, dialkyl peroxides (but not hydroperoxides) can exhibit the negative shift relative to hydrogen peroxide reduction potentials anticipated through inductive effects of alkyl substitution.

In anhydrous media *tert*-butylperoxy esters are remarkably like alkyl hydroperoxides in their polarographic behavior; they reduce in the same potential range and have a similar persistent small shift to more positive $E_{1/2}$ values with increasing size of the acyl group. In aqueous media there is a very strong positive shift, with *tert*-butyl perbenzoate, for example, moving to the *positive* side of 0.0 V. Such reduction behavior is more typical of peroxy acids than hydroperoxides, and suggests that in aqueous media perester reduction may proceed through a peracid intermediate—this despite the generally accepted idea (129) that peroxy esters are derivatives of, and hydrolyze to, acids and hydroperoxides, rather than peroxy acids and alcohols.

In semiaqueous solvent media peroxy esters show more complex behavior. Schulz and Schwarz (164), working with $0.25N$ ammonium nitrate in dioxane–water solutions, found two waves for *tert*-butylperoxy esters—one at −0.85 V, which was diffusion controlled, and an earlier one at −0.1 V, which had a kinetic character; as they increased the water content of the medium this kinetic wave grew at the expense of the diffusion-controlled reduction, suggesting to them that the rate-controlling kinetic step giving rise to the early wave was the heterolytic polar cleavage of the —O—O— bond in the field of the electrical double layer at the dropping mercury electrode, promoted by the increased dielectric constant of the more aqueous media. They implied that the later, diffusion-limited wave resulted from a single-step, homolytic cleavage. Other hypotheses can be advanced, however, involving hydrolysis, adsorption equilibria, and the role of protons in the reduction process. Certainly peroxy esters have a complex and perhaps a controversial polarographic behavior.

The situation is similarly complex for other peroxide classes, whose polarographic behavior can be altered significantly and systematically by factors such as the type of supporting electrolyte or the presence of polymeric additives. For instance, as mentioned earlier, the use of quaternary tetraalkylammonium electrolytes consistently shifts reduction potentials to more negative values (86,159), presumably through a competitive adsorption process at the surface of the mercury drop. Again the presence of a small amount of ethyl cellulose in 1:1 benzene–methanol will allow resolution of a single wave due to fatty ester hydroperoxides into a double wave (99,100) that has been attributed to mono- and dihydroperoxides (149); this effect is maintained if the benzene is replaced by toluene or xylene, but disappears with nonaromatic hydrocarbons and chlorocompounds (99).

In view of the complex effects on both the qualitatively significant half-wave potential and the quantitatively significant diffusion current constant by peroxide structure (both between and within classes), by composition of the solvent medium and selection of supporting electrolyte, and by the presence of modifying additives, there is a need for studies more extensive and penetrating in scope than those conducted heretofore. Such studies should take fuller advantage of the coupling of polarographic analysis with separation techniques (19,114,159) and chemical conversions; this should be a more effective approach with complex mixtures.

3. Analysis by Iodine Liberation Procedures

The variety of iodine reduction procedures for the quantitative determination of organic peroxides is familiar to all who must analyze for these compounds. Perhaps because of this very variety, many workers seemingly unaware of available improvements have clung to old, familiar methods, such as those of Wheeler (205) or Lea (105) for peroxides in oils, long after substantially improved methods (104,145,196) had been developed.

The proliferation of methods has grown partly from the recurring need for improvements in the accuracy, precision, speed, simplicity, or scope of peroxide methods, and partly from different opinions among chemists as to what in a procedure constitutes simplicity rather than complexity, and where the line should be drawn between the practical and impractical.

For instance, if it has been customary to react groups of peroxide samples simply by placing them in a dark cupboard for some time as specified in many reliable procedures (104,145), it may appear impractical to set up facilities for boiling reaction mixtures under reflux in order to gain a shorter reaction period (116,196). One might prefer instead to seek a reagent mixture in which iodide has a greater reducing power at room temperature. On the other hand, if a laboratory regularly carries out analytical reactions at reflux, as in the determination of esters by saponification, for example, it may seem better to use a brief refluxing step (92,116,196) rather than a long room-temperature reaction, particularly if the shorter procedure also promises to end the necessity of excluding atmospheric oxygen.

In this section we consider in depth the basic factors that influence the quantitative determination of peroxides via reduction with iodide. Our intent is to assist analytical chemists in selecting or in tailoring iodide reduction procedures that will best fit the requirements of their specific analytical situations. In the several categories of iodide methods there are often acceptable alternatives in both apparatus and procedural details that are consistent with sound primary principles. The procedure selected should be based on these principles, and should also reflect such practical factors as the degree of analytical difficulty, the availability of equipment, realistically appraised requirements for speed, precision, and sensitivity, and the number and frequency of analyses to be performed. By taking sufficient pains results of the highest quality will usually be obtainable. Judicious corner-cutting, on the other hand, can often produce simpler, quicker, and less expensive analyses that are entirely adequate for a particular application.

For judging the adequacy of a proposed procedure suitable criteria are necessary, unless its performance in the new situation can clearly be extrapolated from past experience. Our laboratory has consistently used criteria (116) that have been cited repeatedly by earlier workers (104,183), namely, that results be substantially independent of sample size, reaction time, and reagent composition, within reasonable limits, and be unaffected by any nonperoxidic components that may be present in the sample.

a. Exclusion of Oxygen

Solvent systems proposed for iodide reduction methods are nearly all based on acetic acid or isopropyl alcohol. Under otherwise equivalent conditions, iodide reduces peroxides more slowly in alcohols than in acetic acid or chloroform–acetic acid. This apparent disadvantage for alcohols is offset, however, by their freedom from discernible interference through air-oxidation of iodide (92,116,196). On the other hand, for the more active acetic acid-based solvents it has become firmly established (104,116) that effective deaeration is necessary to avoid oxygen errors, and all acceptable procedures require this step.

Much variety is evident in the techniques used for deaeration. Almost invariably this step involves purging with an inert gas from an appropriate source: nitrogen or carbon dioxide bubbled through the solution from a cylinder, or carbon dioxide generated in the reaction flask by adding small pieces of Dry Ice or a gram or two of sodium bicarbonate. We have consistently favored the use of gas from a cylinder, feeling that this permits better control of the operation, is independent of water content in the reagent, and seems to be more convenient once a distributing manifold and modified reaction flasks (116) have been constructed; in this preference we have substantial support (75,104,145). Direct addition of Dry Ice is widely used (13,23) in room-temperature reductions, and can function also at higher temperature (24) with assistance from a slit rubber tube (Bunsen) valve. In addition, Wibaut and co-workers' improvement (206) of the sodium bicarbonate technique (108,178) produces excellent results with peroxides that react at room temperature.

Sully obtained complete deaeration without any purge gas by refluxing chloroform–acetic acid or ether–acetic acid mixtures in a flask surmounted with a long tube, the top portion of which was water-cooled (187); boiling solvent vapors physically excluded oxygen from the reaction zone. The fine performance of this clever adaptation is offset, however, by the necessity of employing specialized, somewhat ungainly, glassware and a relatively complex procedure.

The issue of specialized glassware is often present in iodometric peroxide methods, and almost always arises because of deaeration steps. Besides Sully's special condenser, other examples are Lea's flask with a separate compartment for samples and a special plug for sealing (104); Heaton and Uri's train of pear-shaped flasks or special photometer tubes for macro or micro determinations with rigorous oxygen exclusion (75); Mair and Graupner's flask with a side-arm (116); and the exit bubbler of Wibaut and co-workers (206). Our recommendation is that nonstandard apparatus be avoided wherever it is practical to do so because of the added costs in fabrication and maintenance, but that it be employed whenever it will significantly improve the operability or the results of a procedure.

It is surprising how many published methods fail to specify continuance of protection from air oxidation during final titration of the liberated iodine. For results of the highest accuracy such action should be taken.

Deaeration procedures can range from vigorous and complex (75,104,187) to practical and casual (13,23), with the latter approach giving surprisingly good results in appropriate circumstances. Exclusion of oxygen during actual reduction of the peroxide is of primary importance (187), for it is then that reaction between free-radical intermediates and molecular oxygen can occur, to produce additional peroxide in proportion to the original peroxide content. Reasonable precautions at other stages of the procedure also should yield secondary improvements in accuracy and precision; such gains may be of crucial importance at low peroxide concentrations (2,75). The development of improved deaeration techniques requiring little special apparatus or manipulative effort, yet providing rigorous oxygen exclusion both for frequent and occasional peroxide analyses, would be most welcome.

b. Maintenance of an Adequate Iodide/Iodine Ratio

The advantages of using a relatively large excess of iodide have been discussed recently (116). Besides increasing the rate of peroxide reduction, this practice also prevents errors due to volatilization of iodine and to the addition of iodine to unsaturated materials (196). By maintaining the equilibrium $I_2 + I^- \leftrightharpoons I_3^-$ far to the right, the excess iodide holds liberated iodine as triiodide ion, a form in which it differs from molecular iodine by being both nonvolatile and unable to undergo addition to double bonds.

Wagner, Smith, and Peters (196) compared 2 and 7 g of sodium iodide for the reduction of 2 meq of peroxides; this corresponded to initial iodide concentrations of 0.26 and 0.95M, and to final iodide/triiodide molar ratios of 10 and 44. Although they concluded that the amount of iodide is not critical provided a large excess is present, their data indicate consistently higher results with the larger iodide quantity, as shown in Table XX. In

TABLE XX

Effect of Iodide/Triiodide Ratio on Peroxide Analyses by a Sodium Iodide–Isopropyl Alcohol Method (196).

Peroxide	Peroxide analysis,[a] using	
	2 g NaI[b]	7 g NaI[c]
Hydrogen peroxide	29.4 (29.7, 29.6, 29.4)[d]	29.7
Benzoyl peroxide	98.8 (98.8, 98.0, 98.5)[d]	100.2
Tetralin hydroperoxide	96.9 (97.3, 97.1, 97.3)[d]	98.1
tert-Butyl hydroperoxide	96.3 (96.5, 96.5, 96.0)[d]	97.5

[a] Percent recovery, based on 100% purity. Water added before titration to sharpen the iodine–thiosulfate endpoint.

[b] Molar ratio I^-/I_3^- approximately 10, assuming 1 mmole of peroxide used.

[c] Molar ratio I^-/I_3^- approximately 44, assuming 1 mmole of peroxide used.

[d] Parenthetical data obtained independently at reaction times of 2, 5, and 15 min, respectively, show results to be independent of reaction time.

our laboratory (62) an evaluation of iodide–isopropyl alcohol methods (92,196) showed a somewhat different result: Liberation of iodine by hydroperoxides was stoichiometric for final iodide/triiodide ratios as low as 3–5, corresponding to consumption of 40–50% of the added iodide in the reaction,

$$ROOH + 3I^- + 2H^+ \rightarrow ROH + H_2O + I_3^- \tag{19}$$

Results were some 30% low, however, for ratios of 0.3, in which 90% of the iodide was consumed.

Experience has indicated that these low results were due to incomplete peroxide reduction or loss of volatile molecular iodine from the boiling reaction mixture. There must have been a different origin, however, for the consistently lower results obtained by Wagner, Smith, and Peters at ratios near 10 (compared with 44), seen in Table XX. Since these lower results were independent of reaction time, they were probably caused by reaction of molecular iodine with an impurity in the isopropyl alcohol solvent, rather than by volatilization. Hartman and White showed evidence for similar losses of iodine in isopropyl alcohol (71) in a report that recommended the alternative use of *tert*-butyl alcohol. The data in Table XX indicate that this apparent interference was suppressed by use of the larger amount of iodide.

c. Improbability of Iodine Absorption by Olefins

Wagner, Smith, and Peters investigated at some length the possibility of iodine absorption by unsaturated materials. They showed (196) that whereas molecular iodine readily adds to the double bonds in a monoolefin (cyclohexene) and a conjugated diolefin (isoprene), there is no trace of such addition by triiodide. They eliminated the idea that triiodide addition might occur if catalyzed by peroxides, particularly in conjugated systems, by demonstrating that there is no iodine loss during reduction of benzoyl peroxide with sodium iodide in the presence of isoprene. Skellon and Wills similarly reduced benzoyl peroxide in the presence of large excesses of oleic acid (178), without detecting iodine absorption; in addition, Banerjee and Budke studied the analysis for traces of peroxide in many solvents, saturated and unsaturated, and found no indication of interference from olefins (11).

Wagner, Smith, and Peters obtained further evidence on the nonreactivity of triiodide with olefins in the analysis of four autoxidized monoolefins, diisobutylene, 2-pentene, cyclohexene, and tetralin, whose peroxidic components are the easily reduced alkyl hydroperoxides. In each case, the resulting peroxide number was essentially independent of reaction time, sample size, and iodide concentration (196). The identical results obtained with 2 and 7 g of sodium iodide reagent were good evidence that hydroperoxide reduction had been complete and that liberated iodine had not been absorbed by the unsaturated substrates.

In these experiments with olefins the final iodide/triiodide molar ratios were all relatively high, ranging from 20 up to 1100, and above 50 in three-quarters of the cases. The data warrant the conclusion that, at least in media based on isopropyl alcohol, interferences from olefins of all types may be prevented in iodometric peroxide determinations by using sufficiently high proportions of excess iodide reagent. The same is probably true in the other useful media.

It has been suggested (196) that triiodide may add to conjugated diolefins when catalyzed by diolefin peroxides (these are mixtures of cyclic and polymeric alkyl peroxides formed respectively by intra- and intermolecular 1,4-addition of molecular oxygen to conjugated diolefins). The suggestion was based on a single experiment in which the peroxide number obtained for an autoxidized diolefin was somewhat lower in the presence than in the absence of additional peroxide-free diolefins, 309 compared to 341. We consider this evidence weak, inasmuch as the same study also made it clear that the reduction of diolefin peroxides in the sodium iodide–isopropyl alcohol system is variable and only partial. These latter findings, rather than iodine absorption, could more reasonably account for the observed differences.

While interferences due to the volatility of iodine and its addition to olefinic double bonds can be eliminated by using a sufficient excess of alkali iodide reagent, this measure will not, of course, prevent interferences from reaction of iodine with acetone (23,71,92) or reduction with mercaptans and other sulfur compounds.

d. Analysis of Easily Reduced Peroxides

The classes of organic peroxides vary greatly in their ease of reduction by iodide, and a clear and seemingly natural distinction has been recognized between (*a*) the more easily reduced types that respond quickly and quantitatively to relatively mild reagents, and (*b*) those less easily reduced varieties that do not. It is on peroxides in the first category that the majority of iodometric analyses are performed; they are reduced by sodium or potassium iodide in acetic acid or an alcohol at room temperature (although the alcoholic reagents are customarily heated to boiling). It has been proposed that organic peroxides be classified operationally for iodometric analysis into three categories (116) based on response to a refluxing sodium iodide–isopropyl alcohol reagent. These categories, shown in Table XXI, are: easily reduced (rapid response), moderately stable (slow response), and difficultly reduced (no response). The easily reduced subdivision includes principally the peroxy acids, the diacyl peroxides, and the alkyl hydroperoxides; its members are defined by the requirement for rapid quantitative reduction. A special case can be made for the *tert*-butyl

TABLE XXI

Classification of Organic Peroxides by Ease of Reduction with Sodium Iodide–Isopropyl Alcohol at Reflux

Category	Membership	Response to reagent[a]
Easily reduced	Peroxy acids, diacyl peroxides, all hydroperoxides, and certain other compounds	Liberate iodine stoichiometrically in 5 min or less
Moderately stable	Most peresters, aldehyde peroxides and ketone peroxides, di-*n*-alkyl peroxides, and mixed *n*-alkyl peroxides	Liberate iodine, but stoichiometric reduction requires more than 5 min.
Difficultly reduced	Transannular peroxides, diaralkyl peroxides, di-*tert*-alkyl peroxides, and mixed aralkyl-*tert*-alkyl peroxides	Show no iodine liberation with the test reagent even upon prolonged refluxing

[a] Based on reaction of 0.5–1 mmole with 2 g of NaI plus 2.5–5 ml of acetic acid in 25–50 ml isopropyl alcohol boiling under reflux.

peroxy esters, which, in effect, can be promoted from the moderately stable category by catalysis with a trace of ferric ion (172), as will be seen.

Two basic rules for sound quantitative iodometric peroxide analysis developed earlier in this section specify the exclusion of molecular oxygen except in alcoholic media, and the use of an adequate excess of iodide reagent. For the easily reduced peroxides, two other rules complete the basic requirements: the reduction period must be adequate and there must be freedom from interference by other components of the sample. These components may include both substances that liberate iodine, such as the slower reacting, moderately stable peroxides or ferric iron (the latter interference can be controlled by complexing with a small amount of a fluoride salt), and those that absorb it, such as acetone and mercaptans.

Some of the published iodide reduction procedures that we consider satisfactory for quantitative analysis of the easily reduced peroxides are shown in Table XXII. Iodometric procedures alone cannot differentiate mixtures of the several types of these reactive peroxides; there are more broadly based methods for doing so (82) described in Table XI.

The potassium iodide–acetic acid methods in Table XXII specify brief reaction periods at room temperature. With chloroform also present, 1 hr is allowed for complete reaction of lipid peroxide (Lea's cold method)—probably an excessive period for peroxy acids and diacyl peroxides. The sodium iodide–isopropyl alcohol reagent mixtures are generally heated to shorten reaction times in this less active medium; reaction at room temperature is also reasonably fast, however. Buncel and Davies used this reagent at room temperature to determine peroxysilanes (21), allowing the reaction to proceed overnight in the dark.

We call special attention to the method of Abrahamson and Linschitz (2) usually cited for the use of a dead-stop endpoint in the titration of liberated iodine in colored solutions. This method, in addition, has a remarkable combination of sensitivity, precision, and accuracy that places it in a class by itself for iodometric peroxide determination by a volumetric technique. For instance, 4 μeq of Tetralin hydroperoxide was determined with better than 0.5% accuracy and precision, titrations being 16.60 $\pm$ 0.03 ml of 0.00026*N* thiosulfate.

The modified reagent proposed by Hartman and White (71) deserves a broader evaluation than it has received; it was applied only to a few fats and oils and to benzoyl peroxide. This reagent consists of sodium iodide in *tert*-butyl alcohol acidified with citric acid and modified with carbon tetrachloride. *tert*-Butyl alcohol was selected because of its inertness to molecular iodine and good solubility for fats; citric acid was used because it is a fiftyfold more strongly ionized acid than acetic, and hence a better

TABLE XXII

Some Satisfactory Iodometric Methods for Easily Reduced Peroxides

Reagent	Solvent	Temperature	Deaeration	Remarks	Ref.
1. NaI	Isopropyl alcohol (acidified)	Boiling	None	Quantitative reaction in 5 min or less	116, 196
2. NaI or KI	Isopropyl alcohol (acidified)	Boiling	Continuous N_2 or CO_2 flushing during titration	Precise and accurate at trace peroxide levels, in colored solutions; uses electrometric "dead stop" endpoint	2
3. KI	Acetic acid	Room	Several lumps of Dry Ice	Reaction period 5 min; results very precise	13, 23
4. KI	Acetic acid	Room	$NaHCO_3$	Reaction period 10 or 15 min in the dark; very precise results	178, 206
5. KI	Acetic acid–chloroform, 3:2	Boiling	N_2 or CO_2 gas	Reaction period of 1 hr in the dark appears excessive; special apparatus	104, 177
6. KI	Acetic acid–chloroform, 1:1	Boiling	Chloroform vapor	Zero blank; reaction period 3–5 min	187
7. NaI	Acetic acid–chloroform, 3:2 $4 \times 10^{-5} M$ in Fe(III)	Room	N_2 gas	Quantitative reduction of *tert*-butyl peroxy esters in 5–10 min	172

source of hydrogen ions; carbon tetrachloride or chloroform kept the fat (and molecular iodine) in a bottom layer during titration.

This reagent has two noteworthy characteristics: (*a*) Its reducing power is appreciably enhanced over that of sodium iodide–isopropyl alcohol, acidified with acetic acid, and is approximately equivalent to that of sodium (or potassium) iodide–acetic acid–carbon tetrachloride (or chloroform); it can apparently yield quantitative results on easily reduced peroxides in 15 min at room temperature. (*b*) It consistently gives a zero blank, whether in the presence or absence of air. There appears to be a slight oxygen error, with results slightly higher in the presence of air because of induced peroxide formation (although not for the rapidly reacting benzoyl peroxide). Areas needing investigation are behavior of this reagent system on other classes of peroxides, with and without carbon tetrachloride, with and without heating, and with a more adequate amount of iodide, and the effect of substituting citric for acetic acid in the regular sodium iodide–isopropyl alcohol procedure (116,196).

Silbert and Swern's iron(III)-catalyzed modification (172) of Lea's cold method is included in Table XXII as a special case. The *tert*-butylperoxy esters, for which this method is so effective, are categorized ordinarily as moderately stable peroxides, but are promoted to the easily reduced category by its use. This will be discussed further in the following subsection.

There have been two approaches to the continuous removal of iodine as it is formed during peroxide reduction in iodide–alcohol media (30,120). In each case the primary aim was to prevent iodine loss through volatility and side reactions (120), or, specifically, through addition to conjugated diolefins (30). Despite the fact that fears of such loss appear to be generally unwarranted, the idea of continuously consuming liberated iodine is intriguing and deserves consideration in its own right.

Matthews and Patchan (120) describe continuous thiosulfate titration of liberated iodine by means of an automatic potentiometric titrator in the determination of peroxide number in petroleum products. This approach was claimed to allow the determination of both diacyl peroxides and alkyl hydroperoxides, whereas the arsenite method (198) was thought to be restricted to hydroperoxides only. Since the first use of arsenite as a peroxide reagent (170) was for the determination of dibenzoyl peroxide, this advantage does not appear valid. The procedure is too complex and fussy to offer advantage.

In another approach (30), Dahle and Holman have included in the reagent a known (excess) amount of a thioglycolate ester whose thiol group consumes iodine as rapidly as produced; excess thioglycolate is determined by titration with a standard iodine solution. They used a solvent mixture of 50% absolute ethyl alcohol, 30% glacial acetic acid, and 20% chloroform, containing 0.2% (0.01M) octyl thioglycolate. In a semimicro procedure, up to 100 mg of lipid sample was dissolved in 1 ml of the solvent and reacted for 15 min after two drops of saturated aqueous potassium iodide had been added. Evaluation of this new procedure on lipids was clouded by use of the outmoded Wheeler method (205) as a standard of comparison. It would appear worthwhile to investigate the parameters of this method more extensively.

e. Control of Reducing Power in Iodide Reagents

The more stable peroxides described in Table XXI, which include essentially all types except peroxy acids, diacyl peroxides, and hydroperoxides, require more vigorous reaction conditions to bring about their quantitative reduction within a reasonable period of time. Several parameters that may be varied to increase and control the reducing power of iodide reagents have been discussed recently (116); they include the use of a catalyst, choice of solvent medium, choice of reaction temperature, addition of mineral acid, and, most important, the concentration of water. We shall now discuss these factors individually.

(1) Catalyst

For a large, important class of organic peroxides, namely, the *tert*-butylperoxy esters, use of a catalyst produces a dramatic increase in the rate of reduction by iodide (172), as we have noted; Silbert and Swern discovered that traces of iron(III) in an acetic acid–chloroform medium (protected from oxygen) produced a precise and stoichiometrically accurate liberation of iodine by these peroxides within 5 min at room temperature. In the absence of iron, the reduction required more than 4 hr, and even with stronger hydriodic acid reagents it took nearly an hour (116). They proposed a five-step modified Haber-Weiss redox–free radical mechanism to account for the stoichiometry and products of reaction. According to this scheme, one-half of the iodine is produced in a ferric–ferrous cycle wherein iodide reduces iron(III) to iron(II) and the latter is converted back to iron(III) by reducing the peroxy ester to an acid anion and a *tert*-butoxy radical. The other half of the iodine is formed during decomposition of the *tert*-butoxy radicals; these attack water or acetic acid to yield hydroxyl radicals, which oxidize iodide.

They found catalyst concentration to be very important, with an optimum of only 0.001% as ferric chloride hexahydrate, or $4 \times 10^{-5}M$ in iron(III). Higher catalyst levels gave low stoichiometry, perhaps because *tert*-butoxy radicals would be present at higher concentrations and might decompose partially by alternative reactions.

The *tert*-butylperoxy ester analysis is the only application that we have found in which reduction of a peroxide by iodide was significantly improved by a catalyst. Use of iron(III) with diacyl peroxides (173) and of cuprous chloride with aralkyl hydroperoxides (77) appears uncalled for, in view of the inherent high reactivity of these peroxides.

(2) Choice of Solvent Medium

Acetic acid has been virtually the exclusive choice as a reaction medium for iodide reduction of less reactive peroxides, as it is in this solvent that iodide shows its top reducing power. Alcohols (71,92,196) or acetic anhydride (139) are mainly useful with easily reduced peroxides. Acetone has been tried by several workers, but it is a most unsuitable solvent (23,71,92) because it absorbs part of the iodine by some as yet unexplained mechanism.

This acetone-absorbed iodine is released slowly during titration with thiosulfate, causing the endpoint to fade quickly. We tried stepwise titration back to the endpoint on one occasion, and with persistence recovered substantially all of the missing iodine.

(3) Choice of Reaction Temperature

The inherent reducing power of iodide reagent mixtures can be varied widely by modifying composition, as we shall see shortly. In addition, of course, the strength of any such reagent system can be modified by varying its temperature, which would seem to allow great flexibility in tailoring the reducing power of a peroxide reagent. There are some practical considerations in varying temperature, however. Room temperature is the lowest that has been found useful and is the easiest to use. Ice temperature was used on one occasion (183) in the evaluation of a reagent acidified with sulfuric acid, but the resulting procedure was considered of dubious value. The next easiest to use, and the maximum that should be considered normally, is the reflux temperature of the reagent mixture. There is a pitfall in this means of temperature control, however, in that the presence of volatile components may lower the reagent's refluxing temperature significantly below its expected value. For example, acetic acid solutions of sodium iodide boil at nearly 120°C; but the addition of chloroform or carbon tetrachloride as auxiliary solvent, or the dilution of peroxide samples with, e.g., benzene or cyclohexane, would markedly lower the temperature of a boiling iodide–acetic acid reagent and tend to slow the peroxide reduction.

In the range between ambient and reflux, reaction temperatures can be controlled directly with heating baths, a device used in several well-known methods: The "hot" modification of the Lea method (104) employs a boiling water bath and also one maintained at 77°C, and the hydriodic acid method of Vaughan and Rust (41) carries out reduction of di-*tert*-butyl peroxide at 60°C. Such methods of heating have seemed to us unnecessary and undesirable complications in most instances, to be considered only when they are more conveniently available than boiling under reflux. This view would require revision if future studies should show reaction temperature to be an important parameter in tailoring reagent activity.

(4) Addition of Mineral Acid—The Role of Water Concentration

For increased reducing power of iodide-in-acetic acid reagents, an alternative to the use of higher temperatures is the addition of an acid stronger than acetic. Because hydriodic acid is well known to be a more potent reducing agent than iodide salts, it seems clear that the effectiveness of mineral acids in enhancing the reducing power of iodide–acetic acid solutions is due to the *in situ* formation of hydriodic acid. The reason why iodide is a more effective peroxide reducer in acetic acid than in alcohols may well be that in the acidic solvent a small but significant concentration of hydriodic acid is formed from the iodide reagent.

An extreme case is the direct substitution of hydriodic acid for an iodide salt. Vaughan and Rust's hydriodic acid method (41) for di-*tert*-butyl peroxide uses equal volumes of acetic acid and constant boiling (57%) hydriodic acid. Employing such a high proportion of the concentrated hydriodic acid in order to obtain strong reducing conditions is self-defeating, however. In this laboratory (116) we have obtained more vigorous reducing conditions with much lower proportions, the optimum ratio being about 2 ml of 57% hydriodic acid per 50 ml of acetic acid; at room temperature, dicumyl peroxide, although inert to sodium iodide–acetic acid, is quantitatively reduced by this dilute hydriodic acid within 10 min.

Larger proportions of hydriodic acid, e.g., 5 ml per 50 ml of acetic acid, make distinctly weaker reagents. This apparently inverse relation between reagent concentration and effectiveness can be rationalized in terms of a diminution in the reducing power of hydriodic acid through hydration via a series of equilibria that may be represented by the equation:

$$HI + nH_2O \rightleftharpoons HI \cdot H_2O + (n - 1)H_2O \rightleftharpoons HI \cdot 2H_2O + (n - 2)H_2O \rightleftharpoons HI \cdot 3H_2O + (n - 3)H_2O \rightleftharpoons \text{higher complexes} \quad (20)$$

This thesis assumes that the active reducing agent is unhydrated or lightly hydrated hydrogen iodide. The concentration of this component in a hydriodic acid–acetic acid–water reagent will depend very strongly on the acetic acid/water ratio, because acetic acid and water are also associated in the equilibrium $HOAc + H_2O \rightleftharpoons HOAc \cdot H_2O$; hence, the drier the acetic acid the more effectively can this solvent dehydrate the series of hydriodic acid–water complexes. This key factor, the water content of the reagent, is controlled by the proportion of 57% hydriodic acid employed. The observed inversion of reducing power with reagent concentration indicates that there is a higher concentration of unhydrated, i.e., active, hydrogen iodide per 50 ml of acetic acid with the use of 2 ml of 57% hydriodic acid than with 5 ml or even 50 ml of this reagent.

As the proportion of 57% hydriodic acid is decreased in such simple hydriodic acid–acetic acid reagents, in order to obtain more anhydrous (and hence more active) reaction media, it is impossible to maintain adequate iodide/triiodide ratios. This problem can be solved, however, by generating controlled amounts of hydriodic acid *in situ* through addition of controlled amounts of mineral acid to a sodium iodide–acetic acid solution; such a medium can simultaneously be both relatively anhydrous and iodide-rich.

Earlier workers such as Stansby used this approach in a limited way (183); we have studied the parameters more extensively (116) and have described reagent mixtures that provide two useful levels of reducing power above that of the sodium iodide–isopropyl alcohol system, as will be discussed. Again the activity of these reagents is controlled by their water content, which is the compositional variable of primary significance.

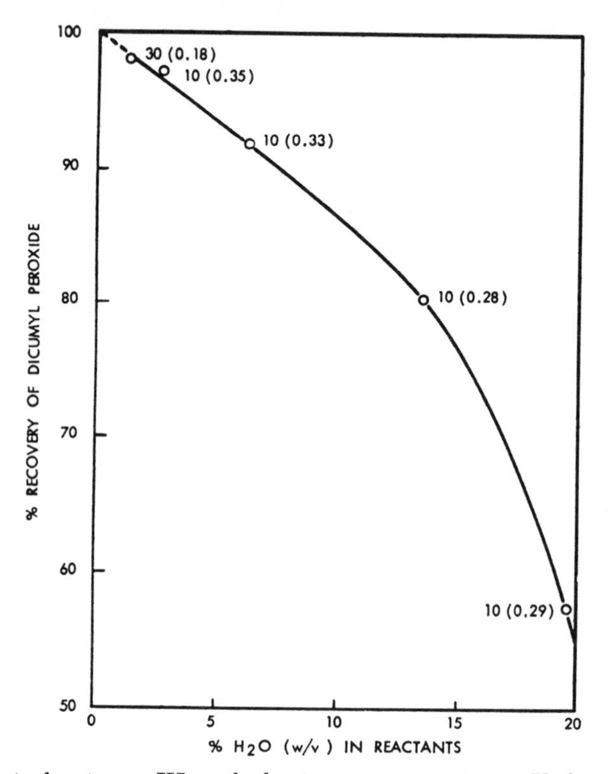

Fig. 4. Effect of water on HI methods at room temperature. Various proportions of acetic acid, 37% HCl, water, and NaI or KI. Reaction time, min: 10, 30; molar concentration of iodide: (0.18), (0.35), (0.33), etc.

Figure 4 illustrates how reactivity at room temperature in typical iodide–acetic acid–hydrochloric acid reagents increases as the water content is decreased. For instance, a composition containing 3% water will quantitatively reduce dicumyl peroxide and all but the most intractable peroxides at room temperature. Coupling low water content with higher reaction temperatures further enhances reducing powers to a remarkable degree; for example, di-*tert*-butyl peroxide, which is inert to the reagent with 3% water at room temperature, is quantitatively reduced in less than 5 min when heated to reflux.

Reagent strength is a liability when it causes reduction of sample components other than peroxides (with accompanying liberation of iodine). These interfering reductions are often stoichiometrically quantitative; it has been shown, for instance (116), that the above reagent (3% water) at its refluxing temperature, about 120°, will liberate 1 mole of iodine per mole of many cyclic alcohols or of olefins produced by dehydration of alcohols. These may include peroxide reduction products, such as the cumyl alcohol formed by reduction of cumene hydroperoxide or dicumyl peroxide: Both this compound and its dehydration product, α-methyl styrene, produce a mole of iodine in the refluxing reagent within 15 min. This reaction causes no difficulty in the analysis of samples for which either of these specific peroxides is known to be the only reducible component initially present, for then a quantitative correction can be made. When unknown amounts of the above nonperoxides are initial components, however, substantial errors will occur if so potent a reagent is used. In general, to avoid errors of this type in an iodine liberation procedure, one must appropriately decrease the reducing power of the reagent system.

The data in Fig. 5 show that water plays the same role in moderating reducing power at reflux that it does at ambient temperature. In these experiments, cumyl alcohol was chosen as substrate, to provide a moderate reaction rate. The top point on the curve represents the iodide–acetic acid–hydrochloric acid reagent containing 3% water, which we have been discussing. The others were simple mixtures of acetic acid and 57% hydriodic acid, with the highest water content corresponding to the composition of Vaughan and Rust's hydriodic acid reagent (41). Note that at 60°C cumyl alcohol is extensively reduced. Consequently, this well-known iodine liberation method for determining di-*tert*-butylperoxide is unsuitable for use with dicumyl and other aralkyl peroxides (or hydroperoxides) because their reduction produces interfering alcohols.

Concentrated (37%) hydrochloric acid is the most useful mineral acid for preparing hydriodic acid reagents. Its 63% water content presents no practical difficulty, as suitable reagents must not be too dry; sufficient

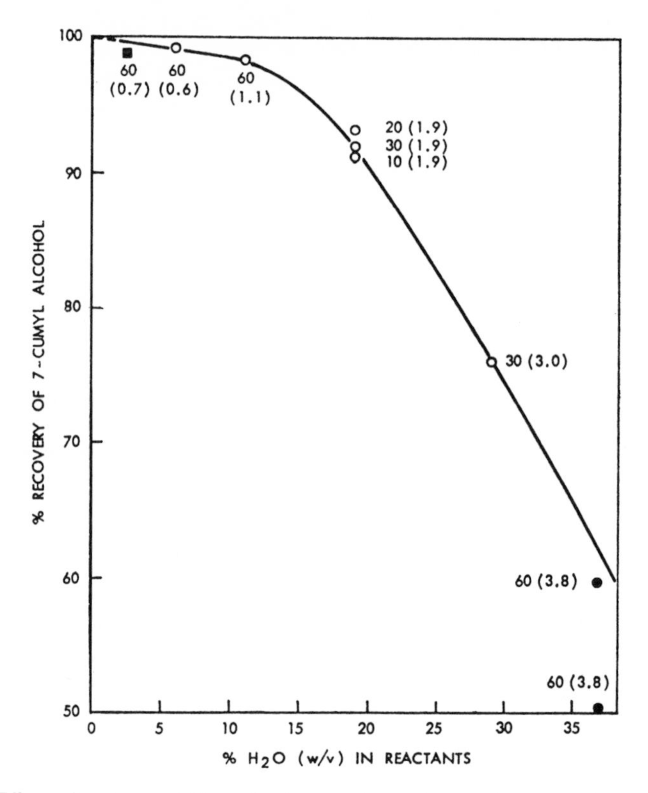

Fig. 5. Effect of water on HI methods at reflux. (○) Acetic acid–57% HI, refluxed at about 120°C. (●) Acetic acid–57% HI, reacted at 60°C, as in reference 41. (■) Acetic acid–NaI–37% HCl, refluxed at about 120°C; reaction time, minutes: 10, 20, 30, 60; molar concentration of iodide: (0.7), (1.1), (3.0), etc.

water must be present to prevent excessive volatilization of hydrogen iodide (a phenomenon observed (116) when anhydrous reagents have been prepared by adding acetic anhydride or by using hydrogen chloride gas in lieu of concentrated hydrochloric acid).

Some water is needed also to mitigate the extremely potent reducing power of anhydrous hydrogen iodide. This factor was emphasized (116) by the result of adding concentrated sulfuric acid to a sodium iodide–acetic acid solution in order to produce anhydrous hydrogen iodide; at room temperature rapid reduction to H_2S occurred, with copious liberation of iodine.

It appears that the 3% (1.5M) water content obtained with 2 ml of 37% hydrochloric acid per 50 ml of acetic acid is close to a practical minimum limit for the water concentration in hydriodic acid reagents. Susceptibility to reduction by hydriodic acid makes sulfuric an unwise choice for a mineral acid, even if appropriately diluted with water.

A brief evaluation has been made of acidification with 85% phosphoric acid (116). This mineral acid produced hydriodic acid reagents that were distinctly lower in reducing power than those from hydrochloric or sulfuric acids, in spite of their relatively anhydrous nature. This would appear to mean that phosphoric acid–hydrogen iodide or phosphoric acid–hydrogen iodide–water complexes were formed that suppressed the hydriodic acid reducing power. The possibility of obtaining usefully modified reagents by acidification with phosphoric acid has not been adequately explored yet.

As might be anticipated, we have found that quite satisfactory reagents can be obtained by acidifying sodium iodide–acetic acid solutions with small amounts of concentrated hydriodic rather than concentrated hydrochloric acid (116). The discovery that concentrated hydriodic acid can be freed of iodine simply by shaking with mercury before use (70,117) has made this substitution practical. Water contents are equivalent as the two acids contain equal concentrations of water (75%) on a volume basis.

Data on blanks gave evidence that the use of hydriodic rather than hydrochloric acid produces a slightly stronger reagent. These blank values appear to arise from traces of difficultly reduced impurities in the acetic acid solvent. With 1, 2, or 3 ml of 57% hydriodic acid per 50 ml of acetic acid, and either 30 or 60 min reaction at reflux, blanks were consistently in the range of 0.10–0.12 meq of iodine, which implies quantitative reduction of the impurities. On the other hand, with 2 ml of 37% hydrochloric acid under similar circumstance, blanks were typically 0.05–0.07 meq, and were proportional to the reaction time, indicating that reduction had been only partial. In a limited comparison of applications, these two alternative types of reagent otherwise appeared to be equivalent.

f. Analysis of Moderately Stable Peroxides

Above the level of reducing power adequate for easily reduced peroxides, two additional useful levels have been defined, one with intermediate and one with strong reducing properties. Examples are described in Table XXIII. The intermediate methods are a compromise between the ability to reduce a large group of moderately stable peroxides (Table XXI) and the need to be relatively free from interference by nonperoxidic components. Methods in the strong category are more seriously affected by such interference; reduction of nonperoxides is apt to be severe and removable only

TABLE XXIII

Composition of Useful Iodimetric Reagents for Moderately Stable[a] and Difficultly Reduced[a] Peroxides

Reagent[b]: Acetic acid solution of	HI molarity	Total I^- molarity	H_2O concentration		Reaction temp.	Remarks	Ref.
			Molarity	%(w/v)			
A. Methods of Intermediate Reducing Power							
1. NaI–HCl[c]	0.5	0.75	1.7	3.1	Room temp.	Quantitative reaction of dicumyl peroxide in 10 min; di-*tert*-butyl peroxide inert	116
2. NaI–H_2O[d]	$\sim 10^{-5}$	0.75	3.1	5.7	Boiling[e]	Quantitative reduction of dicumyl peroxide in 20 min; di-*tert*-butyl peroxide inert; little interference from nonperoxides	116
B. Methods of Strong Reducing Power							
1. NaI–HCl[c,f]	0.5	0.75	1.7	3.1	Boiling[e]	Quantitative reduction of di-*tert*-butyl peroxide in 5 min, frequent interference from nonperoxides, e.g., quantitative reduction of cumyl alcohol	116
2. KI–HCl[c]	0.17	0.17	6.8	12.3	Boiling[e]	Weaker reagent than previous entry; only 90% reduction of di-*tert*-butyl peroxide, and no attack on cumyl alcohol	82

[a] As defined in Table XXI.
[b] All reagents deaerated with N_2 or CO_2 gas.
[c] Forms HI.
[d] Method II in reference 116.
[e] $\sim$120°C.
[f] Method III in reference 116.

by destroying the effectiveness of the reagent in the intended peroxide reduction.

Reagents for the two intermediate-level methods in Table XXIII are closely related, both stemming from a solution of 6 g of sodium iodide in 50 ml of acetic acid. In the first case, 2 ml of 37% hydrochloric acid is added and the reagent used at room temperature. In the second, the solution is used at the boiling point after adding 3 ml of water to suppress reduction of nonperoxides.

We have used these methods extensively for the determination of dicumyl peroxide in autoxidized cumene and for similar analysis in related aralkyl systems. Frequently, the samples contained potentially reducible by-products, typified by cumyl alcohol and α-methyl styrene in autoxidized cumene. Our experience has shown that the second method, the boiling sodium iodide–acetic acid–water reagent, is more reliable for such samples than is the first, the room-temperature hydriodic acid method. With the latter, neither the alcohol nor olefin by-products showed the slightest trace of iodine liberation in the absence of dicumyl peroxide. In its presence, however, they were both partially reduced, causing peroxide analyses to be proportionately high.

This interference appears to have been specifically due to peroxide reversal of Markownikoff's rule during addition of hydriodic acid to the side-chain double bond of α-methyl styrene (116); the resulting primary organic iodide group should be reduced readily by excess hydriodic acid, unlike the normally expected tertiary iodide. Such reversed addition in the presence of peroxides is well known for hydrobromic acid, although it does not occur for hydrochloric acid and had not been observed previously for hydriodic acid; Royals had predicted such a possibility, however (160).

The preferred method was reached by two modifications of the room temperature method. First its hydriodic acid concentration was drastically reduced (from $0.5M$ to 10^{-4} or $10^{-5}M$) by eliminating the addition of mineral acid; the attendant reduced reactivity was restored by heating the reagent to boiling. It was found that reduction of cumyl alcohol and α-methyl styrene occurred with this reagent also, although now in the absence of peroxides as well as in their presence.

The second modification was to add an appropriate small amount of water. This has an effect summarized in Fig. 6. The shaded areas indicate that the addition of increasing amounts of water causes the interference to decrease progressively. Note that when interference was eliminated (3.0 ml water added), the necessary reaction period for dicumyl peroxide reduction increased from 10 to 20 min.

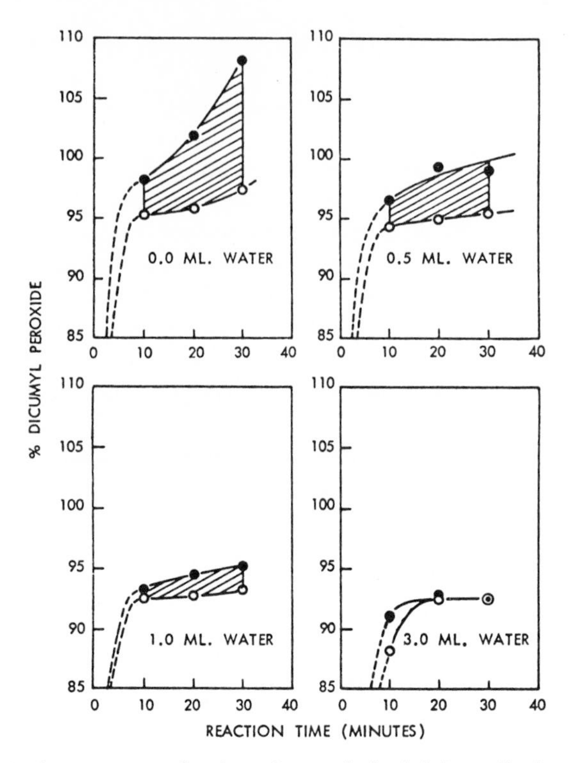

Fig. 6. Effect of water on reduction of cumyl alcohol by refluxing sodium iodide–acetic acid. (○) 75 mg of dicumyl peroxide. (●) 75 mg of dicumyl peroxide + 250 mg of cumyl alcohol. 3 g NaI (generous measure) in 50 ml of acetic acid, reaction at reflux.

Note also the stoichiometric deficiency in iodine liberation by the dicumyl peroxide, which was a pure specimen. As has been reported (116), this behavior is not well understood, but it is remarkably precise. The following are typical reproducible recoveries obtained for several pure bis(aralkyl) peroxides by the recommended procedure (30 min reaction at reflux under inert gas with 6 g of sodium iodide and 3 ml of water in 50 ml of acetic acid):

Dicumyl peroxide	92.5%
Di(*p*-cymene)peroxide	87.5%
Di(*p*-isopropylbenzene)peroxide	89.0%

When results are corrected for these conversions, they appear to be both accurate and precise. The same method is applicable to other similar peroxides after calibration with pure specimens. Such analyses have been found to be independent of sample size and reaction time, within broad limits (116).

A successful application of this method to the determination of 0.25–0.5% dicumyl peroxide in fire-resistant polystyrene materials (containing other peroxides) recently has been described by Brammer, Frost, and Reid (17); the dicumyl peroxide was recovered by acetone extraction, separated from other peroxides by thin-layer chromatography, by the method of Knappe and Peteri (91), and determined directly on the silica gel substrate of the TLC spot by a tenfold scaledown of the original iodometric procedure (116). Only about 1-mg portions of dicumyl peroxide were used; at the 0.5% level, the standard deviation was 0.03% absolute, corresponding to 95% confidence limits of ±12% relative for single determinations.

An alternative and much simpler technique for the final analysis in this case would be ultraviolet spectrophotometry, on an extract of the TLC spot.

Whether controlled additions of water could have similarly suppressed the previously discussed interferences in the room temperature hydriodic acid method (2 ml of 37% hydrochloric acid per 50 ml of iodide solution) has not been investigated. There is a good possibility, however, that such would prove to be the case, although reaction periods might have to be substantially lengthened.

g. Analysis of Difficultly Reduced Peroxides

The final two methods in Table XXIII bracket the practical upper and lower limits for reducing power of reagents that can determine the difficultly reduced peroxides. The first, our so-called method III, has the most powerful reagent for peroxide analysis that has been described (116). The second is the method used by Horner and Jürgens (82) to determine dialkyl peroxides when resolving mixtures of peroxide classes by the procedural combinations described in Table XI. The reasons why these methods should differ in reducing power are clear in Table XXIII, which shows that the second has four times the water content and less than two-fifths the hydriodic acid concentration of the first.

A limited comparison of these two methods is summarized in Table XXIV. The Horner and Jürgens method apparently reduced both dicumyl and di-*tert*-butyl peroxides quantitatively; the much stronger method III performed as previously reported (116), quantitatively reducing di-*tert*-butyl peroxide and producing 3 moles rather than 1 mole of iodine per mole of dicumyl peroxide. The relatively low results for

TABLE XXIV

Comparison of Two Methods for Difficultly Reduced Peroxides

Sample	Reaction time, min	Apparent percent peroxide	
		Mair and Graupner Method III[a]	Horner and Jürgens ROOR method[b]
Dicumyl peroxide (commercial product)	50	271.5 (90.5% of 3 moles I_2/mole)	86.5
	30	—	89.0, 90.2
Di-*tert*-butyl peroxide (nominally 97% pure)	50	96.0	87.0
	30	—	86.5

[a] Procedure as in reference 116; reagent is 6 g NaI + 2 ml conc. HCl in 50 ml acetic acid.

[b] Reagent as in reference 82: 1 g KI (as 1 ml saturated aqueous solution) + 5 ml conc. HCl in 30 ml acetic acid; apparatus and procedural steps the same as for method III in reference 116.

di-*tert*-butyl peroxide by the Horner and Jürgens method probably indicate that significant losses due to sample volatility occurred under the conditions of the analysis. This loss does not occur in method III, apparently because reduction is much more rapid—complete in less than 5 min.

In view of the effectiveness of water concentration in controlling the reducing power of a sodium iodide–acetic acid system, as clearly shown in Fig. 6, the few results in Table XXIV by the Horner and Jürgens reagent, whose composition is detailed in section B2 of Table XXIII, suggest similar effectiveness for this technique with hydriodic acid–acetic acid systems. We recommend that this lead be evaluated systematically and that the effects of using intermediate reaction temperatures be explored in tailoring iodometric peroxide reagents.

As indicated above in Section IV-B-3-e-(4), the practical upper limit on reducing power of hydriodic acid reagents is set by the minimum water content necessary to prevent evolution of gaseous hydrogen iodide from the boiling reaction mixture. The third method in Table XXIII, the so-called method III, is about at this limit. The reagent for this method consists of 6 g of sodium iodide and 2 ml of 37% hydrochloric acid per 50 ml of acetic acid (116). Because it is employed at the boiling point, an important step in the procedure is the avoidance of even a modest inert gas purge of the reaction flask during boiling in order to prevent what can otherwise be substantial losses of hydrogen iodide and, in the case of volatile compounds such as di-*tert*-butyl peroxide, substantial losses of the sample also.

In peroxide analysis, there appears to be no need for a further increase in reducing power, for method III quantitatively reduces the stable di-*tert*-alkyl peroxides within 5 min, and also can be expected to reduce satisfactorily the other family of relatively inert peroxides, the cyclic trimeric ketone peroxides (we have not, however, tested this point).

A modified approach to the analysis of difficultly reduced peroxides is the procedure of Braithwaite and Penketh for the determination of butadiene polyperoxide (16a). They have used a boiling reagent solution of lithium iodide in isooctanol, acidified with phosphoric acid. This interesting combination allows the use of an alcohol at a high enough temperature to promote rapid reaction, along with good iodide solubility. This new system has been applied to no other peroxides; it would appear to deserve a broader evaluation.

h. Hydriodic Acid Reduction of Nonperoxides

A study of reactivity by several classes of potentially reducible non-peroxides in method III (116) has turned up two apparent generalizations: It is clear that alcohols and olefins that result from dehydration of the alcohols are equally reducible; it seems also that nonperoxides, to be reducible, must contain a cyclic structure.

Among terpenes and related compounds, there is a clean division of the monocyclics into reactive and unreactive types. Thus, all difunctional monocyclics, i.e., menthadienes, menthenols, and menthadiols, liberate exactly 1 mole of iodine within 15 min, whereas the corresponding mono-functional menthenes or menthols are virtually inert even after prolonged reaction periods. Bicyclic terpenes are also virtually inert unless they can isomerize to difunctional monocyclics.

Rate of reduction of exocyclic double bonds conjugated with aromatic rings, and the equivalent alcohols, depends on the level of alkyl substitution on the α-carbon. For example, α-methyl styrene and cumyl (α,α-dimethyl benzyl) alcohol are reduced promptly and quantitatively; styrene and α-methyl benzyl alcohol, on the other hand, are reduced more slowly and incompletely. This phenomenon is illustrated in Fig. 7, in which reaction rates are compared for primary, secondary, and tertiary benzyl alcohols.

A double bond conjugated with two benzene rings, as in *trans*-stilbene, is almost inert. So is the nonaromatic conjugated double bond in 1-phenyl-cyclohexane.

Most noncyclic unsaturated structures are unaffected by the refluxing hydriodic acid–acetic acid reagent; for example, oleic, linoleic, and con-jugated linoleic acid were all inert. We have found, however, that 2,5-

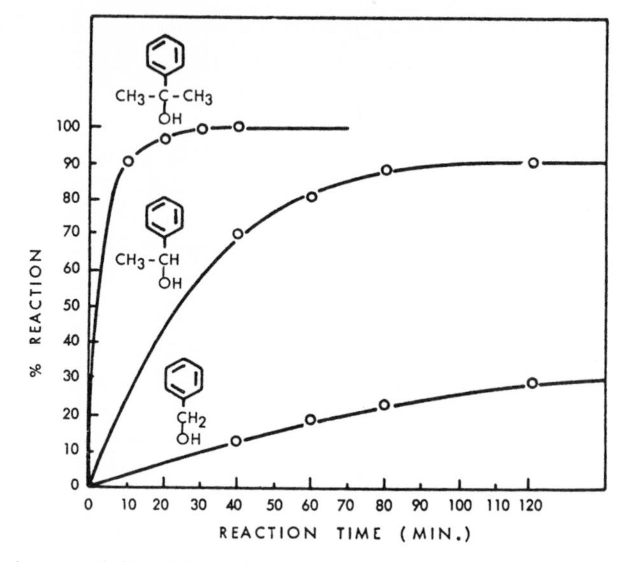

Fig. 7. Relative reactivity of benzyl alcohols in a vigorously reducing HI–acetic acid method.

dimethyl-2,5-(di-*tert*-butylperoxy) hexane produces almost 3 rather than the expected 2 moles of iodine per mole by method III—presumably by partial reduction of the initial product, 2,5-dimethyl-2,5-hexanediol. The behavior of other diols and polyols should be investigated.

4. Colorimetric Analysis with Leuco Methylene Blue

The strong points and weak points of Ueberreiter and Sorge's colorimetric method for peroxides (181,192) have been surveyed earlier in Section IV-A-3-d. This method would be outstanding if difficulties in preparing pure reagent and in preventing deterioration of its dilute benzene solutions could be diminished. Some steps have been taken in this direction. Thus Martin (117) and Banderet and co-workers (9) describe nearly identical methylene blue reduction processes that improve yields of the leuco product by precipitating its ammonium salt. In manipulation of the pure product and its benzene solutions the latter workers made extensive use of flasks and ampules fitted with break-seals. In this way, although with considerable effort, they prepared very satisfactory reagent solutions.

A more practical answer to the problem of reagent supply, it seems to us, is for producers of specialty analytical reagents to prepare and market a colorimetric reagent grade of leuco methylene blue, perhaps in sealed glass ampules containing enough of the material to prepare a liter of reagent.

Another approach to the problem of storing benzene solutions of leuco methylene blue in a suitably decolorized state has been described by Dulog (47). Following a suggestion by Ueberreiter and Sorge (181), he stored the reagent solution in the reservoir of an automatic buret over platinized asbestos in an atmosphere of hydrogen rather than nitrogen. Filter paper fastened over the inlet prevented passage of catalyst to the buret. With this technique, use of pure, light yellow leuco methylene blue was not necessary; greenish, surface-oxidized preparations served just as well, as the solution was reduced before use. It was necessary to keep the solution well protected from light.

It is interesting to contrast the general behavior of leuco methylene blue with that reported for *N*-benzoyl leuco methylene blue by Eiss and Giesecke (52). The latter reagent was obtained commercially in a suitably pure grade, and its benzene solutions were unaffected by air so that they could be stored satisfactorily in brown bottles at room temperature. This reagent was thus apparently free from the two chief drawbacks of leuco methylene blue. It had an important defect of its own, however, in that reaction with peroxides was relatively slow.

Attempts to accelerate this reaction by heating produced erratic results, even at 30°C. Nine metallic naphthenates were investigated as promoters, on the theory that they would accelerate decomposition of the peroxides into free radicals that oxidize the leuco reagent. Among these, cobalt and calcium had no effect, and cerium and manganese gave excessive reagent blanks; some acceleration was produced by zinc and lead naphthenates, however, and zirconium naphthenate was found to be excellent, reducing the reaction period for several hydroperoxides from about 40 hr to less than 1 hr. For benzoyl peroxide the time decrease was from 120 to 30 hr; this is still an unreasonably slow determination for such a reactive peroxide.

The benzoyl leuco methylene blue method was also very sensitive to the effects of light, apparently more so than the leuco methylene blue method; for the latter, no protection from light was specified originally (117,181, 192), although later workers took general precautions (9,47).

Another important difference between these two reagents lies in their relative sensitivities for determining peroxide, the sensitivity of the benzoyl derivative being about twice as great as that of leuco methylene blue. This can be seen in Table XXV, where the absorbance data of Eiss and Giesecke have been expressed as molar absorptivities; the average for the peroxides studied was 155,000 liters/mole-cm, compared to 77,200 for leuco methylene blue (181). As with leuco methylene blue, response of the benzoyl derivative was largely independent of peroxide structure, and precision appears highly satisfactory.

TABLE XXV

Molar Absorptivities at 662 mμ for Peroxides Reacted with *N*-Benzoyl Leuco Methylene Blue[a]

Peroxide	Period for complete color development	Molar absorptivity[b]
Cumene hydroperoxide	40 min (38 hr[c])	148,000 ± 4000[d]
p-Menthane hydroperoxide	2 hr	153,000 ± 2500[d]
tert-Butyl hydroperoxide	30 min (36 hr[c])	144,000 approx.[e]
Benzoyl peroxide	30 hr (120 hr[c])	150,000 ± 3000[d]
Lauroyl peroxide	5 hr	183,000 ± 4000[d]
	Average =	155,000

[a] Calculated from the data of Eiss and Giesecke (52).

[b] In units of liters/mole-cm, at 662 mμ. Comparable value for leuco methylene blue (Ueberreiter and Sorge (181)), based on 7 peroxides, was 77,200 ± 3900.

[c] Without use of zirconium naphthenate.

[d] 95% confidence interval.

[e] Did not obey Beer's law; required working curve.

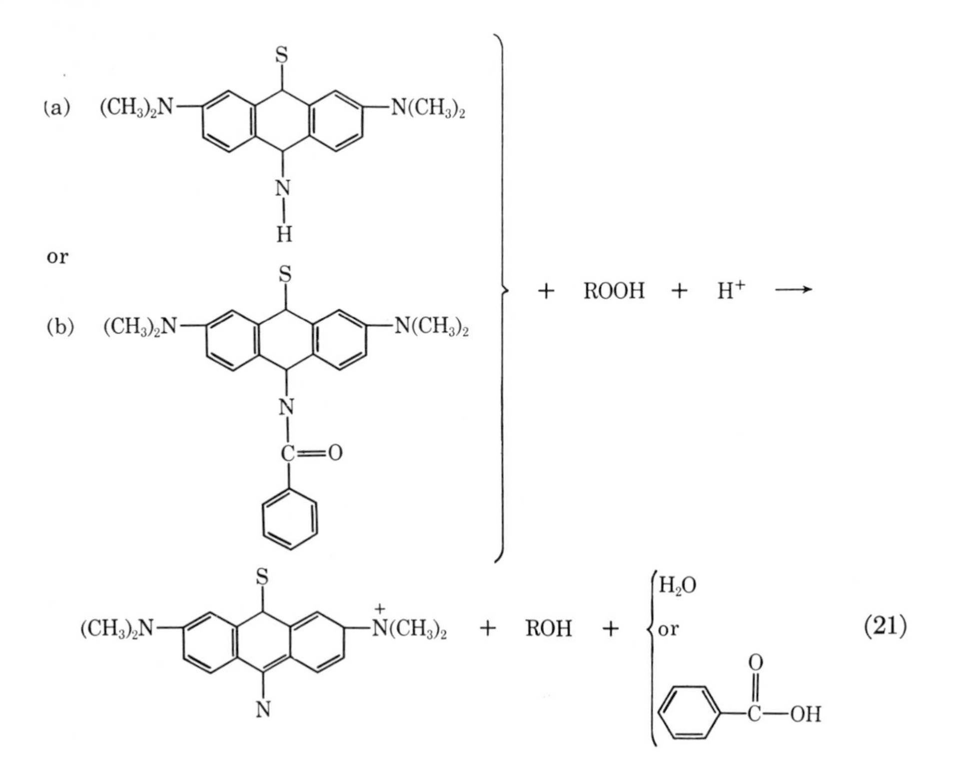

It can be assumed that both reagents produce the same color body, namely, the methylene blue cation, according to the nominal reactions represented by reaction (21). The difference in absorptivities thus provides a valuable clue to the reaction mechanisms in these colorimetric systems, which will now be discussed.

Pure methylene blue trichloroacetate has been shown to have a molar absorptivity of 88,100 (181), which should thus represent the theoretically maximum molar response of peroxides to either leuco reagent by the above reactions. Ueberreiter and Sorge recognized that the oxidation of leuco methylene blue by organic peroxides must be more complicated. They pointed out that the observed spectra of authentic methylene blue and of peroxide-oxidized leuco methylene blue were different; both had a main peak at 655 mμ and a shoulder at about 600 mμ, but in oxidized leuco methylene blue the molar absorptivity of the main peak was only 42,700, less than half that of authentic methylene blue, and the relative size of the shoulder had increased (an effect attributed to formation of thiazine derivatives).

These differences are indicated in Table XXVI, where we see also that

TABLE XXVI

Molar Absorptivities of Methylene Blue and Oxidized Leuco Methylene Blue[a]

Color body	Molar absorptivities at		Ratio
	655 mμ	600 mμ	a_{600}/a_{655}
Methylene blue[b]	88,100	31,600	0.36
Completely oxidized[c] leuco methylene blue	42,700	27,500	0.64
Partially oxidized[d] leuco methylene blue	—	—	0.59

[a] Calculated from the spectra of Ueberreiter and Sorge (181), Fig. 3.

[b] Methylene blue (Merck) in benzene containing 0.025% trichloroacetic acid.

[c] Leuco methylene blue oxidized in benzene–trichloroacetic acid medium with a 100-fold excess of 1,1′-bis-hydroperoxydicyclohexyl peroxide.

[d] Leuco methylene blue oxidized in benzene–trichloroacetic acid medium with 10% of a stoichiometric amount of fluorenone peroxide.

the spectra of *partially* and *completely* peroxide-oxidized leuco methylene blue are similar, each having a shoulder with about 60% of the main peak absorptivity, compared to only 36% for pure methylene blue.

It was recognized (181) that a complex scheme of equilibria among a variety of oxidized forms could exist, and that the observed spectrum of oxidized leuco methylene blue could not be assigned to a definite oxidation state. It was evident, however, that this spectrum was qualitatively independent of the nature of the peroxide and of the ratio of leuco methylene blue to peroxide. Therefore, Ueberreiter and Sorge concluded that there was *in effect* an unequivocal reaction between the reagent and any of the peroxides, as implied in the above equation, even though the stoichiometry was not exact and was different from that indicated by the nominal reaction.

It should now be emphasized that the average molar absorptivity in Ueberreiter and Sorge's method, 77,200 ± 3900 liters/mole-cm (based on peroxide concentration but due to absorption by peroxide-oxidized leuco methylene blue), was well *above* the value of 42,700 shown in Table XXVI for the conversion of a mole of leuco methylene blue. In fact, this response was some 180% of the expected maximum for the existing (although unknown) reaction paths. This consideration suggests that again in the case of leuco methylene blue, as seen earlier with the ferrous thiocyanate and the leuco chlorophenolindophenol methods (Sections IV-A-3-a and -b), and as suggested in the case of aromatic diamines (Section IV-A-3-c), there is an oxygen error—large positive deviations, proportional to the peroxide content, which can be attributed to induced formation of hydroperoxides by oxygen dissolved in the reaction system. With leuco methylene blue there must be the significant difference, however, that this effect is independent of many factors, including peroxide structure, reagent concentration, and reaction temperature.

It seems certain that the unusually intense molar response of benzoyl leuco methylene blue to peroxides, indicated by the data of Eiss and Giesecke (Table XXV) is due similarly to an oxygen error. Note that the average molar absorptivity with benzoyl leuco methylene blue (155,000 liters/mole-cm based on peroxide concentration) is 175% of the value of 88,100 reported for authentic methylene blue, a ratio almost identical with that for leuco methylene blue in Ueberreiter and Sorge's method (again based on peroxide concentration), namely, 180% of the 42,700 found for oxidized leuco methylene blue.

In addition to the implied oxygen error, the higher molar absorptivity with benzoyl leuco methylene blue suggests strongly that peroxide oxidation of this reagent proceeds by a different and more efficient reaction path, one that produces methylene blue from the benzoyl leuco derivative with almost exactly the stoichiometry shown above in the equation for the nominal reaction. This hypothesis implies that the spectrum of peroxide-oxidized *N*-benzoyl leuco methylene blue should be virtually the same as that of authentic methylene blue, without the modified shoulder-band intensity found characteristic of the oxidized unsubstituted leuco derivative (Table XXVI). That such is actually the case can be seen in a spectrum of the oxidized *N*-benzoyl leuco reagent reported by Eiss and Giesecke (52), on which we have calculated that the shoulder band has 33% of the intensity of the main peak, compared with 36% shown in Table XXV for pure methylene blue (181).

As in the case of the aromatic diamine reagents, it appears that no one has examined the implied oxygen effect for leuco methylene blue and the *N*-benzoyl derivative by studying their behavior as peroxide reagents in oxygen-free systems. These studies should be done so that the relative merits of ferrous thiocyanate, aromatic diamines, and leuco methylene blue as colorimetric reagents for determining traces of peroxides can be assessed objectively.

V. RECOMMENDED LABORATORY PROCEDURES

A. ANALYSIS BY GAS CHROMATOGRAPHY

Many organic peroxides have sufficient volatility and thermal stability to permit ready analysis by gas chromatography; where applicable, this is frequently the preferred method.

Apparatus

Any conventional gas chromatograph (thermal conductivity detector) that can be fitted with a glass column and operated with on-column sample loading to minimize decomposition. Use a 3- to 10-ft glass column with a packing such as didecyl phthalate or silicone grease on diatomaceous earth or powdered poly(tetrafluoroethylene).

Procedure

Use conventional gas chromatographic techniques to inject a sample solution containing an appropriate internal standard. Elute the components either isothermally at suitable temperature or by temperature programming over an appropriate range.

Precision and Accuracy

Bukata and co-workers reported relative standard deviations of 0.3% for di-*tert*-butyl peroxide and of about 1.0% for several other peroxides and mixtures. Duplicate results accordingly should agree within 1% relative for di-*tert*-butyl peroxide and within about 3% relative for less ideal peroxides.

B. ANALYSIS BY POLAROGRAPHY

Virtually all organic peroxides can be reduced and quantitatively determined polarographically, down to relatively low concentrations, in the absence of interference from reducible nonperoxides; such interference can generally be eliminated by an appropriate preliminary separation by thin-layer or column chromatography. The following general procedure covers the use of standard aqueous and nonaqueous (1:1 benzene–methanol) solvents, and also *N*,*N*-dimethylformamide (DMF) saturated with tetramethylammonium chloride, a system particularly useful for determining more difficultly reduced peroxides.

Apparatus

Use a Sargent Model XV or XXI polarograph, or the equivalent, with a polarographic H-cell, a dropping mercury electrode and a saturated calomel electrode (S.C.E.). Use a single-compartment cell and a mercury pool anode with DMF; the relatively low conductivity in this system makes resistance too high to use an H-cell.

Reagents

Lithium chloride, for general use.

Lithium sulfate, for peroxy acids and diacyl peroxides reducing on the positive side of 0 V.

Tetramethylammonium hydroxide, 10% aqueous solution, for dialkyl peroxides in a 60% ethanol system.

Tetramethylammonium chloride.

Ethanol (2 BA), *methanol* (anhydrous), *benzene* (reagent grade).

N,N-Dimethylformamide, anhydrous (Note a); to prepare a suitable polarographic medium, add 2% water and saturate with tetramethylammonium chloride.

Procedure

(1) *Selection of Polarographic System*

Carry out a variety of preliminary experiments with the peroxide system whose polarographic analysis is desired, to determine the extent of possible interferences and how to resolve them (e.g., by a preliminary separation or chemical modification), and to select optimum (or at least suitable) values for the many polarographic parameters: solvent composition, nature and concentration of supporting electrolyte, nature and concentration of maximum suppressors or other additives, dropping rate for the mercury capillary electrode, appropriate peroxide concentration range in the polarographic medium, and appropriate potentials between which to measure the diffusion current wave height.

(2) *General Preparation of Solutions for Polarography*

Add the several components to a 25-ml volumetric flask in such amounts and in such an order that upon diluting to volume with the last (usually the principal solvent), one has a homogeneous solution containing the desired concentration of reducible species in the desired polarographic medium.

(3) *Calibration*

Prepare a stock solution of the standardizing peroxide (a pure specimen or one that has been independently assayed) in the solvent medium or a suitable solvent component, at an appropriate concentration. From this solution prepare a series of dilutions to give at least three standards having the selected overall composition and peroxide contents covering the range of interest.

If using a polarographic H-cell, fill its cathode compartment to a suitable level with one of the standard solutions, and add a few milliliters of an aqueous chloride salt to the anode compartment. Immerse a saturated calomel electrode in the anode compartment and a dropping mercury electrode in the cathode compartment. Depth of solution should be slightly less on the anode than on the cathode side.

If using the single-compartment cell, as with the DMF polarographic system, pour into it a suitable volume of the standard solution, add just enough mercury to cover the bottom of the cell, and immerse in the cell an electrode assembly consisting of a dropping mercury electrode, a contact with the mercury pool, and a glass frit for deaeration.

With either type of cell, deaerate the solution with nitrogen and obtain a polarogram over the selected potential range. Repeat this procedure for each standard.

From the polarogram, read the increase in diffusion current between selected potentials below the foot and above the top of the polarographic reduction wave. Plot this increase against some function of peroxide concentration, conveniently the milligrams of peroxide in the volumetric flask. The resulting plot should be a straight line with origin very near zero (Note b).

(4) *Analysis of Samples*

Weigh an appropriate amount of the unknown sample into a 25-ml volumetric flask and make it up to the selected medium composition exactly as for the standards. Obtain a polarogram and read the diffusion current increase from it just as for the standards. To guard against unsuspected changes in electrode characteristics, measure the mercury drop time at a reference potential, such as the top of the reduction wave, in each series of analyses. Calculate the amount of peroxide and its concentration in the sample from the calibration curve.

Precision and Accuracy

With careful calibration, absence of interferences, and optimum selection of the polarographic parameters, analysts can expect precision and accuracy to about 2–3% of the amount present.

Notes. a. If commercially available DMF is not sufficiently pure, as determined by running a blank polarogram, it may be purified as follows: Charge a 5-liter flask with 3 liters of DMF. Attach an insulated distillation column, 18 in. × 1.5 in. diameter, filled with an extruded metal (Cannon) packing, and distil at total reflux until the head temperature is constant. Then take off and discard some 200 ml at a reflux ratio of about 20:1 (head temperature should then equal the boiling point of pure DMF), and distil about 1500 ml of good quality DMF at a reflux ratio of about 10:1 (a lower ratio may yield poor quality DMF).

b. Since the calibration will not generally pass exactly through zero, it is better to use a calibration curve than to calculate a proportionality constant. An alternative graphical construction with advantages for calibrations extending over wide concentration ranges has been described by Meites (122): this is to plot C/i_d vs log i_d, using semilog paper. Once i_d has been measured for a sample solution, such a graph yields the proportionality constant by which it must be multiplied to obtain the concentration or amount of the peroxide in question. The semilog plot enables a single graph to be accurate at both low and high concentrations.

C. ANALYSIS BY IODIDE REDUCTION

The literature on organic peroxide analysis contains numerous procedures based on iodide reduction, some of which are excellent, many acceptable in limited areas of application, and a few rather poor—the last should be readily recognizable on the basis of the earlier discussions. We have selected for this section two broadly applicable procedures with which we have had much experience (methods V-C-1 and V-C-2), and two others that appear very useful, although we are personally less acquainted with them (methods V-C-3 and V-C-4).

1. Determination of Easily Reduced Peroxides, with Sodium Iodide in Isopropyl Alcohol (116)

This is an adaptation of the procedure by Wagner, Smith, and Peters (196) based on the original alcohol solvent method of Kokatnur and Jelling (92). Also acceptable for these peroxide classes are the iodide procedures of Wibaut and co-workers (206), Lea (102, 104), Paschke and Wheeler (145), and Cass (23).

Reagents

(*1*) *Acidified isopropyl alcohol.* Add 100 ml of glacial acetic acid per liter of anhydrous isopropyl alcohol.

(*2*) *Sodium iodide solution in isopropyl alcohol.* Dissolve 200 g of sodium iodide in a liter of warm anhydrous isopropyl alcohol by stirring; store in a brown bottle.

Procedure

Add 10 ml of the sodium iodide–isopropyl alcohol solution to 25–50 ml of acidified isopropyl alcohol (Note a) in a 250-ml conical flask with a ground-glass joint. Add up to 2.5 meq of sample, weighed directly or as an aliquot of solution in a suitable solvent; hydrocarbons, chlorinated solvents, esters, ethers, acetic acid, and water may be used, but not acetone.

Reflux (or simmer) for some 5 min (Note b). Add 5–10 ml of water and titrate with standard sodium thiosulfate solution to a yellow-to-colorless endpoint. Run blanks, for highest accuracy; these should not exceed 0.002 meq of triiodide.

Calculations

(*1*) Content of a specific peroxide, P:

$$\text{Wt \% P} = \frac{V \times N \times M_P}{S \times n \times 20} = \frac{V \times N \times F_p}{S}$$

where V is the net volume of the titration with sodium thiosulfate of normality N; S is the sample weight, in grams; M_P is the molecular weight of peroxide, P, which has n peroxide groups per molecule; and F_P is a combined calculation factor for P.

(*2*) Active oxygen content:

$$\text{Wt \% active oxygen} = \frac{V \times N \times 0.8}{S}$$

(*3*) Peroxide number:

$$\text{Peroxide number} = \text{meq peroxide/kg} = \frac{V \times N \times 1000}{S}$$

Precision and Accuracy

The relative standard deviation of this method on samples of 1–3 meq, estimated from 21 degrees of freedom, was 0.23% (116); at the 95% confidence level, duplicate results should agree within 0.6% relative.

Accuracy is equivalent to the precision: There is neither a positive nor a negative bias.

Notes. a. Do not reverse this order of addition; it precludes the possibility of contacting the sample directly with unacidified iodide, which we have observed to cause low results, at least for hydroperoxides.

b. Many reactive peroxides will have been quantitatively reduced in much less than a 5-min reaction period, particularly if they have been added to a preheated reagent—such shorter reactions may be used if properly defined; on the other hand, extended reaction periods, of an hour or more, normally have no adverse effect. If the presence of a moderately stable peroxide is suspected, which would cause interference in a 5-min reaction through partial reduction, a range of reaction periods should be applied; it may be possible to achieve total reduction, and a resolution of both types, or at least to separate the response of the easily reduced peroxides by extrapolation to zero time.

2. Determination of Moderately Stable Peroxides, with Sodium Iodide in Acetic Acid (6% Water) (116)

This method was devised specifically for the quantitative assay of dicumyl and related bis(aralkyl) peroxides without interference from the corresponding *tert*-alcohols or their α-methyl styrene dehydration products. It is applicable to the moderately stable peroxides, in general, and quantitatively includes easily reduced peroxides when they are present; it is essentially without effect on di-*tert*-alkyl peroxides.

For bis(aralkyl) peroxides there is a stoichiometric deficiency for reduction by this iodide reagent. Its basis is not clear, but it is appreciable and, fortunately, very precise; hence by analysis of pure specimens one can determine the fraction of a mole of iodine produced per mole of peroxide to provide a correction factor C.

Apparatus

(*1*) *Special reaction flasks,* like that illustrated in Fig. 8.

(*2*) *Electrical heaters and reflux condensers;* for repetitive analyses, a six-place bank of individually controlled hot plates and cool-joint Liebig condensers is recommended.

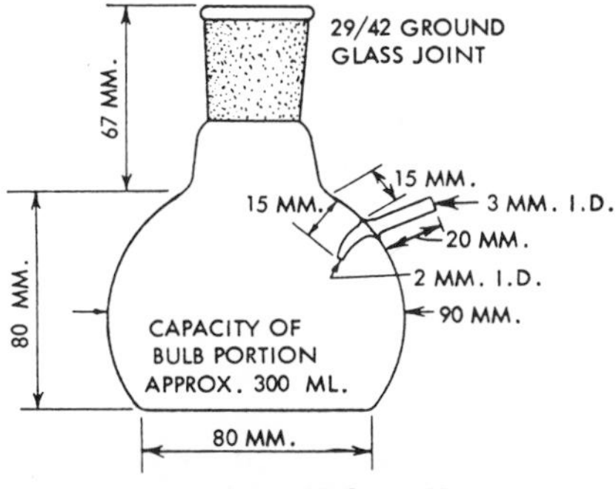

Fig. 8. Special reaction flask for iodometric peroxide analysis.

(*3*) *Inert gas cylinder*, either carbon dioxide or nitrogen; for the six-place setup, a distributing manifold can be constructed from pipe tees, close nipples, Hoke valves, and tubing connectors, as shown in Fig. 9. Location in a laboratory fume hood is convenient, to remove acetic acid fumes and to protect the reactants from bright sunlight; too much illumination produces high and erratic blanks (116).

(*4*) *Cooling bath;* a trough of rectangular cross section long enough to accommodate a row of six reaction flasks, and containing static or running water. This is a convenience for repetitive analyses. Use lead collars (truncated cones) on the flasks to prevent tipping from buoyancy.

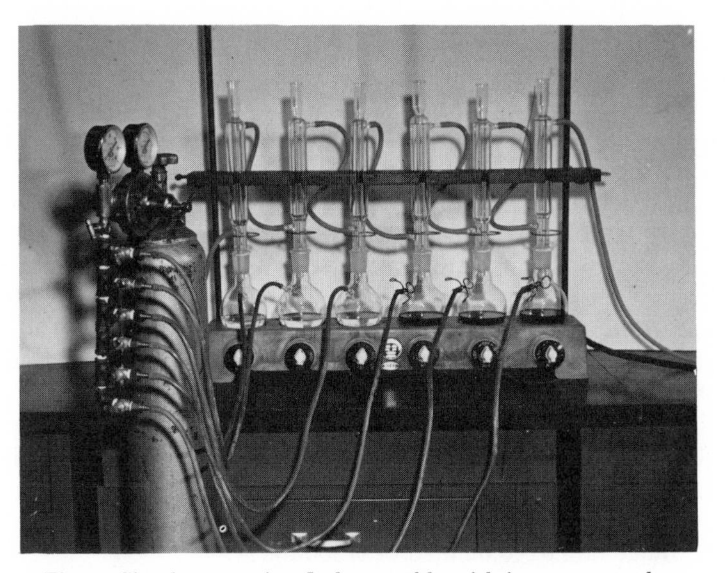

Fig. 9. Six-place reaction flask assembly with inert gas supply.

Reagents

(*1*) *Sodium iodide,* reagent grade, granular (the powdered form is excessively hygroscopic). Use a grade that does not contribute significantly to the blank.

(*2*) *Acetic acid, glacial,* reagent grade.

Procedure

Deaerate portions of acetic acid solvent as follows: Place 50 ml of acetic acid and several boiling chips in each of the special reaction flasks, attach rubber-tubing leads from the inert gas manifold, and adjust its valves to give a suitable flow, one that just perceptibly dimples the surface of the acetic acid solvent. Place the flasks on hot plates under reflux condensers, close off the inert gas flow by placing a pinch clamp close to the side arm on the flasks (Note a), and bring to reflux briefly; then take off the pinch clamps, remove the flasks from the hot plates and allow them to cool (Note b).

Add 6 g of sodium iodide (Note c) and 3.0 ml of water to each flask, swirling the contents briefly to dissolve most of the solid reagent.

If a sample is expected to contain little peroxide, then at this point carefully pipet into its reaction flask and a reagent blank 5.00-ml portions of a 0.1–0.2M solution of triiodide in acetic acid, allowing reproducible draining periods of 45–60 sec. This addition prevents interference from compounds like cumyl alcohol and α-methyl styrene in one situation not otherwise guarded against by the reagent composition: when the sample contains *no peroxide* and should give a result of nil. Partial reduction of these nonperoxides, which otherwise would occur, is prevented by the presence of the added iodine (116).

To the prepared reagents now add sample containing up to 2.5 meq of peroxide, by pipetting a 5.00- or 10.00-ml portion of a known solution in acetic acid or peroxide-free (alumina-treated) xylene; allow reproducible pipet draining periods of 10–15 sec for xylene and 45–60 sec for acetic acid solutions. Add a similar amount of the same solvent to the blanks. Reflux these reaction mixtures for 30 min, or for another reaction period that has been found appropriate, and remove from the hot plates, manipulating the pinch clamps as in the initial deaeration step (Note b).

Add 100 ml of water and titrate immediately with 0.1N standard sodium thiosulfate to the yellow-to-colorless triiodide self-indicator endpoint; starch is useful with samples that have color interference. Blanks should be in the range of 0.01–0.02 meq of triiodide.

Calculation

(*1*) *Assay of a specific peroxide*, P:

$$\text{Wt \% P} = \frac{V \times N \times M_P}{S \times n \times C \times 20} = \frac{V \times N \times F_P}{S}$$

where V is the net volume of the titration with sodium thiosulfate of normality N; S is the sample weight in the solution aliquot, in grams; M_P is the molecular weight of the peroxide, P, which has n peroxide groups per molecule and a stoichiometry correction factor, C (Note d); and F_P is a combined calculation factor for P.

(*2*) *Assay of dicumyl peroxide:*

$$\text{Wt \% dicumyl peroxide} = \frac{V \times N \times 270.4}{S \times 0.925 \times 20} = \frac{V \times N \times 14.62}{S}$$

(*3*) *Active oxygen content, uncorrected:*

$$\text{Wt \% apparent active oxygen} = \frac{V \times N \times 0.8}{S}$$

Precision and Accuracy

The relative standard deviation of this method on 1- to 3-meq samples, estimated from 26 degrees of freedom, was 0.30% (116); at the 95% confidence level, duplicate results should agree within 0.9%, relative.

Since there is a negative bias for some peroxides, accuracy depends on the care with which the correction factor has been determined.

Notes. **a.** The use of a pinch clamp in this location prevents the condensation of acetic acid in the inert gas line that occurs if flow is stopped at the needle valves.

b. We have found it desirable to interrupt the inert gas flow (by applying a pinch clamp) whenever a flask has been placed on a hot plate, and to resume this purge again just prior to removing the flask. Such a deaeration procedure is as effective as continuous purging, while avoiding the possibility of excessive and uncertain solvent loss—and the accompanying uncertainty in reagent composition—through inadvertently refluxing for too long a period.

c. Measure the sodium iodide portions in a test tube marked at the appropriate level, and deliver them rapidly into the flasks through a powder funnel; add the water via pipet.

d. The following correction factors have been established (116):

Peroxide	C, moles I_3^- per mole peroxide
Dicumyl	0.925
Di(*p*-cymyene)	0.875
Di(*p*-isopropylbenzene)	0.89

When the stoichiometry correction factor is unknown, it should be determined by analysis of a pure specimen.

3. Determination of Difficultly Reduced Peroxides, with Hydriodic Acid in Acetic Acid (7% Water) (82)

This method has been proposed by Horner and Jürgens for the determination of dialkyl peroxides (82). (See Table XI-D1.) It has been discussed in Section IV-B-3-g, with information in Tables XXIII and XXIV. It appears to reduce bis(aralkyl) peroxides quantitatively without interference from their associated nonperoxides (although this point has not been fully established) and also to reduce di-*tert*-alkyl peroxides.

We recommend that the following procedure summary, applied with suitable apparatus and deaeration procedures (e.g., those described in method V-C-2) be used as the basis for developing optimum variations of this water-modified hydriodic acid–acetic acid reagent for particular dialkyl peroxide analysis systems.

Procedure

React an aliquot part of the sample in 2–5 ml of benzene or chloroform with 1 ml of saturated aqueous potassium iodide and 5 ml of concentrated hydrochloric acid in 30 ml of glacial acetic acid under carbon dioxide or nitrogen by boiling for 30 min. After cooling, dilute with 30–50 ml of water and titrate the liberated iodine with $0.1N$ standard sodium thiosulfate.

4. Determination of *tert*-Butylperoxy Esters, with Fe(III)-Catalyzed Sodium Iodide in Chloroform–Acetic Acid (172).

This procedure was devised by Silbert and Swern for the more convenient analysis of *tert*-butyl peresters; as has been discussed in Section IV-B-3-e-(1), the use of an appropriate small iron(III) concentration dramatically accelerates the iodide reduction of this peroxide subclass.

Apparatus

Iodine flasks, or 250-ml glass-stoppered conical flasks.

Reagents

(*1*) Use commercial reagents of high purity, including: *acetic acid, chloroform, ferric chloride hexahydrate,* and *sodium iodide.*

(*2*) *Ferric chloride solution,* $7.5 \times 10^{-5}M$, in acetic acid. Dissolve 0.25 g of ferric chloride hexahydrate in 250 ml of acetic acid (0.1% w/v); make a further 50-fold dilution in acetic acid (0.002% w/v), which solution will be $7.5 \times 10^{-5}M$ in Fe(III).

(3) *Sodium iodide*, saturated aqueous. Add 1.8–2 g of sodium iodide per ml water, and equilibrate.

Procedure

Weigh a portion of sample containing up to 1–1.5 meq of *tert*-butyl perester into an iodine flask and dissolve in 10 ml of chloroform; flush the flask with nitrogen for 20 sec; add 2 ml of saturated sodium iodide, followed by 15 ml of $7.5 \times 10^{-5}M$ Fe(III) in acetic acid (Note a); then place the flask in the dark for 5–10 min. Add 50 ml of water, and titrate the solution to a starch endpoint with $0.1N$ sodium thiosulfate. Run a blank on the reagents.

Calculations

Calculate wt % peroxide or wt % active oxygen by the equations given in method V-C-1, above.

Precision and Accuracy

Silbert and Swern reported a relative standard deviation of 0.42%; at the 95% confidence level, duplicate results should agree within about 1.3% relative. The accuracy, expressed as the average percent difference between calculated and found values, was stated to be 0.29% (172).

Note. a. The procedure in the reference indicates that flushing the reaction flask with nitrogen is normally discontinued prior to addition of the reagents. Even if this is not the case, it appears certain that traces of molecular oxygen will enter the flask from the Fe(III)–acetic acid solution, which is not deaerated before use. Hence the subsequent peroxide reduction can scarcely occur in a totally oxygen-free system.

D. ANALYSIS OF PEROXIDE MIXTURES ACCORDING TO CLASS (82)

Mixtures containing peroxides from a number of classes can be analyzed quantitatively as to class without prior separation by a group of chemical methods described by Horner and Jürgens (82) (see Section IV-A-2-c). By appropriate use of reductions with diphenyl sulfide, triethyl arsine, potassium iodide, and hydriodic acid, along with acidimetry, these procedures can quantitatively determine in the same sample any or all of the chemical classes: carboxylic acid, peroxy acid, diacyl peroxide, alkyl hydroperoxide, and dialkyl peroxide.

Procedures and Calculations

Eleven procedures covering all possible combinations of the applicable classes have been listed in Table XI, along with formulas for calculating the results.

Precision and Accuracy

Horner and Jürgens list titration errors ranging from ±0.1 to ±0.2 ml of standard titrant for the various steps in a five-component analysis. These correspond to variations in the calculated results ranging from ±1.5% to ±5% relative, for samples where molar concentrations of the components are all of the same order of magnitude; the relative precision will be poorer for minor components.

There appeared to be no positive or negative bias in the various procedures.

E. ANALYSIS BY REDUCTION WITH SODIUM ARSENITE (198)

This method was devised by Walker and Conway for determining peroxides in petroleum products. These components, primarily hydroperoxides, are extracted into a hot aqueous phase where they are reduced by a sodium arsenite reagent; consumption of the latter is the basis for a quantitative estimate.

The principal merit of the method is its avoidance of interference by sulfur compounds that are present in most gasolines and fuel oils, and cause low results by iodide reduction methods. Another advantage is the ability to use relatively large samples, so that very low hydroperoxide concentrations can be accurately determined in these products.

Although we have no personal experience with the procedure, it appears very sound.

Apparatus

Tetraethyllead extraction apparatus (3), commercially available, modified by the substitution of a three-way for a two-way stopcock at the bottom, to allow more convenient purging with nitrogen during the reduction procedures.

Reagents

Sodium arsenite, 0.1*N*. Dissolve 4.9450 g of pure, dry arsenious oxide in 50 ml of 1*N* sodium hydroxide. Make neutral or very slightly acid to litmus paper with 1*N* sulfuric acid, and dilute to 1 liter.

Iodine, 0.05*N*. Dissolve 12.7 g of iodine and 40 g of potassium iodide in 25 ml of water, and dilute to 2 liters. Standardize against the sodium arsenite solution.

Procedure

Into the additions funnel of the extraction apparatus introduce 75 ml of water, 2.0 ml of 1*N* sodium hydroxide, and 2 drops of phenolphthalein indicator solution. In accordance with the anticipated hydroperoxide content, add the amounts of sample and of $0.1N$ sodium arsenite given in Table XXVII. If the amount of sample used is less than 50 ml, add enough petroleum ether to bring the volume of the top phase to between 50 and 100 ml.

TABLE XXVII

Sample Size and Reagent Volume for the Sodium Arsenite Method[a]

Anticipated hydroperoxide number[b]	Ml of sample	Ml of $0.1N$ sodium arsenite
0–1	100.0	5.00
1–5	50.0	7.00
5–10	25.0	7.00
10–50	10.0	15.00
50–100	5.0	15.00
100–1000	0.5	15.00

[a] Method of Walker and Conway (198).
[b] Milliequivalents of active oxygen per liter.

Drain the mixture into the lower part of the apparatus and rinse the funnel with 25 ml of 95% ethyl alcohol. Bubble nitrogen through the solution at a moderate rate for 5 min, then turn on the heater. When refluxing begins, adjust the flow of nitrogen and heat for maximum reflux. If the pink phenolphthalein color disappears during reflux, restore the alkalinity with $1N$ sodium hydroxide and add 2 ml in excess. Reflux for 45 min, then acidify the mixture dropwise with $10N$ sulfuric acid. Turn off the heater and cool the extractor to approximately room temperature. Shut off the nitrogen and allow the phases to separate.

Withdraw the lower phase into a 250-ml separatory funnel; extract three times with 25-ml portions of chloroform, and discard the chloroform washings. Transfer the solution to a 250-ml beaker and chill in ice. Add 1 g of sodium bicarbonate and titrate the solution with the standard iodine solution. Determine the endpoint electrometrically or with starch indicator.

Run a blank determination with each sample or series of samples, using the same amount of reagent and solvents.

Calculations

(*1*) Hydroperoxide number, milliequivalents active oxygen per liter:

$$\text{Hydroperoxide number} = \frac{1000N(b - s)}{v}$$

where b and s are the milliliters of iodine solution, of normality N, used to titrate blank and sample, respectively, and v is the milliliters of sample used. If more than one-third of the arsenite has been consumed by the sample, i.e., if $b/(b - s)$ is less than 3, then repeat the determination with adjusted proportions of sample and reagent chosen from Table XXVII.

(*2*) Active oxygen content, for weighed samples:

$$\text{Wt \% active oxygen} = \frac{0.8N(b - s)}{w}$$

where w is the grams of sample used.

Precision and Accuracy

Walker and Conway indicate that the accuracy of the method is 0.03 absolute for hydroperoxide numbers less than 1, and within 2% relative for numbers greater than 1 (198). Although data were not given, the precision is apparently similar to the accuracy.

F. ANALYSIS BY COLORIMETRIC METHODS

For determining trace amounts of easily reduced organic peroxides, this section presents four colorimetric procedures: two based on reduction with ferrous salts (with and without deaeration), one on iodide, and one on leuco methylene blue. They have been selected on the basis of our literature search and the above discussion in Sections IV-A-3 and IV-B-6; we have had little direct experience with colorimetric estimation of peroxides.

1. Reduction with Ferrous Thiocyanate in Benzene–Methanol, in the Presence of Oxygen (180)

This method has been cited by Lea (102) as the most precise and sensitive available for the colorimetric determination of traces of hydroperoxide in lipids by means of ferrous thiocyanate, although its absolute accuracy is poor. It has been shown unequivocally, with several peroxidized fatty methyl esters, to give peroxide estimates that are high by a factor of almost 2. Accuracy may be improved by calibrating with an independent method. This method should be applicable to traces of hydroperoxide or other easily reduced peroxide in any substrate soluble in 7:3 benzene–methanol.

Apparatus

(*1*) A photoelectric photometer with a filter centered in the wavelength range of 460–490 mμ, or a spectrophotometer.

(*2*) Reaction tubes. Photometer tubes with a 10.0-ml volume calibration marking, or 10-ml volumetric flasks. Clean glassware with chromic–sulfuric acid and rinse successively with tap water, distilled water, and 95% ethanol.

Reagents

(*1*) *Benzene and methanol*, reagent grade. Dry and purify by fractional distillation, if necessary.

(*2*) *Distilled water*, freed from residual trace metals by mixed-bed ion exchange.

(*3*) *Ferrous chloride solution*, 0.44% (w/v). To 1.0 g of hydrated ferrous sulfate dissolved in 50 ml of water slowly add, with stirring, a solution of 0.8 g of hydrated barium chloride in 50 ml of water and 2 ml of 10*N* hydrochloric acid. (All reagents should be free of ferric iron.) Allow the precipitated barium sulfate to settle and decant the clear solution into a bottle protected from light. The solution is sufficiently stable to be usable for about a week.

(*4*) *Ammonium thiocyanate solution*, 30% (w/v).

(*5*) *Standard ferric iron solution.* Dissolve 0.500 g bright iron wire in iron-free hydrochloric acid and oxidize with hydrogen peroxide. Boil off the excess peroxide and dilute the solution to 500 ml with water.

Procedure

To 3.0-ml portions of anhydrous methanol in 10-ml calibrated tubes or volumetric flasks, add aliquots of benzene solutions of the sample (typically, fat-bearing extracts) containing up to 3 μmoles apparent peroxide, and dilute to 10.0 ml with benzene. Analyze not more than about 15 samples at a time and include one control tube containing only solvents and reagents for every five unknowns.

Mix equal volumes of the ferrous chloride and ammonium thiocyanate reagents; add 0.1 ml to each tube and immediately distribute by shaking. Place all the tubes for 5 min in a bath at 50 $\pm$ 2°C and then for about 10 min in a bath at room temperature, gently agitating them for the first half-minute in each bath.

Measure the absorbance of the tube contents in a 1-cm cell against that of the control tube (Note a), at an appropriate wavelength (Note b) in a suitable photometer.

When absorbance readings are higher than 0.85, repeat the analyses with smaller aliquots of the benzene solution.

Calibration

Prepare a 1:100 dilution of the standard ferric chloride solution (1 mg Fe(III) per ml) with 7:3 benzene–methanol; also dilute 0.05 ml of the ferrous chloride reagent in 100 ml of the same solvent. Dilute exact 1-, 3-, and 5-ml aliquots of the ferric chloride solution to 10.0 ml with the ferrous chloride solution (to approximate the total iron content of the reaction mixture during analysis of samples), add 0.05 ml of ammonium thiocyanate reagent, and mix the contents immediately. Develop the colors as described in the procedure, and measure the absorbances in a 1-cm cell against pure solvent. This should give a linear plot of absorbance vs. Fe(III) concentration. Calculate C, the micrograms of Fe(III) per tube for which the absorbance is 1.00.

Alternatively, calibrate by analyzing samples of the pertinent substrate whose peroxide content has been determined by a reliable iodide reduction method. Calculate R, the ratio of μmoles of peroxide per sample aliquot (iodimetric) per μmoles of apparent peroxide per sample aliquot (Fe(III) calibration).

Calculation

(*1*) Apparent peroxide content of sample aliquot, based on calibration with standard Fe(III):

$$\mu\text{moles apparent peroxide} = A \times C/112 = A \times F$$

where A is the absorbance of the sample tube contents, C is the micrograms of Fe(III) per tube for which the absorbance is 1.00 (from the calibration), and 112 is the micrograms of Fe(III) nominally equivalent to 1.00 μmole of peroxide.

(*2*) Peroxide content of sample aliquot, based on iodimetric analysis of a comparable sample:

$$\mu\text{moles peroxide} = A \times F \times R$$

where R is as defined above.

Notes. a. The intensity of the developed color is stable for at least an hour in subdued light.

Measure the control itself against pure solvent to check the condition of the ferrous chloride reagent; freshly prepared, the control absorbance should be about 0.01; discard the reagent when this value exceeds 0.03.

b. Smith used a photoelectric photometer with a filter at 490 mμ (180); Kolthoff and Medalia measured ferric thiocyanate absorbance at 460 mμ (93).

2. Reduction with Ferrous Perchlorate in Acetone, in the Absence of Oxygen (93)

For cumene hydroperoxide, this procedure has been shown to give results that are reasonably precise and independent of sample size; it is also quite accurate, yielding 2 moles of ferric ion per mole of hydroperoxide; with fat and fatty acid hydroperoxides, however, the stoichiometry appeared severely and unpredictably low, although precision remained satisfactory. (See discussion in Section IV-A-3-a.)

Reagents

Acetone, purified. Distil drum-lot C.P. acetone at about 1 liter/h through a short unpacked column; to 4 liters of this distillate add 10 ml of $0.5M$ ferrous perchlorate in $0.4N$ perchloric acid, and redistil at the same rate. The product may be kept in a full bottle for several days. It should give a blank corresponding to less than $3 \times 10^{-7}M$ ferric iron.

Ferrous perchlorate, $0.1M$, in $2M$ perchloric acid. Place 0.28 g of pure iron wire in 50 ml of 20% perchloric acid, in a loosely stoppered 2-ounce bottle; keep at 50°C overnight, to dissolve; cool, and store under nitrogen.

Ammonium thiocyanate solution. Dissolve 14 g of ammonium thiocyanate in 100 ml of acetone or water.

Procedure

Dissolve the sample, containing up to 3 μmoles peroxide, in 80 ml of acetone in a 100-ml volumetric flask. Add 1.75 ml of water and a few glass beads, then heat to boiling, and continue to boil gently for 5 min. (Although the reference is not specific as to method or rate of heating (93), it is clearly necessary to take normal precautions against flammability hazards and undue decrease in solvent volume.)

Add 0.25 ml of the $0.1M$ ferrous perchlorate reagent, and continue boiling for a further 5 min. Then cool the solution, add 2 ml of the 14% ammonium thiocyanate solution, dilute to volume and read the absorbance at 460 mμ in a suitable photometer.

Run a blank on the reagents for each series of samples.

Calibration

Prepare a series of solutions from a ferric perchlorate solution obtained by dissolving iron wire in perchloric acid and oxidizing with excess hydrogen peroxide, which is then destroyed by boiling for several minutes; these acetone solutions should contain 0.28% ammonium thiocyanate, 2%

water, and $0.001M$ perchloric acid, with Fe(III) ranging from $0.6\text{–}6 \times 10^{-5}M$. Measure their absorbance at 460 mμ after standing 5 min at room temperature, and construct a calibration curve.

Alternatively, calibrate by analyzing samples containing known concentrations of the specific peroxide to be determined.

Calculations

Subtract the absorbance of the blank from that of the samples, and estimate from the calibration curves either:

(*1*) Apparent peroxide number (meq peroxide per kg) from the calibration with ferric perchlorate, assuming a 2:1 ratio between Fe(III) and peroxide; or

(*2*) Actual concentration of a specific peroxide, from its calibration with a standard sample.

Precision and Accuracy

Kolthoff and Medalia's study permits an estimation that duplicate results should probably agree within 20% relative. Absolute accuracy, except for cumene hydroperoxide, will be poor and unknown. Accuracy based on calibration through an iodometric method will equal the combined precision of the two methods.

3. Reduction with Iodide in Acetic Acid–Chloroform (10)

This method was developed by Banerjee and Budke for determining traces of peroxides in organic solvents. We recommend it because of its simplicity and broad applicability in the concentration range down to 10 ppm active oxygen, along with adequate precision and absolute accuracy. (The reference contains directions also for a more sensitive variation that performs well in the 1–10 ppm active oxygen range; it employs a special combined reaction flask–spectrophotometer cell.)

Procedure

Pipet a 5-ml portion of the sample (if a solvent) or a 5-ml aliquot of a solution (if the sample is a more concentrated peroxide) into a 25-ml volumetric flask and dilute to volume with 2:1 acetic acid–chloroform. (At least 80% of this mixture can be replaced with isopropyl alcohol without altering peroxide analyses.) Include a blank with each set of sample analyses.

Insert a hypodermic needle or glass capillary to the bottom of each flask and purge with a fine stream of nitrogen for 1.5 min. Add 1 ml of fresh 50% aqueous potassium iodide solution and continue purging for an additional minute. Then remove the hypodermic, stopper the flasks, and let them stand in the dark for 1 hr (or such shorter period as may have been found adequate).

At the end of this time, measure the absorbance of the solutions as rapidly as possible (to minimize the effect of air oxidation) at 470 mμ in matched 1-cm cells against water as a reference in a suitable spectrophotometer. Subtract the absorbance of the blank from that of the samples.

Calibration

Dissolve 0.1270 g of iodine in 2:1 acetic acid–chloroform and dilute to 100 ml in a volumetric flask. (This is equivalent to 80.0 μg of active oxygen per ml.) Pipet 0-, 1-, 2-, 3-, 4-, and 5-ml aliquots of this solution into 25-ml volumetric flasks, and dilute to volume with the acetic acid–chloroform solution. Purge with nitrogen for 1 min (Note a), then add 1 ml of 50% potassium iodide, and again purge for 1 min. Measure the absorbance immediately at 470 mμ, using 1-cm cells and a water reference. Plot absorbance against micrograms of active oxygen equivalent to the iodine in each flask (Note b).

Calculation

From the calibration curve, obtain the micrograms of active oxygen equivalent to the net absorbances of the sample solutions. Convert these to ppm active oxygen in the sample.

Precision and Accuracy

Data of Banerjee and Budke on the analysis of 16 different peroxides and on determining traces of peroxide in 9 different solvents show a relative standard deviation of about 1%. Hence, at the 95% confidence limit, duplicate analyses should agree within 3.2% relative. Accuracy appears equal to the precision.

Notes. **a.** Avoid excessive purging at this point, as the iodine is volatile until after the potassium iodide addition.

b. This calibration is said to be identical to that obtained using aliquots of a solution containing known amounts of hydrogen peroxide (10); its use avoids a hydrogen peroxide assay step.

4. Reduction with Leuco Methylene Blue Trichloroacetate in Benzene (47)

This method by Dulog is a modification of the original Ueberreiter and Sorge procedure based on suggestions made by them (181). It appears to have overcome the principal difficulties in preparing and storing high quality reagent solutions that hampered wider utilization of the original method.

Since its response on a molar basis appears independent of peroxide structure, it can be calibrated with any convenient reactive peroxide—a feature shared with this method only by those based on iodide reduction. The leuco methylene blue method also has the greatest sensitivity of the colorimetric peroxide methods, outperforming ferrous thiocyanate and iodide methods in this regard by about a factor of 3.

The peroxide reduction is best carried out in benzene, so that samples should be soluble in this solvent. The method is applicable to all easily reduced peroxides; its utility with more stable types has not been adequately explored.

Reagents

(*1*) *Leuco methylene blue,* pure, crystalline. Prepare from methylene blue by the phenylhydrazine reduction of Landauer and Weil (101) as described by Ueberreiter and Sorge (181), or as modified by Martin (117) and by Banderet and co-workers (9).

Reagent purity is less critical with this method than with earlier procedures. Since the reagent solution is catalytically hydrogenated just before use, it has been found (47) that pure light-yellow crystals are not necessary, and that greenish preparations that have suffered some surface oxidation can be used just as well.

(*2*) *Leuco methylene blue solution,* 0.3 g/liter, in benzene, stored over platinized asbestos under hydrogen. Carry out reduction of the reagent in the storage vessel of an automatic buret from which the reagent is dispensed (Note a). Place a liter of benzene in the reagent flask, deaerate by bubbling with nitrogen for a few minutes, and then add 0.3 g of solid leuco methylene blue and a quantity of platinized asbestos (Note b), while continuing the nitrogen flow. Purge the automatic buret with nitrogen and insert it in the storage flask. For 5–15 min before use of this reagent solution, bubble a gentle stream of hydrogen first through a gas-washing bottle of benzene and then through the stored solution. Protect the leuco methylene blue solution from light.

(*3*) *Trichloroacetic acid solution,* 10 g/liter, in benzene.

Procedure

To a small conical flask add about 10 ml of the trichloroacetic acid solution, 1 ml of the freshly reduced leuco methylene blue solution, and 2 ml of a sample solution containing 20–200 μmoles of peroxide. (Use 2-ml portions of suitable liquid substrates directly; prepare benzene solutions of solid or too highly peroxidized samples.) Heat to reflux for 5–15 min (Note c), avoiding direct light as far as possible; cool in the dark, and transfer to a 20–25-ml volumetric flask, rinsing and diluting to volume with trichloroacetic acid solution. Measure the absorbance at 643 mμ against benzene in 1-cm cells, using a suitable spectrophotometer. Run a blank on the leuco methylene blue reagent solution and subtract its absorbance.

Calibration

Prepare a calibration curve of absorbance vs peroxide content for each peroxide to be analyzed, at least until this is shown by experience to be unnecessary (Note d).

Calculation

Calculate the peroxide content of the sample from the measured net absorbance of its reaction solution, by referring to the calibration curve, or multiplying by the appropriate calibration factor.

Precision and Accuracy

Based on Ueberreiter and Sorge's data (181), discussed in Section IV-A-3-d, this method has a relative standard deviation of 1.85%, so that at the 95% confidence level, duplicate determinations should agree within 5.2% relative, down to a peroxide number of 0.025.

Notes. a. There are clearly a number of equivalent arrangements. Dulog recommended the following (47): As storage vessel use a 1-liter round-bottomed flask mounting an automatic buret equipped for the exclusion of air and having a connection through a side arm with a hydrogen cylinder. Cover the bottom part of the buret, which dips into the stored solution, with filter paper held in place by copper wire bands to prevent entrance of platinized asbestos to the buret.

b. Use a commercially available product containing 5% platinum; suspend a suitable quantity in a little ethanol under nitrogen and replace the nitrogen with hydrogen for a few minutes to reduce the catalyst; then flush with nitrogen again and remove the ethanol under nitrogen by evaporation or by washing and decanting with deaerated benzene.

c. Dulog prefers a standard heating step to Ueberreiter and Sorge's practice of heating carefully (bringing just to boiling) only when necessary (181); this indicates that heating apparently has no deleterious effects on color development.

d. Without comment, Dulog recommends drawing up a calibration curve in each case (47); this differs from the parent Ueberreiter and Sorge procedure (181), where a calibration curve was required only for old reagents for which blank absorbances had increased to 0.05—in other cases a simple calibration factor was used, the same for each peroxide on a molar basis.

GENERAL PEROXIDE REFERENCES

Organic Peroxides, A. G. Davies, Butterworths, London, 1961. A lucid guide to the chemistry of the many peroxide classes. Detection and analysis section brief and concise.

Organic Peroxides, E. G. E. Hawkins, Van Nostrand, New York, 1961. An excellent and timely complement to Davies' book. Has a more extensive and informative section on analysis.

"Organic Peroxides and Peroxy Compounds," N. A. Milas, in *Encyclopedia of Chemical Technology,* Vol. 10, pp. 58–84, R. E. Kirk and D. F. Othmer, Editors, Interscience, New York, 1953. Concise yet comprehensive summary of the field of organic peroxides. Has useful tables of properties for many peroxide classes and listing of commercially available peroxy compounds.

"Determination of Organic Peroxides," A. J. Martin, in *Organic Analysis,* Vol. 4, pp. 1–64, John Mitchell, Jr., Ed., Interscience, New York, 1960. Broad selective coverage of available methods, with particularly good sections on colorimetric and polarographic techniques. Many detailed procedures.

Organische Peroxyde, W. Eggersglüss, Monographien zu *"Angewandte Chemie"* und *"Chemie-Ingenieur-Technik,"* Nr. 61, Verlag Chemie, Weinheim, 1951. Studies of the chromatography, distribution coefficients, micro gas analysis, reactions, and preparation of organic peroxides.

REFERENCES

1. Abraham, M. H., A. G. Davies, D. R. Llewellyn, and E. M. Thain, *Anal. Chim. Acta,* **17,** 499 (1957).
2. Abrahamson, E. W., and H. Linschitz, *Anal. Chem.,* **24,** 1355 (1952).
3. American Society for Testing and Materials, Philadelphia, *1967 Book of ASTM Standards,* Part 17, Designation D526-66.
4. Amundsen, L. H., and L. S. Nelson, *J. Am. Chem. Soc.,* **73,** 242 (1951).
5. Anton, A., *J. Appl. Polymer Sci.,* **9,** 1631 (1965).
6. Arrhenius, S., *Acta Chem. Scand.,* **9,** 715 (1955).
7. Bailey, P. S., *Chem. Rev.,* **58,** 925 (1958).
8. Balandin, A. A., L. Kh. Freidlin, and N. V. Nikiforova, *Bull. Acad. Sci. USSR Div. Chem. Sci. (English Transl.),* **1958,** 126; **1959,** 1138.
9. Banderet, A., M. Brendle, and G. Riess, *Bull. Soc. Chim. France,* **1965,** 626.
10. Banerjee, D. K., and C. C. Budke, *Anal. Chem.,* **36,** 792 (1964).
11. Banerjee, D. K., and C. C. Budke, *Anal. Chem.,* **36,** 2367 (1964).
12. Barnard, D., and K. R. Hargrave, *Anal. Chim. Acta,* **5,** 476 (1951).
13. Bartlett, P. D., and R. Altschul, *J. Am. Chem. Soc.,* **67,** 812 (1945).
14. Bartlett, P. D., and R. E. Pincock, *J. Am. Chem. Soc.,* **82,** 1769 (1960).

15. Bellamy, L., *The Infrared Spectra of Complex Molecules*, 2nd ed., Wiley, New York, 1958, Chapters 7 and 8.
16. Bernard, M. L. J., *Ann. Chim. (Paris)*, **10,** 315 (1955).
16a. Braithwaite, B., and G. E. Penketh, *Anal. Chem.*, **39,** 1470 (1967).
17. Brammer, J. A., S. Frost, and V. W. Reid, *Analyst*, **92,** 91 (1967).
18. Brill, W. F., *J. Am. Chem. Soc.*, **87,** 3286 (1965).
19. Bruschweiler, H., and G. J. Minkoff, *Anal. Chim. Acta*, **12,** 186 (1955).
20. Bukata, S. W., L. L. Zabrocki, and M. F. McLaughlin, *Anal. Chem.*, **35,** 885 (1963).
21. Buncel, E., and A. G. Davies, *J. Chem. Soc.*, **1958,** 1550.
21a. Buzlanova, M. M., V. F. Stepanovskaya, and V. L. Antonovskii, *J. Anal. Chem. USSR (English Transl.)*, **21,** 1324 (1966).
21b. Buzlanova, M. M., V. F. Stepanovskaya, A. F. Nesterov, and V. L. Antonovskii, *J. Anal. Chem. USSR (English Transl.)*, **21,** 454 (1966).
22. Cartlidge, J., and C. F. H. Tipper, *Anal. Chim. Acta*, **22,** 106 (1960).
23. Cass, W. E., *J. Am. Chem. Soc.*, **68,** 1976 (1946).
24. Cass, W. E., *J. Am. Chem. Soc.*, **72,** 4915 (1950).
25. Castrantas, H. M., D. K. Banerjee, and D. C. Noller, *Fire and Explosion Hazards of Peroxy Compounds*, ASTM STP 394, Am. Soc. Testing Materials, 1965.
26. Chapman, R. A., and K. Mackay, *J. Am. Oil Chemists' Soc.*, **26,** 360 (1949).
26a. Cohen, I. R., T. C. Purcell, and A. P. Altshuller, *Environ. Sci. Technol.*, **1,** 247 (1967).
26b. Courtier, J. C., *Methodes Phys. Anal.*, **1966,** 23; through *Chem. Abstr.*, **65,** 17125g (1966).
27. Criegee, R., *Record Chem. Progr. (Kresge-Hooker Sci. Lib.)*, **18,** 111 (1957).
28. Criegee, R., and G. Paulig, *Chem. Ber.*, **88,** 712 (1955).
29. Criegee, R., W. Schorrenberg, and J. Becke, *Ann. Chem.*, **565,** 7 (1949).
30. Dahle, L. K., and R. T. Holman, *Anal. Chem.*, **33,** 1960 (1961).
31. Darley, E. F., K. A. Kettner, and E. R. Stephens, *Anal. Chem.*, **35,** 589 (1963).
32. Dasler, W., and C. D. Bauer, *Ind. Eng. Chem., Anal. Ed.*, **18,** 52 (1946).
33. Dastur, N. N., and C. H. Lea, *Analyst*, **66,** 90 (1941).
34. Davies, A. G., *J. Chem. Soc.*, **1958,** 3474.
35. Davies, A. G., *Organic Peroxides*, Butterworths, London, 1961.
36. Davies, A. G., and R. Feld, *J. Chem. Soc.*, **1956,** 665.
37. Davies, A. G., R. V. Foster, and R. Nery, *J. Chem. Soc.*, **1954,** 2204.
38. Davies, A. G., D. G. Hare, and R. F. M. White, *J. Chem. Soc.*, **1960,** 1040.
39. Davies, A. G., D. G. Hare, and R. F. M. White, *J. Chem. Soc.*, **1961,** 341.
40. Davison, W. H. T., *J. Chem. Soc.*, **1951,** 2456.
41. Dickey, F. H., J. H. Raley, F. F. Rust, R. S. Treseder, and W. E. Vaughan, *Ind. Eng. Chem.*, **41,** 1673 (1949); U.S. Pat. 2,403,771 (July 9, 1946), W. E. Vaughan and F. F. Rust (to Shell Development Co.).
42. Dobrinskaja, A., and M. Neumann, *Acta Physicochim. URSS*, **10,** 297 (1939).
43. Dobson, G., and G. Hughes, *J. Chromatog.*, **16,** 416 (1964).
44. Doehnert, D. F., and O. L. Mageli, *Mod. Plastics*, **36** (No. 6), 142 (1959).
44a. Driver, M. G., R. B. Koch, and H. Salwin, *J. Am. Oil Chemists' Soc.*, **40,** 504 (1963).
45. Dubouloz, P., J. Fondarai, J. Laurent, and R. Marville, *Anal. Chim. Acta*, **15,** 84 (1956).
46. Dugan, P. R., *Anal. Chem.*, **33,** 696, 1630 (1961).

47. Dulog, L., *Z. Anal. Chem.*, **202,** 192 (1964).
48. Dulog, L., and K. H. Burg, *Z. Anal. Chem.* **203,** 184 (1964).
49. Dykstra, S., and H. S. Mosher, *J. Am. Chem. Soc.*, **79,** 3474 (1957).
50. Egerton, A. C., A. J. Everett, G. J. Minkoff, S. Rudrakanchana, and K. C. Salooja, *Anal. Chim. Acta*, **10,** 422 (1954).
51. Eggersglüss, W., *Organische Peroxyde*, Verlag Chemie, Weinheim, 1951.
52. Eiss, M. I., and P. Giesecke, *Anal. Chem.*, **31,** 1558 (1959).
53. Everett, A. J., and G. J. Minkoff, *Trans. Faraday Soc.*, **49,** 410 (1953).
54. Farbenfabriken Bayer Aktiengesellschaft, Trade Publication AN 565 from Technical Applications Group K, *Commercially Available Organic Peroxides as of October, 1964.*
55. Feinstein, R. N., *J. Org. Chem.*, **24,** 1172 (1959).
56. Fierz-David, H. E., *Chimia*, **1,** 246 (1947); through *Chem. Abstr.*, **42,** 2228 (1948).
57. Forman, E. J., and C. M. Wright, Hercules Incorporated, Research Center, Wilmington, Delaware, private communication (1959).
58. Foxley, G. H., *Analyst*, **86,** 348 (1961).
59. Fujiwara, S., M. Katayama, and S. Kamio, *Bull. Chem. Soc. Japan*, **32,** 657 (1959).
60. Giguere, P. A., and D. Lamontagne, *Can. J. Chem.*, **29,** 54 (1951).
61. Goddu, R. F., "Near-Infrared Spectrophotometry," in *Advances in Analytical Chemistry and Instrumentation*, C. N. Reilley, Ed., Interscience, New York, 1963, Vol. 1, p. 406.
62. Goddu, R. F., Hercules Incorporated, Research Center, Wilmington, Delaware, private communication, (1952).
63. Gosman, B. A., *Collection Czech. Chem. Commun.*, **7,** 467 (1935).
64. Graupner, A. J., Hercules Incorporated, Research Center, Wilmington, Delaware, private communication (1954).
65. Gribov, L. A., *Opt. Spectry. (USSR) (English Transl.)*, **9,** 93 (1960).
66. Gribov, L. A., and A. V. Karyakin, *Opt. Spectry. (USSR) (English Transl.)*, **9,** 350 (1960).
67. Guillet, J. E., and M. F. Meyer, *Ind. Eng. Chem., Prod. Res. Develop.*, **1,** 226 (1962).
68. Guillet, J. E., T. R. Walker, M. F. Meyer, J. P. Hawk, and E. B. Towne, *Ind. Eng. Chem., Prod. Res. Develop.*, **3,** 257 (1964).
69. Hahto, M. P., and J. E. Beaulieu, *J. Chromatog.*, **25,** 472 (1966).
69a. Hamm, D. L., and E. G. Hammond, *J. Dairy Sci.*, **50,** 1166 (1967).
69b. Hamm, D. L., E. G. Hammond, V. Parvanah, and H. E. Snyder, *J. Am. Oil Chemists' Soc.*, **42,** 920 (1965).
70. Handy, C. T., and H. S. Rothrock, *J. Am. Chem. Soc.*, **80,** 5306 (1958).
71. Hartman, L., and M. D. L. White, *Anal. Chem.*, **24,** 527 (1952).
72. Hartman, L., and M. D. L. White, *J. Sci. Food Agr.*, **3,** 112 (1952).
73. Hartmann, S., and J. Glavind, *Acta Chem. Scand.*, **3,** 954 (1949).
74. Hawkins, E. G. E., *Organic Peroxides*, Van Nostrand, New York, 1961.
74a. Hayano, S., T. Ota, and Y. Fukushima, *Bunseki Kagaku*, **15,** 365 (1966); through *Chem. Abstr.*, **67,** 39941r (1967).
75. Heaton, F. W., and N. Uri, *J. Sci. Food Agr.*, **9,** 781 (1958).
76. Hochstein, F. A., *J. Am. Chem. Soc.*, **71,** 305 (1949).
77. Hock, H., and H. Kropf, *Chem. Ber.*, **92,** 1115 (1959).
78. Hoffman, J., *J. Org. Chem.*, **22,** 1747 (1957).
79. Hoffmann, H., *Angew. Chem.*, **72,** 77 (1960).

80. Holman, R. T., C. Nickell, O. S. Privett, and P. R. Edmondson, *J. Am. Oil Chemists' Soc.*, **35**, 422 (1958).
81. Horner, L., and W. Jurgeleit, *Ann. Chem.*, **591**, 138 (1955).
82. Horner, L., and E. Jürgens, *Angew. Chem.*, **70**, 266 (1958).
83. Horner, L., and E. Jürgens, *Ann. Chem.*, **602**, 135 (1957).
84. Hyden, S., *Anal. Chem.*, **35**, 113 (1963).
85. Japan Electron Optics Laboratory Co., Ltd. (JEOL), Trade Literature on Model JLC-2A "Universal" automatic recording liquid chromatograph.
86. Johnson, R. M., and I. W. Siddiqi, *J. Polarog. Soc.*, **11**, 72 (1965).
87. Kalbag, S. S., K. A. Narayan, S. S. Chang, and F. A. Kummerow, *J. Am. Oil Chemists' Soc.*, **32**, 271 (1955).
88. Karyakin, A. V., *Russ. Chem. Rev.* (*English Transl.*), **30**, 460 (1961).
89. Kenyon, W. C., J. E. McCarley, E. G. Boucher, A. E. Robinson, and A. K. Wiebe, *Anal. Chem.*, **27**, 1888 (1955).
90. Kirk, A. D., and J. H. Knox, *Trans. Faraday Soc.*, **56**, 1296 (1960).
90a. Klyueva, N. D., G. E. Muratova, A. I. Kashlinskii, and A. V. Sokolov, *J. Anal. Chem. USSR* (*English Transl.*), **22**, 253 (1967).
91. Knappe, E., and D. Peteri, *Z. Anal. Chem.*, **190**, 386 (1962).
91a. Kober, R., Ger. (East) Pat. 53,354, Jan. 5, 1967; through *Chem. Abstr.*, **67**, 122070b (1967).
92. Kokatnur, V. R., and M. Jelling, *J. Am. Chem. Soc.*, **63**, 1432 (1941).
93. Kolthoff, I. M., and A. I. Medalia, *Anal. Chem.*, **23**, 595 (1951).
94. Kolthoff, I. M., and A. I. Medalia, *J. Am. Chem. Soc.*, **71**, 3777, 3784, 3789 (1949).
95. Kolthoff, I. M., and E. B. Sandell, *Textbook of Quantitative Inorganic Analysis*, 3rd ed., Macmillan, New York, 1952, pp. 574, 600, 705.
96. Kornblum, N., and H. E. DeLaMare, *J. Am. Chem. Soc.*, **73**, 880 (1951).
97. Kornblum, N., and H. E. DeLaMare, *J. Am. Chem. Soc.*, **74**, 3079 (1952).
98. Kovner, M. A., A. V. Karyakin, and A. P. Efimov, *Opt. Spectry.* (*USSR*) (*English Transl.*), **8**, 64 (1960).
99. Kuta, E. J., and F. W. Quackenbush, *Anal. Chem.*, **32**, 1069 (1960).
100. Kuta, E. J., and F. W. Quackenbush, *J. Am. Oil Chemists' Soc.*, **37**, 148 (1960).
101. Landauer, P., and H. Weil, *Ber.*, **43**, 198 (1910).
102. Lea, C. H., *J. Sci. Food Agr.*, **3**, 586 (1952).
103. Lea, C. H., *J. Soc. Chem. Ind.* (*London*), **64**, 106 (1945).
104. Lea, C. H., *J. Soc. Chem. Ind.* (*London*), **65**, 286 (1946).
105. Lea, C. H., *Proc. Roy. Soc.* (*London*), *Ser. B*, **108**, 175 (1931).
106. Lewis, W. R., and F. W. Quackenbush, *J. Am. Oil Chemists' Soc.*, **26**, 53 (1949).
107. Lewis, W. R., F. W. Quackenbush, and T. DeVries, *Anal. Chem.*, **21**, 762 (1949).
108. Liebhafsky, H. A., and W. H. Sharkey, *J. Am. Chem. Soc.*, **62**, 190 (1940).
109. Linschitz, H., J. Rennert, and T. M. Korn, *J. Am. Chem. Soc.*, **76**, 5839 (1954).
110. Lobunez, W., J. R. Rittenhouse, and J. G. Miller, *J. Am. Chem. Soc.*, **80**, 3505 (1958).
111. Loftus-Hills, G., and C. C. Thiel, *J. Dairy Res.*, **14**, 340 (1946).
112. Lorenz, O., *Anal. Chem.*, **37**, 101 (1965).
112a. Lueck, H., and R. Maack, *Fette, Seifen, Anstrichmittel*, **68**, 81, 305, 609 (1966).
113. Maack, R., and H. Lück, *Experientia*, **19**, 466 (1963).
114. MacNevin, W. M., and P. F. Urone, *Anal. Chem.*, **25**, 1760 (1953).
115. Mageli, O. L., S. D. Stengel, and D. F. Doehnert, *Mod. Plastics*, **36** (No. 7), 135 (1959).

116. Mair, R. D., and A. J. Graupner, *Anal. Chem.*, **36**, 194 (1964).
117. Martin, A. J., "Determination of Organic Peroxides," in *Organic Analysis*, John Mitchell, Jr., Ed., Interscience, New York, 1960, Vol. IV, pp. 1–64.
118. Matic, M., and D. A. Sutton, *Chem. Ind. (London)*, **1953**, 666.
119. Matic, M., and D. A. Sutton, *J. Chem. Soc.*, **1952**, 2679.
120. Matthews, J. S., and Joan F. Patchan, *Anal. Chem.*, **31**, 1003 (1959).
121. McMillan, G. R., *J. Am. Chem. Soc.*, **83**, 3018 (1961).
122. Meites, L., *Polarographic Techniques*, 2nd ed., Interscience, New York, 1965, p. 396.
123. Merz, J. H., and W. A. Waters, *J. Chem. Soc.*, **1949**, S15.
124. Middleton, J. T., *Ann. Rev. Plant Physiol.*, **12**, 431 (1961).
125. Milas, N. A., "Organic Peroxides and Peroxy Compounds," in *Encyclopedia of Chemical Technology*, Vol. 10, R. E. Kirk and D. F. Othmer, Ed., Interscience, New York, 1953, p. 58.
126. Milas, N., and I. Belic, *J. Am. Chem. Soc.*, **81**, 3358 (1959).
127. Milas, N., and A. Golubovic, *J. Am. Chem. Soc.*, **81**, 3361 (1959).
128. Milas, N. A., and D. M. Surgenor, *J. Am. Chem. Soc.*, **68**, 205 (1946).
129. Milas, N. A., and D. M. Surgenor, *J. Am. Chem. Soc.*, **68**, 642 (1946).
130. Minkoff, G. J., *Discussions Faraday Soc.*, **9**, 320 (1950).
131. Minkoff, G. J., *Proc. Roy. Soc. (London), Ser. A.*, **228**, 287 (1951).
132. Mitchell, A. G., and W. F. K. Wynne-Jones, *Discussions Faraday Soc.*, **15**, 161 (1953).
133. Mitchell, J., Jr., and L. R. Perkins, *Mod. Plast.*, **44** (8), 158 (1967).
134. Mitchell, J., Jr., and D. M. Smith, *Aquametry*, Interscience, New York, 1948, p. 142.
135. Molyneux, P., *Tetrahedron*, **22**, 2929 (1966).
136. Noller, D. C., and D. J. Bolton, *Anal. Chem.*, **35**, 887 (1963).
137. Noller, D. C., S. J. Mazurowski, G. F. Linden, F. J. G. DeLeeuw, and O. L. Mageli, *Ind. Eng. Chem.*, **56** (12), 18 (1964).
138. Norris, T. O., and A. L. Ryland, unpublished work, E. I. du Pont de Nemours & Co., Wilmington, Delaware.
139. Nozaki, K., *Ind. Eng. Chem., Anal. Ed.*, **18**, 583 (1946).
140. Orr, A. A., Hercules Incorporated, Research Center, Wilmington, Delaware, private communication, 1960.
141. Parker, W. E., C. Ricciuti, C. L. Ogg, and D. Swern, *J. Am. Chem. Soc.*, **77**, 4037 (1955).
142. Parker, W. E., L. P. Witnauer, and D. Swern, *J. Am. Chem. Soc.*, **79**, 1929 (1957).
143. Parks, T. D., and K. A. Hansen, *Anal. Chem.*, **22**, 1268 (1950).
144. Parsons, R., "The Structure of the Electrical Double Layer and Its Influence on the Rates of Electrode Reactions," in *Advances in Electrochemistry and Electrochemical Engineering*, Vol. 1, P. Delahay, Ed., Interscience, New York, 1961, p. 1.
145. Paschke, R. F., and D. H. Wheeler, *Oil Soap*, **21**, 52 (1944).
146. Philpotts, A. R., and W. Thain, *Anal. Chem.*, **24**, 638 (1952).
147. Pobiner, H., *Anal. Chem.*, **33**, 1423 (1961).
148. Proske, G., *Anal. Chem.*, **24**, 1834 (1952).
148a. Purcell, T. C., and I. R. Cohen, *Environ. Sci. Technol.*, **1**, 431 (1967).
149. Quackenbush, F. W., and E. J. Kuta, Div. Anal. Chem., 142nd ACS Meeting, Atlantic City, N. J., Sept. 1962.
150. Ramsey, J. B., and F. T. Aldridge, *J. Am. Chem. Soc.*, **77**, 2561 (1955).
151. Redington, R. L., W. B. Olson, and P. C. Cross, *J. Chem. Phys.*, **36**, 1311 (1962).

152. Reimers, F., *Sb. Mezinarad. Polarog. Sjezdu. Praze, 1st Congr., 1951*, Pt. 1, 203; through *Chem. Abstr.*, **46**, 6998d (1952).
153. Ricciuti, C., C. O. Willits, C. L. Ogg, S. G. Morris, and R. W. Riemenschneider, *J. Am. Oil Chemists' Soc.*, **31**, 456 (1954).
154. Rieche, A., E. Schmitz, and P. Dietrich, *Chem. Ber.*, **92**, 2239 (1959).
155. Rieche, A., and M. Schulz, *Angew. Chem.*, **70**, 694 (1958).
156. Rieche, A., M. Schulz, H. E. Seyfarth, and G. Gottschalk, *Fette, Seifen, Anstrichmittel*, **64**, 198 (1962).
157. Roberts, E. R., and J. S. Meek, *Analyst*, **77**, 43 (1952).
158. Rogers, M. T., and T. W. Campbell, *J. Am. Chem. Soc.*, **74**, 4742 (1952).
159. Romantsev, M. F., and E. S. Levin, *J. Anal. Chem. USSR* (*English Transl.*) **18**, 957 (1963).
160. Royals, E. E., *Advanced Organic Chemistry*, Prentice-Hall, Englewood Cliffs, N. J., 1954, p. 369.
161. Ryland, A. L., Div. Anal. Chem., 142nd ACS Meeting, Atlantic City, N. J., Sept. 1962.
162. Sambale, E., *Plaste Kautschuk*, **10**(1), 33 (1963).
163. Satterfield, C. N., and A. H. Bonnell, *Anal. Chem.*, **27**, 1174 (1955).
164. Schulz, M., and K. H. Schwarz, *Monatsber. Deut. Akad. Wiss. Berlin*, **6**(7), 515 (1964).
164a. Schulz, M., and K. H. Schwarz, *Z. Chem.*, **7**, 176 (1967).
165. Schulz, M., H. Seeboth, and W. Wieker, *Z. Chem.*, **2**, 279 (1962).
166. Serif, G. S., C. F. Hunt, and A. N. Bourns, *Can. J. Chem.*, **31**, 1229 (1953).
167. Sheppard, N., *Discussions Faraday Soc.*, **9**, 322 (1950).
168. Siddiqi, A. M., and A. L. Tappel, *Chemist-Analyst*, **44**, 52 (1955).
169. Siemens, A. M. E., *Brit. Plastics*, **35**, 357 (1962).
170. Siggia, S., *Anal. Chem.*, **19**, 872 (1947).
171. Silbert, L. S., *J. Am. Oil Chemists' Soc.*, **39**, 480 (1962).
172. Silbert, L. S., and D. Swern, *Anal. Chem.*, **30**, 385 (1958).
173. Silbert, L. S., and D. Swern, *J. Am. Chem. Soc.*, **81**, 2364 (1959).
174. Silbert, L. S., L. P. Witnauer, D. Swern, and C. Riccuiti, *J. Am. Chem. Soc.*, **81**, 3244 (1959).
175. Simon, W., J. T. Clerc, and R. E. Dohner, *Microchem. J.*, **10**, 495 (1966).
176. Simon, W., and H. Giacobbo, *Chem. Ingr.-Tech.*, **37**, 709 (1965).
177. Skellon, J. H., and M. H. Thurston, *Analyst*, **73**, 97 (1948).
178. Skellon, J. H., and E. D. Wills, *Analyst*, **73**, 78 (1948).
179. Skoog, D. A., and A. B. H. Lauwzecha, *Anal. Chem.*, **28**, 825 (1956).
180. Smith, G. H., *J. Sci. Food Agr.*, **3**, 26 (1952).
181. Sorge, G., and K. Ueberreiter, *Angew. Chem.*, **68**, 486 (1956).
182. Stahl, E., *Chemiker-Ztg.*, **82**, 323 (1958).
182a. Stamm, J., *Bull. Soc. Pharm. Esthonia*, **5**, 181 (1925); *J. Pharm. Chim.*, **118**, 214 (1926); *Analyst*, **51**, 416 (1926).
183. Stansby, M. E., *Ind. Eng. Chem., Anal. Ed.* **13**, 627 (1941).
184. Stephens, E. R., E. F. Darley, O. C. Taylor, and W. E. Scott, *Intern. J. Air Pollution*, **4**, 79 (1961); through *Chem. Abstr.*, **55**, 19082h (1961).
184a. Stein, R. A., and V. Slawson, *Anal. Chem.*, **35**, 1008 (1963).
185. Stern, V., and S. Polak, *Acta Physicochim. URSS*, **11**, 797 (1939).
186. Stouffer, J. E., *Barber-Colman Chromatogram*, **5**, No. 1 (1965); Stouffer, J. E., T. E. Kersten, and P. M. Krueger, *Biochim. Biophys. Acta*, **93**, 191 (1964).
187. Sully, B. T. D., *Analyst*, **79**, 86 (1954).

188. Sutton, D.A., *Chem. Ind. (London)*, **1951**, 272.
189. Swern, D., J. E. Coleman, H. G. Knight, C. Ricciuti, C. O. Willits, and C. R. Eddy, *J. Am. Chem. Soc.*, **75**, 3135 (1953).
190. Swern, D., and L. S. Silbert, *Anal. Chem.*, **35**, 880 (1963).
191. Swern, D., L. P. Witnauer, C. R. Eddy, and W. E. Parker, *J. Am. Chem. Soc.*, **77**, 5537 (1955).
191a. Szobor, A., and I. Back, *Tetrahedron Letters*, **1966**, 3985.
191b. Terent'ev, A. P., G. C. Larikova, and E. A. Bondarevskaya, *J. Anal. Chem. USSR (English Transl.)*, **21**, 312 (1966).
191c. Terent'ev, V. A., N. Kh. Shtivel, and V. L. Antonovskii, *Zh. Prikl. Spektrosk.*, **5**, 463 (1966); through *Chem. Abstr.*, **66**, 52055d (1967).
192. Ueberreiter, K., and G. Sorge, *Angew. Chem.*, **68**, 352 (1956).
193. Verderame, F. D., and J. G. Miller, *J. Phys. Chem.*, **66**, 2185 (1962).
194. Voellmin, J., P. Kriemler, I. Omura, J. Seibl, and W. Simon, *Microchem. J.*, **11**, 73 (1966).
195. Wagner, C. D., H. L. Clever, and E. D. Peters, *Anal. Chem.*, **19**, 980 (1947).
196. Wagner, C. D., R. H. Smith, and E. D. Peters, *Anal. Chem.*, **19**, 976 (1947).
197. Wagner, C. D., R. H. Smith, and E. D. Peters, *Anal. Chem.*, **19**, 982 (1947).
198. Walker, D. C., and H. S. Conway, *Anal. Chem.*, **25**, 923 (1953).
199. Ward, G. A., and R. D. Mair, *Anal. Chem.*, **41**, 538 (1967).
200. Waters Associates, Inc., Framingham, Mass., Trade Literature on Model R-4 Liquid Chromatograph.
201. Watson, E. S., M. J. O'Neill, J. Justin, and N. Brenner, *Anal. Chem.*, **36**, 1233 (1964).
202. Wawzonek, S., E. W. Blaha, R. Berkey, and M. E. Runner, *J. Electrochem. Soc.*, **102**, 235 (1955).
203. Weissberger, A., E. S. Proskauer, J. A. Riddick, and E. E. Toops, Jr., *Organic Solvents*, 2nd ed., Interscience, New York, 1955.
204. Welch, F. J., H. R. Williams, and H. S. Mosher, *J. Am. Chem. Soc.*, **77**, 551 (1955).
205. Wheeler, D. H., *Oil Soap*, **9**, 89 (1932).
206. Wibaut, J. P., H. B. van Leeuwen, and B. van der Wal, *Rec. Trav. Chim.*, **73**, 1033 (1954).
207. Williams, H. R., and H. S. Mosher, *Anal. Chem.*, **27**, 517 (1955).
208. Williams, H. R., and H. S. Mosher, *J. Am. Chem. Soc.*, **76**, 2984, 2987 (1954).
209. Williams, H. R., and H. S. Mosher, *J. Am. Chem. Soc.*, **76**, 3495 (1954).
210. Willits, C. O., C. Ricciuti, H. B. Knight, and D. Swern, *Anal. Chem.*, **24**, 785 (1952).
211. Willits, C. O., C. Ricciuti, C. L. Ogg, S. G. Morris, and R. W. Riemenschneider, *J. Am. Oil Chemists' Soc.*, **30**, 420 (1953).
212. Wolf, R., *Bull. Soc. Chim. France*, **1954**, 644.
213. Wolfe, W. C., *Anal. Chem.*, **34**, 1328 (1962).
214. Wurster, C. F., Jr., L. J. Durham, and H. S. Mosher, *J. Am. Chem. Soc.*, **80**, 327 (1958).

INDEX